German-English and English-German

Dictionary of Technological Terms Used in Electrical Communication

By

O. Sattelberg
of the Telegraphentechnische Reichsamt
Berlin

Part Second
German-English

Springer-Verlag Berlin Heidelberg GmbH
1926

Englisch-Deutsches und Deutsch-Englisches

Wörterbuch der Elektrischen Nachrichtentechnik

Von

O. Sattelberg
im Telegraphentechnischen Reichsamt
Berlin

Zweiter Teil
Deutsch-Englisch

Springer-Verlag Berlin Heidelberg GmbH
1926

Alle Rechte vorbehalten
© Springer-Verlag Berlin Heidelberg 1926
Originally published by Julius Springer in Berlin in 1926
Softcover reprint of the hardcover 1st edition 1926

ISBN 978-3-662-31746-4 ISBN 978-3-662-32572-8 (eBook)
DOI 10.1007/978-3-662-32572-8

Vorwort zum englisch-deutschen Teil.

Das letzte Jahrzehnt hat nur wenigen Gebieten einen solchen Aufschwung gebracht wie der Elektrischen Nachrichtentechnik. Es erscheint daher gerechtfertigt, zumal im Hinblick auf die im Werden begriffene Übereinkunft über ein zwischenstaatliches europäisches Fernsprechnetz und auf andere internationale Vereinheitlichungsbestrebungen, zur Verringerung der sprachlichen Schwierigkeiten ein Wörterbuch der Nachrichtentechnik zu schaffen.

Vollständigkeit, sprachliche und sachliche Richtigkeit hoffe ich in so weitem Maß erreicht zu haben, daß das vorliegende Werk in den meisten vorkommenden Fällen ein brauchbares Hilfsmittel zu sein verspricht. Wenn die bezeichneten Ziele nicht erreicht wurden, ja, wohl nie ganz erreicht werden können, so liegt das vielleicht weniger an einem Mangel an Sorgfalt, als an mancherlei Hindernissen, die sich der Bearbeitung entgegengestellt haben. Nicht das kleinste dieser Hindernisse liegt in der in der Fernmeldetechnik als einem in schnellem Fortschritt befindlichen Gebiet ganz besonders fühlbaren Uneinheitlichkeit der Terminologie.

Die englischen Ausdrücke sind so gut wie ausschließlich englisch-amerikanischen Fachschriften entnommen worden; Übersetzungen aus der deutschen in die englische Sprache kommen nur in ganz vereinzelten Fällen vor. Zur Zusammenstellung des Stoffes habe ich den verschiedensten Arbeitsgebieten entnommene Bücher, Broschüren, Zeit- und Patentschriften englischen und amerikanischen Ursprungs von insgesamt gegen 9000 Druckseiten durchgearbeitet. In ausgedehntem Maße dienten deutsche Fachschriften zum Vergleich. In dankenswerter Weise haben in einzelnen Fällen Fachgenossen ihren Rat zur Verfügung gestellt.

Trotz aller aufgewendeten Sorgfalt kann ich nicht hoffen, daß nicht gewisse Lücken und einzelne Unrichtigkeiten in dem Werk enthalten sind. Für deren Mitteilung zwecks Verwertung bei einer späteren Auflage werde ich den Benutzern dankbar sein.

Berlin, im Januar 1925.

O. Sattelberg.

Vorwort zum deutsch-englischen Teil.

Der zweite Teil ist nicht lediglich eine Umstellung des englisch-deutschen Teils. Es wurde vielmehr auf die deutsche Ausdrucksweise viel weitergehend Rücksicht genommen, als dies nur durch eine Umkehrung möglich gewesen wäre. Dementsprechend sind auch recht zahlreiche Übersetzungen aus dem Deutschen in das Englische eingefügt worden. Der Stoff wurde um mehrere hundert Stichwörter vermehrt. — So darf ich hoffen, daß der zweite Teil eine ebenso freundliche Aufnahme finden wird, wie dies beim ersten Teil der Fall gewesen ist.

Berlin, im Januar 1926.

O. Sattelberg.

Abkürzungen.

A	Selbstanschlußwesen	
B	Bau von Leitungen	
F	Fernsprechwesen	
K	Fernkabel	
L	Leitungstheorie	
R	Funkwesen	
T	Telegraphie	
V	Verstärkertechnik	
am.	amerikanisch	
engl.	englisch	
ab:	abgekürzt	
cf.	vergleiche	
v.	siehe	

Abbreviations.

	Automatic Telephony.
	Construction of Outdoor Plant.
	Telephony.
	Long-distance Cables.
	Line Theory.
	Radio.
	Telegraphy.
	Valves, Amplifiers.
	American.
	English.
	abbreviated.
	confer.
	see.

Quellenverzeichnis.

Steinmetz, The Theory and Calculation of Alternating Current Phenomena, New York 1897.
H. W. Malcolm, The Theory of the Submarine Telegraph and Telephone Cable, London 1916.
J. G. Hill, Telephonic Transmission, London 1920.
J. H. Morecroft, Principles of Radio Communication, New York 1921.
T. E. Herbert, Telegraphy, London 1921.
— Telephony, London 1923.
H. H. Harrison, Printing Telegraph Systems, London 1923.
A. B. Crotch, The Elements of Automatic Telephony, London 1924.
F. Anson, The W. E. Co.'s Automatic Telephone System, London 1916.
H. G. White, Electric Bells, Alarms and Signalling Systems, London 1921.
H. Viard, Vocabulaire en cinq Langues, Paris 1920.
W. L. Weber, Handy Electrical Dictionary, London.
Marconi Yearbooks, London 1916—1924.
Henley's Workable Radio Receivers, New York 1924.
British Engineering Standards Association, London,
    Radio Communication, Dec. 1923.
    Telegraphs and Telephones, July 1924.
D. Murray, Press-the-Button Telegraphy, London.
The Post Office Electrical Engineers' Journal, London.
The Telegraph and Telephone Journal, London.
The Journal of the Institution of Electrical Engineers, London.
The Radio Review, London (1920—22).
The Electrician, London.
Electrical Communication (W. E. C.), New York.
The Telegraph and Telephone Age, New York.
The Journal of the American Institute of Electrical Engineers.
Broschüren von: Peters, Heising, Andres, Colpitts, Blackwell, Rhoads, Campbell, Jewett, Craft, Hartley, Hill, King, v. d. Bijl, Bureau of Standards (Washington).
Englische und amerikanische Patentschriften.
Muret-Sanders, Enzyklopädisches Wörterbuch der englischen und deutschen Sprache, Berlin.

A.

Abändern, to alter, to modify, to change.

Abänderung f, alteration, modification, change.

abbinden, to bind; to set (Mörtel, mortar); to sew, to lace (Kabel, cables).

abbrechen, to break down (eine Leitung, a line).

abbrühen, to boil out, B.

Abbrühen n, boiling-out (von Kabeln, of cables), B.

ABC-Telegraph m, ABC telegraph;

Geber m des –-–en, communicator.

Abdachung f, slope, roofing, dip;

Stangen- –, pole roof.

abdämpfen, to boil out, B.

Abdämpfen n, boiling-out, B.

abdecken, to cover.

Abdeckplatte f, cover, der Kabelbrunnen: manhole cover, box-lid.

Abdeckung f, capping, cover.

abdrosseln, to choke;

vollständig –, to choke out.

Abdruck m, print(ing), impression (einer Type, of a type).

unsauberer –, smudgy impression;

– im Fluge, flying print;

– durch Abwälzen, rolling print;

– auf Blättern (Streifen), page (tape) printing.

abdrucken, to print, imprint.

aberregen, to de-energize.

Aberregung f, de-energization.

Abfall m, Metall usw.: chip waste; Sinken: fall, drop, descent;

vollständiger –, decay;

steiler –, steep fall, precipitation;

Spannungs- –, drop of potential, potential fall.

abfallen, to fall (off), to drop, to descend, to droop; Relais: to release, to restore; vollständig: to decay; Kurve: to slope (down);

steil –, to precipitate.

Abfallen n, fall, drop; des Relais: release, releasing.

abfallend, falling off; decaying;

langsam –es Relais, slow to release relay, slowreleasing relay.

Abfallzeit f, releasing time.

Abfederung f, padding.

abfertigen, to dispatch.

Abfertiger, dispatcher.

Abfertigung f, dispatch;

– -zeit f, handling time.

abflachen, to flatten, to smooth (down).

Abflachung f, flattening.

Abflachungsdrossel f, smoothing choke;

– -kondensator m, smoothing condenser.

Sattelberg, Wörterbuch: Deutsch-Englisch.

abfließen, to leak off.
Abfrage-apparat *m*, service instrument set, operator's telephone, operator's phone set. service apparatus, answering equipment;
— **-beamtin** *f*, answering operator;
— **-betrieb** *m*, direct trunking *F*;
— **-einrichtung** *f*, = Abfrageapparat;
— **-garnitur** *f*, operator's set, speaking set, operator's telephone;
— **-klinke** *f*, answering jack, home jack;
— **-platz** *m* für Teilnehmer, home position, answering position;
— **-schnur** *f*, answering cord;
— **-stöpsel** *m*, answering plug.
abfragen, to answer, to enquire (the number), *F*.
Abfragen *n*, answering, enquiry, *F*.
Abfühlnadel *f*, pecker, selecting needle, *T*.
Abgabe *f*, Gebühr: charge, fee, rate; von Telegrammen: sending, transmission.
Abgang *m*, departure.
Abgangs-amt *n*, departure station;
— **-verkehr**, *m*, outgoing traffic.
abgeben, to deliver (Strom usw.); Telegramme: to transmit.
abgedeckte Stelle *f*, radio pocket, radio shadow, dead spot, *R*.
abgeflacht, flat(tened), oben: flat-topped.
abgegebene Leistung *f*, output.
abgeglichen, balanced.
abgegrenzt, defined, definite.
abgehen, to depart, to leave, to start.
abgehend, outgoing (Strom, Verkehr usw.);
— **es Gespräch** *n*, out call;
— **e Verbindung** *f*, outgoing trunk, outlet, *A*.
abgekürzte Morsezahlen *pl*, contracted morse figure signals, *T*.
abgelaufen, run down (Feder); expired (Patent usw.).
abgeleitete Einheit *f*, derived unit.
abgeschirmt, screened;
— **e Stelle** *f*, radio shadow, radio pocket, dead spot, *R*. [off.
abgeschaltet, disconnected, cut
abgeschlossen, closed, terminated (durch einen Widerstand, by a resistance); isoliert: sealed;
in sich —, self-contained.
abgeschrägt, bevelled, skewed;
— **e Polränder**, skewed pole tips.
abgesetzt, reduced, recessed.
abgestimmt, tuned, in tune (auf to); syntonic;
auf Resonanz —, tuned to resonance;
gleich (verschieden) —, tuned alike (differently tuned);
nicht —, untuned;
scharf (unscharf) —, sharply (flatly) tuned;
— **er Anruf** *m*, tuned ringing, harmonic ringing;
— **er Kreis** *m*, tuned circuit.
abgestuft, stepped, graduate.
abgewickelt, rolled out;
— **gezeichnet**, represented as rolled out;
— **e Linie** *f*, evolute.
abgezapft, tapped (-off).
Abgleich *m*, balance, balancing.
abgleichen, to balance, to equilibrate.
Abgleichen *n*, balancing.
Abgleichfehler *m*, unbalance, balance error.
Abgleichung *f*, balance;
schlechte —, unbalance, want of balance.

abgreifen, to tap.
Abgreifpunkt *m*, tap, tapping point.
Abgrenzung *f*, definition.
abgrenzen, to define, to mark.
Abgriff *m*, tap.
Abhang *m*, slope.
abhängen, to depend (von, on, upon); den Hörer: to lift off (the receiver).
abhängig, dependent;
voneinander —, interdependent;
— e Veränderliche *f*, dependent variable.
Abhängigkeit *f*, dependence;
gegenseitige —, interdependence (zwischen, between);
lineare —, linear dependence;
einen Wert in — von einem anderen darstellen, to plot a value against another value.
abheben, to lift off, to remove (den Hörer, the telephone).
Abheben *n* des Hörers, removal of the receiver.
abhelfen, to remedy.
abhorchen, abhören, to listen to.
Abhörende(r), listener(-in).
abisolieren, to bare, to skin, to strip (einen Draht, a wire).
abklingeln, to ring off.
abklingen, to fade out, to decay, to die down, to die away, to die out.
Abklingen *n*, decay, fading (-out).
abkühlen, to cool, sich — to cool (down).
Abkühlung *f*, cooling.
ablagern, to deposit (Schlamm, mud);
— (lassen), to season, to superseason (Faser, Holz, fibre, wood.
Ablagerung *f*, deposition; das Abgelagerte: deposit.

Ablauf *m*, expiration.
ablaufen, to run down, to return (Nummernscheibe, dial switch); Recht, Patent: to expire;
schnell —, to bolt (Uhrwerk);
— lassen, to release, to let go (die Nummernscheibe, the dial switch).
Ablaufen *n*, running-down, return(ing); expiration;
schnelles — eines Uhrwerks, bolting.
Ablaufzeit *f* des Nummernschalters, time of running down of the dial, *d*.
abläuten, to ring off.
Abläutezeichen *n*, ring-off signal.
ableiten, to derive (von, from), to deduce *M*; Strom: to leak off, to drain off.
Ableitung *f*, derivation (nach, with respect to), erste (zweite) —, first (second) derivative, *M*; Leitungskonstante: leakance, line shunt conductance, leak conductance;
Nebenschluß: leak, leakage;
dielektrische —, dielectric leakance;
mit — behaftet, leaky;
mit erhöhter — laden, to leakload;
— — — belastet, leak loaded;
— — —, Ladung *f*, leak loading.
Ableitungs-dämpfung *f*, leakance loss;
— -glied *n*, shunt element (eines Kettenleiters, of a network);
— -strom *m*, leak(age) current, leakance current, stray current;
— -verlust *m*, leakance loss, leakage loss;
— -widerstand, leak(age) resistance; der Röhre: resistance leak.

1*

ablenken, to deflect, to inflect, to deviate, Lichtstrahlen: to diffract.

Ablenkung *f,* deflection (im Galvanometer, on the galvanometer), deviation, inflection, Licht: diffraction;

größte —, amplitude of deflection;

Winkel= —, angular deflection;

Ablenkungs=amplitude *f,* amplitude of deflection;

— =elektroden, deflecting plates *pl* (eines Kathodenstrahlenojzillographen, of a cathode ray oscillograph);

— =kraft *f,* deflecting force;

— =messer *m,* deflectometer;

— =platten, deflecting plates pl.

Ablesefernrohr *n,* reading telescope.

ablesen, to read (off), to take a reading.

Ablesung *f,* reading, test reading; eine — nehmen, to take a reading.

(mit) direkte(r) —, direct reading;

Spiegel= —, mirror reading;

Zähler= —, meter reading.

abliefern, to deliver.

Ablieferung *f,* delivery.

ablösen, to loosen, eine Schicht: to peel, to strip (off); im Dienst: to relieve.

abloten, to plumb.

ablöten, to unsolder.

Abmaß *n,* tolerance.

abmessen, to measure; nach der Zeit: to time.

Abmessung *f,* measurement; Größe: dimension, size;

größte —, Gesamt= —, overall dimension(s *pl*).

Abnahme *f,* fall, diminution, decrease; von Waren: acceptance;

fortschreitende —, progressive diminution;

Spannungs= —, decrease or fall of potential;

— =beamter *m,* testing officer;

— =messung *f,* — =prüfung *f,* acceptance test, factory test.

abnehmbar, removable, detachable.

abnehmen, sinken: to fall (off), to decrease, to diminish; entfernen: to remove, to detach; auf Null —, to die away, to decay.

abnormal, abnormal.

abnutzen, (sich), to wear (out).

Abnützung *f,* wearing, wear and tear.

Abonnement *n,* subscription (auf, to).

Abonnementsgebühr *f,* subscription rate.

Abonnent *m,* subscriber.

abonnieren, to subscribe (auf, to).

abpfählen, to peg out, to stake out (eine Linie, a line), *B.*

abprallen, to rebound.

Abprallen *n,* rebounding, resiliency.

abrechnen, to settle accounts.

Abrechnung, *f,* account; accounting.

abreißen, to tear off; to break down; to break.

Abreißfeder *f,* antagonistic spring, opposing spring, reacting spring, retracting spring, retractile spring.

abrollen, Kabel: to pay out; (sich) — to uncoil.

abrunden, to round off.

Absatzweg *m,* channel, outlet.

absatzweise, intermittent, at intervals.

absaugen, to drain off.

abschaffen, to supersede.

Abschaffung f, supersession.
abschalten, to disconnect, to switch off, to cut off.
Abschalten n, **Abschaltung** f, disconnection, disconnecting, switching-off.
Abschalttaste f, cut-out key.
abschätzen, to estimate.
Abschätzung f, estimation.
abscheren, to shear.
abscheuern, to fray, to chafe.
Abscheuerung f, abrasion.
abschirmen, to shield (gegen, from); to screen (out).
Abschirmung f, screening, shielding.
abschließen, to close, to terminate; dicht —, to seal (off); eine Leitung durch ihren Wellenwiderstand —, to terminate a line in its own impedance.
Abschluß m, termination; sealing (-off);
— durch ein halbes Längs-(Quer-)glied, mid-series (midshunt) termination, L;
Leitungs- —, circuit termination;
— -impedanz f, end impedance, terminal impedance;
— -kabel n, terminal cable, B;
— -kunstschaltung f, terminal network;
— -muffe f (cable) pothead, cable head, cable distribution plug, B;
— -transformator m, — -übertrager m, terminal transformer
— -widerstand m, terminal resistance.
Abschmelz-dauer f, fusing time;
— -strom m, fusing current.
abschmelzen, to fuse, to melt, to blow.
Abschmelzen n einer Sicherung, blowing, fusing, melting of a fuse.

Abschmelz-sicherung f, fuse, safety fuse, safety cut-out;
— -streifen m, fuse strip;
— -stromstärke f, fusing current; blowing or striking point (einer Sicherung, of a fuse);
— -zeit f, time of fusing.
abschmirgeln, to polish, to sandpaper.
abschneiden, to cut.
Abschnitt m, section; stretch;
Untersuchungs- —, testing section (einer Leitung, of a line);
Verstärker- —, repeater section.
Abschnitts-.... sectional;
— -gestänge n, transposition pole B.
abschrägen, to skew, to scarf, to bevel; senken: to slope.
abschrauben, to screw off, to unscrew.
abschrecken, to quench (Stahl, steel).
abschreiben, to copy, to write up (Telegramme, messages).
Abschreiben n, writing-up, copying (von Streifen, of slips).
Abschreibung f, depreciation.
absenden, to forward, to dispatch.
Absender m, sender.
Absendung f, dispatch.
absetzen, niederschlagen: to deposit; mit einem Absatz versehen: to reduce, to recess.
absieben, to screen out.
absolut, absolute;
— e Einheit f, absolute unit.
absondern, to separate, to eliminate.
absorbieren, to absorb.
Absorption, f, absorption.
Absorptions-faktor m, absorption factor;
— -strom, absorption current;
— -(strom)kreis m, absorbing circuit.

Abspann=dachständer *m*, roof end standard, *B*;
— =**draht** *m*, span wire;
— =**isolator** *m*, terminal insulator;
— =**pfahl** *m*, span pole;
— =**seil** *n*, span rope;
—= **stange** *f*, stay pole, terminal pole.

abspannen, to strain back to, to stay, to dead-end, to terminate (einen Draht, a wire).

Abspannung *f*, dead-ending, termination; span.

Absperrhahn *m*, stop cock.

Abspringen *n* (des Blitzes), side-flashing (of lightning).

abspulen, to uncoil, to wind off.

Abstand *m*, distance, space, spacing, interval, clearance, separation; in Abständen von... km, on ... km spacing; in (un)gleichen Abständen, (un-)evenly spaced;
geringer —, proximity;
Kontakt= —, contact clearance;
Pol= —, pole clearance;
Verstärker= —, repeater spacing;

Abstands-leiste *f*, spacing strip;
— =**stück** *n*, spacing piece, distance piece;
— =**taste** *f*, blank key, spacing key, *T*;
— =**zeichen** *n*, spacing signal, *T*.

abstecken, to peg out, to stake out (eine Linie, a line), *B*.

Absteckstange *f*, stake.

absteigen, to descend.

abstellen, to remove (Fehler faults); Apparate: to stop.

Abstimm=einrichtung *f*, tuning device;
— =**fähigkeit** *f*, tuning property;
— =**kondensator** *m*, tuning condenser;
— =**mittel**, tuning means pl;

— =**schärfe** *f*, sharpness of resonance, sharpness of tuning;
— =**spule** *f*, tuner, tuning coil, tuning inductance, syntonising coil;
— =**vorrichtung** *f*, (Vielfach=), (multiple) tuner.

abstimmen, to syntonise, to tune (auf, to).

Abstimmen *n*, **Abstimmung** *f*, selection (auf, to), tuning, syntonisation, accordance.

Abstimmung, Vorhandensein *n* der, syntony;
falsche —, mistuning;
feine —, fine tuning;
genaue —, accurate tuning;
grobe —, rough tuning;
scharfe —, sharp tuning;
schlechte —, mistuning;
unscharfe —, flat or imperfect tuning;
Gitterkreis= —, grid tuning;
Ton= —, note tuning, tone tuning.

abstoßen, to repel.

abstoßend, repulsive.

Abstoßung *f*, repulsion.

Abstoßungskraft *f*, force of repulsion, repulsive force.

abstufen, to graduate; mit Schattieren: to shade.

Abstufung *f*, gradation.

abstützen, to support.

Abszisse *f*, abscissa, *pl*: abscissae.

Abteilung *f*, section, division; eines Gestells: bay;
Betriebs= — *f*, service section, *F*, *T*.

abtelegraphieren, to transmit.

abtrennen, to cut off; to dissociate; to separate, to sever.

Abtrennrelais *n*, cut-off relay.

Abtrennung *f*, separation; cutting-off; dissociation.

abtreten, to assign (over).

Abtretender, assignor.
Abtretung f, assignment (an, to); assignation.
Abtrieb, Zuschlag m für, slack (Seekabel, submarine cables).
Abtropf=kante f, drip(ping) edge;
— =ring m, drip ring.
A-B-Verkehr m, junction service *(engl.)*, trunk service *(am.)*.
Abwärts=bewegung f, descent, downward motion;
— =transformator m, step-down transformer, reducing transformer;
— =transformierung f, step(-ping=) down.
abwärts transformieren, to step down.
abwechseln, to alternate.
abwechselnd, alternate, alternating.
abweichen, to depart, to differ; to decline, to deflect, to deviate, to incline; nach oben oder unten — um, to vary up or down by...
abweichend, divergent.
Abweichung f, deviation, departure; declination, deflection, margin;
zulässige —, tolerance, allowance;
— von der wahren Richtung, deviation from the true bearing;
— des Nullpunkts, zero error.
abweisen, to reject.
Abweisung f, rejection.
Abwerfeinrichtung f, arrangement to release switches F.
abwerfen, to release.
Abwerfen n, release (des Ankers, of the armature).
Abwerfsystem n, operator aid system, F.

abwickeln, to wind off, to unwind, to unroll; M: to develop; Verkehr: to handle.
Abwicklung f, unwinding; development M; handling;
— s=zeit f, handling time.
abziehen, to subtract M.
Abzugs=kanal m, (main) sewer;
— =magnet m, trigger magnet.
— =öffnung f (der Trockenelemente usw.) vent.
Abzweig m, branch, leak, shunt, spur, tap;
Abzweig=...., branch(ing)....;
Spulen= — m, coil tap;
— =dose f, connector box;
— =kasten m, coupling box, B;
— = —, Kabel=, cable joint box, B;
— =klinke f, branching jack;
— =leitung f, branch circuit, kurze —: spur line;
— =muffe f, parallel joint(ing) sleeve, cable distributing plug, B;
— =punkt m, distributing point B;
— =spleißstelle f, Y-splice, B;
— =spule f, tapped coil; bridging coil;
— =übertrager m, Vierer=, combining transformer, F;
— =widerstand m, leak resistance, leak coil.
abzweigen, to branch (off), to shunt off, to tee (von, to).
Abzweigung f, branching (-off), tap.
Achat m, agate.
— =hütchen n, agate cup, jewel cup.
Achse f, spindle, shaft(ing), axle (tree), arbor; M: axis;
Antriebs= —, driving shaft;
Null= —, zero axis;
Typenrad= —, type wheel shaft.

Achsenkreuz *n*, coordinate system.
Achs=lager *n*, bearing;
— =**schenkel** *m*, journal.
Achteck *n*, octagon.
achteckig, octagonal.
Achter *m*, **Achterbündel** *n*, eight wire core; quadruple twin.
achterverseilt, quadruple pair....
achtfach, octuple.
Achtfachtelegraph *m*, octuplex telegraph.
abdieren, to add.
Addition *f*, addition.
Additions(rechen)maschine *f*, adding machine.
Ader *f*, wire, core;
  a- —, tip wire, T-wire, *F*;
  b- —, ring wire, R-wire, *F*;
  c- —, c-wire;
  an Erde liegende —, positive wire *A*;
  an Spannung — —, negative wire *A*;
  Doppel= —, pair, two wire core.
Adern=kreuzung *f*, crossing of wires;
  — =**kreuzungsverfahren** *n* **für Kapazitätsausgleich** test-splicing method, *K*;
  — =**paar** *n*, pair;
  — = —, **Trensen=**, (zum Ausfüllen der Lücken der Kabelseele) worming pair;
  — = —, **Zähl=**, key pair.
Admittanz *f*, admittance.
Adressat *m*, addressee.
Adreßbuch *n*, directory.
...=**abrig**, ... wire, ... way;
  doppel= —, pair..., two-wire
  drei= —, threewire..., triplet...;
  ein= —, single core..., single wire, single way...;
  zwei= —, pair..., two-wire....
Aeroplan *m*, aeroplane, airplane.

Affinität *f*, affinity.
Agens *n*, agent.
Aggregat *n*, aggregate.
A h = **Ampèrestunde** *f*, ampere hour.
Ahorn(holz *n*) *m*, maple.
Akazie(nholz *n*) *f*, acacia.
Akkumulator *m*, accumulator, storage cell;
— =**säure** *f*, accumulator acid.
Aktenzeichen *n*, file number (eines Patentes: of a patent).
aktinisch, actinic.
Aktionär *m*, shareholder, stockholder.
aktive Masse *f*, active material, der Elemente: active paste.
Akustik *f*, acoustics *pl*.
akustisch, acoustic(al).
Alaun *m*, alum;
Algebra *f*, algebra.
algebraisch, algebraic(al).
Alkali *n*, alkali.
Alkali= ..., alkaline.
— =**metall** *n*, **Erd=**, alkaline earth metal;
— = — =**gruppe**, **Oxyde** *pl* der, oxides of the alkaline earth group.
Alkohol *m*, alcohol.
alkoholisch, alcoholic(al).
allmählich, gradual.
Alphabet *n*, alphabet, code;
  Dreier= —, ternary or three-unit code, *T*;
  Fünfer= —, **Fünfströme=**, five-unit code, *T*;
  Morse= —, Morse code;
  — = —, **amerikanisches**, American Morse code;
  — = —, **internationales**, Continental Morse code;
  Telegraphen= —, **Telegraphier= —**, telegraph code;
  — = — **mit (un)gleich langen Zeichen**, (un)equal letter code.

Alt, *m* alto, counter tenor;
tiefer —, contralto.
altern, to mature, to age, to (super)season.
Altern *n*, ageing, (super)seasoning.
Alternator *m*, alternator.
alternieren, to alternate.
Alterung *f*, ageing, (super)seasoning.
Alterungseinfluß *m*, ageing effect.
Amalgam *n*, amalgam;
Zink= —, zinc amalgam.
amalgamieren, to amalgamate.
Amalgamierung *f*, amalgamation.
Amateur *m*, amateur;
— =lizenz *f*, amateur licence.
Amboß *m*, anvil.
— =kontakt *m*, anvil contact, buffer contact.
Ambroin *n*, ambroin.
Ammoniak *n*, ammonia (NH_3);
— =gas *n*, ammonia gas (NH_3).
Ammonium *n*, ammonium (NH_4);
Chlor= —, ammonium chloride (NH_4Cl);
— =hydroxyd *n*, ammonium hydrate (NH_5O).
amorph, amorphous.
amortisieren, to amortize.
Amortisation *f*, Amortisierung *f*, amortization.
Ampere *n*, ampere;
— sche Schwimmerregel *f*, Ampere's rule;
— =meter *n*, am(pere)meter, *v*. Strommesser;
— =sekunde *f*, ampere-second;
— =stunde *f*, ampere-hour;
— =stunden-wirkungsgrad *m*, ampere-hour efficiency;
— =—=zahl *f*, ampere-hour capacity;
— =windung *f*, ampere-turn;
— =zahl *f*, amperage.

Amplitude *f*, amplitude, crest;
Maximal= —, maximum amplitude;
Minimal= —, minimum amplitude;
Schwebungs= —, surging amplitude; amplitude of beats;
Schwing(ungs)= —, vibrational amplitude, amplitude of oscillation;
Amplituden=abnahme *f*, reduction in amplitude;
— =entzerrung *f*, correction of amplitudes;
— =verhältnis *n*, ratio of amplitudes;
—=— der Zeichen zu den Luftstörungen, signal-to-static ratio, R;
— =verzerrung *f*, amplitude distortion.
Amt *n*, Tätigkeit: function; office, station; Fernsprech=amt: exchange, central office;
das — anrufen, to ring the exchange, F;
eigenes —, home station;
fernes —, distant station;
unüberwachtes —, unattended office, A;
Annahme= —, accepting or collecting office T;
End= —, terminal station;
Fern= —, trunk exchange, F;
Hand= —, manual exchange, F;
Hilfs= —, satellite exchange, sub-office, A;
Knoten= —, main centre T;
Meß= —, testing office;
Nahverkehrs= —, toll exchange;
Neben= —, minor exchange;
O. B.= —, l. b. exchange, magneto exchange;
Orts= —, local exchange, local central office;
Sammel= —, smaller centre, T;

Amt

Schnellverkehrs- —, no-delay traffic exchange *F;*
Tandem= —, tandem office, *F, A;*
Teil- —, satellite exchange, *A;*
Telegraphen= —, telegraph office, telegraph station;
Übertragungs- —, repeater station, repeating station, *T;*
Unter= —, sub-office;
Untersuchungs- —, testing office;
Verstärker= —, repeater station, amplifying relay station, *V;*
Z. B.= —, c. b. exchange;
Zwischen- —, intermediate station, waystation.
Amts=anlage *f*, internal plant;
— =**anruf** *m*, exchange call, *F;*
— =**batterie** *f*, exchange battery;
— =**bezeichnung** *f*, office code, office prefix, *A;*
— =**bezeichnungssystem** *n*, office code system, *A;*
— =**bezirk** *m*, exchange area;
— =**einheit** *f* mit 10000 Anschlüssen, unit of 10000 lines;
— =**einrichtung** *f*, office equipment, exchange apparatus;
— =**freizeichen** *n*, dialling tone, *A;*
— =**kabel** *n*, office cable;
— =**klinke** *f*, exchange jack (des Nebenstellenschranks, of p. b. x.);
— =**leitung** *f*, exchange line, junction from p. b. x. to exchange, trunk, *F;*
— =**namen**, *pl*, office code, *A;*
— = — =**speicher** *m*, office code register, *A;*
— = — =**wähler** *m*, code switch, *A;*
— =**nebenstelle** *f*, exchange extension set, *F;*
— =**nummer** *f*, exchange number (des Teilnehmers, of the subscriber);
— =**pflege** *f*, exchange maintenance work;
— =**schlüssel** *m*, office code, *A;*
— =**störungspersonal** *n*, exchange fault staff;
— =**seite** *f* (des Hauptverteilers), exchange side, vertical side (of main distributing frame);
— =**summerzeichen** *n*. dialling tone *A;*
— =**taste** *f*, office key, *F;*
— =**telegramm** *n*, service message;
— =**verbindung** *f*, exchange call, *F;*
— =**verdrahtung** *f*, internal wiring;
— =**verkabelung** *f*, office wiring;
— =**zeichen** *n*, dialling tone, *A*.
Analyse *f*, **Analysis** *f*, analysis, *M*.
analysieren, to analyze.
analytisch, analytic(al).
anbringen, to mount, to fit, (an, auf, on), to attach (to).
Anbringung *f*, mounting, attachment.
ändern, (sich), to alter, to vary, to change.
Änderung *f*, variation, alteration, change;
— **mit der Frequenz**, variation with frequency;
prozentuale —, percentage change.
Andrang *m*, rush.
anfänglich, initial.
Anfangs=geschwindigkeit *f*, initial velocity;
— =**permeabilität** *f*, initial permeability, permeability at low magnetizing forces;
— =**punkt** *m*, origin;
— =**spannung** *f*, initial voltage;
— =**zustand** *m*, initial state, initial conditions *pl*.
anfeuchten, to moisten.

Anforderung f, requirement (an, to).

Anfrage f, enquiry.

anfressen, to corrode, to decay.

Anfressung f, corrosion.

Angabe f, datum (pl: data); indication.

Angel f, pivot, hinge.

angerufener Teilnehmer m, called or wanted or required subscriber, called party.

angeschärft, scarfed, thinned.

angeschuhte Stange f, shoed pole, B.

Angestellte(r m) f, employe(e).

angrenzen, to join (an, to).

anhaken, to hook over.

anhalten, to stop, to arrest; wieder —, to rearrest.

Anhalten n, stoppage, stopping.

anhängen, to hang up; den Fernhörer: to restore.

anhäufen, sich, to (ac)cumulate.

Anhäufung f, (ac)cumulation; Telegramme: congestion, accumulation of traffic.

Anhydrid n, anhydrid.

Anilinfarbe f, aniline dye; blaue —, aniline blue.

Anion n, anion.

Anker m, B: stay, guy; des Elektromagnets, der Dynamo: armature; des Relais: tongue; Schiffs- —: anchor;
feststehender (umlaufender) —, fixed (rotating) armature;
Doppel- —, V-stay, double stay, B;
Doppel-T- —, shuttle armature, H-armature;
Draht- —, guy wire; stay, stranded wire stay;
I- —, shuttle armature, H-armature;
Käfig- —, squirrel cage rotor;
Linien- —, longitudinal stay B;
Magnet- —, armature; keeper;
Ring- —, ring armature;
Scheiben- —, disc armature;
Schleifring- —, slip-ring rotor;
Schwächungs- —, adjusting slide, T;
Seiten- —, side guy, lateral stay, B;
Trommel- —, drum armature;
V- —, V-stay, B.
Weicheisen- —, soft iron armature;
Kleben n des —s, sticking or freezing of the armature;
— -bewegung f, armature travel;
— -bohrung f, armature bore;
— -bolzen m, tie bolt, B;
— -draht m, stay wire, B;
— -gesperre n, anchor escapement;
— -haken m, guy wire hook, stay hook, B;
— -halteschraube f, armature holding screw;
— -hub m, armature stroke; play of tongue;
— -klotz m, stay block, B;
— -nut f, armature slot;
— -nutenwellen pl. des Gleichstroms, slot ripple;
— -platte f, anchor plate, B;
— -pfahl m, stay block; stay rod, B;
— -rückwirkung f, armature reaction;
— -schutzpfahl m, stay guard B;
— -spannschraube f, stay tightener, swivel, B;
— -spiel n, play of tongue, armature stroke;
— -spule f, armature coil;
— -stab m, armature bar;
— -stellung f, position of armature;
— -strom m, armature current;
— -umlegefeder f **(des Weckers)** biasing spring (of bell);

Anker
— **-umschlag** *m*, armature travel, transit of the armature;
— **-weg** *m*, armature travel;
— **-windung** *f*, armature coil;
— **mit Gegengewicht**, gravity-controlled armature;
anklammern (an), to clamp (to), to cleat (on).
ankohlen, to char, *B*.
Ankohlen *n* (**der Stangen**), charring (of poles).
ankommend, Strom, Leitung: incoming. Welle: oncoming.
— **es Gespräch** *n*, in call;
— **e Verbindungsleitung** *f*, in junction.
Ankunftsverkehr *m*, incoming traffic.
ankünd(ig)en, to announce, *R*.
Anlage *f*, plant, establishment; Kapital: investment;
fliegende —, temporary plant;
vorläufige —, provisional plant;
zeitweilige —, temporary plant;
Außen- —, external plant;
Innen- —, internal plant;
Not- —, provisional plant; emergency plant;
— **-kosten** *pl*, first cost, purchasing cost.
anlassen, Stahl: to anneal; Motoren: to start.
Anlassen *n*, annealing; starting.
Anlasser *m*, starter;
Flüssigkeits- —, liquid starter;
Selbst- —, automatic starter.
Anlaß-magnet *m*, start(ing) magnet;
— **-schalter** *m*, starting switch;
— **-stromstoß** *m*, starting impulse.
Anlauf-drehmoment *n*, starting torque;
— **-strecke** *f* eines Pupinkabels, end section, first section.

anlaufen, to start;
unter (ohne) Last —, to start under (without) load.
Anlaufen *n*, start(ing).
anlegen, to establish; to lay out (Lager, stores); **an die Leitung usw.**: to put (to), to switch on (to), to apply (to), to join (to), to throw (a line) on to.
Anmelde-abteilung *f*, record section;
— **-beamtin** *f*, record (table) operator, *F*;
— **-leitung** *f*, recording trunk, *F*;
— **-platz** *m*, record position, *F*;
— **-spitzenplatz** *m*, record transfer position, *F*;
— **- —, Beamtin** *f* **am**, record transfer operator, *F*;
— **-stelle** *f*, recording section, *F*;
— **-tisch** *m*, record table, *F*;
— **-zeit** *f*, booking time, *F*.
anmelden, to book, to file, to place, (ein Gespräch a toll call); ein Patent: to apply for a patent; **Patente angemeldet**: patents pending.
Anmelder *m*, applicant.
Anmeldung *f*, application; booking, recording;
Gesprächs- —, toll recording, booking of a call.
Annäherung *f*, approximation, approach;
in erster —, to a first approximation.
Annahme *f*, acceptance; assumption, *M*; von Telegrammen: acceptance, collection;
— **-amt** *n*, accepting office, *T*;
— **-schalter** *m*, counter, *T*.
— **-stelle** *f*, collecting office, *T*;
— **-verweigerung** *f*, refusal, *T*.
annehmen, to accept; to assume; to collect, to accept, (Telegramme messages).

annieten, to rivet.
anodal, anodal.
Anode *f*, anode, positive electrode; von Röhren: plate, output electrode, wing, *V*.
Anoden-..., anodal, anode;
— **-batterie** *f*, anode or plate battery, B-battery;
— **-kreis** *m*, anode- or plate- (to-) filament circuit, discharge circuit, *V*;
— **-licht** *n*, anodal light;
— **-ruhestrom** *m*, feed current, *V*;
— **-schutznetz** *n*, anode-screening grid, *V*;
— **-spannung** *f*, plate voltage, discharge voltage, anode potential, *V*;
— **-speisung** *f*, plate supply, *V*;
— **-strom** *m*, space current, plate current, anode current, discharge current, *V*;
—*— **-änderung** *f*, plate current variation, *V*;
—*— **-kreis** *m*, plate circuit, *V*.
anodenlose Röhre, plateless valve.
anodisch, anodic.
anordnen, to arrange; to group, to design.
Anordnung *f*, arrangement; design; grouping; allocation.
anpassen, to proportion, to adapt; to fit, to accommodate.
Anpassung *f*, proportioning, adaptation, accommodation.
anpassungsfähig, flexible, adaptable.
Anpassungsfähigkeit *f*, adaptability, flexibility.
Anruf *m*, Gespräch: call; Rufen: ring(ing), calling;
abgestimmter —, harmonic ringing;
dringender —, express call;
selbsttätiger —, ohne Rufschlüssel: keyless ringing, mit Maschinenstrom: machine ringing;
selektiver —, **wahlweiser** —, selective ringing;
Verteilung *f* **der —e**, distribution of calls *F*;
— **mit Maschinenstrom**, machine ringing, power ringing;
— **zwischen Teilnehmern einer Gesellschaftsleitung**, reverting call;
Amts- —, exchange call;
Batterie- —, battery ringing;
Fehl- —, lost call;
Induktor- —, (hand) generator ringing;
Wahl- —, selective ringing; selector calling *T*;
—*— **mit abgestimmten Einrichtungen**, harmonic selective ringing, tuned ringing.
— **-anzeiger** *m*, call indicator;
— **-einrichtung** *f* für Übertragungen, calling device, silencer, *T*;
— **-klappe** *f*, calling indicator, *F*;
— **-lampe** *f*, calling lamp, line lamp, *F*; calling-in lamp, *T*;
— **-suchen** *n*, finding action, *A*;
— **-sucher** *m*, (line) finder, finder switch, *A*;
—*—, **erster**, subscriber's line finder, *A*;
—*—, **zweiter**, trunk line finder, *A*;
—*—, **Relais-**, relay line finder, *A*;
—*—* **gestell** *n*, finder rack;
— **-wecker** *m*, call bell;
— **-zeichen** *n*, line signal, indicator, beim Vielfachschrank: home indicator.
anrufen, to ring (up), to call.
Anrufen *n*, ringing, calling.
anrufender Teilnehmer *m*, calling subscriber, calling party, caller.

Anrufer *m*, calling device;
Einzel= —, telegraph selector, *T*;
— **=magnet** *m*, silencer magnet, *T*;
— **=schränkchen**, silencer cabinet *T*.
ansagen, to announce.
Ansager *m*, announcer (des Rundfunksenders, of broadcast transmitter).
ansammeln, to aggregate: (**sich**) —, to accumulate.
Ansammlung *f*, aggregate; accumulation.
Ansatz *m*, extension, prolongation; lug, tail.
ansäuern, to acidulate.
Anschaffungskosten *pl*, prime cost, first cost.
Anschalte=klinke *f*, service jack (für das Abfragegerät, for the operator's set), operator's jack, *F*.
anschalten, to switch on .... (an, to); mit Stöpseln —, to plug up (an, to); to wire up.
anschärfen, to scarf, to thin.
Anschlag *m*, Widerlager: stop, detent, latch; einer Taste, touch, depression of a key; der Kosten: estimate; des Weckers: striking, stroke; des Klopfers: clicking;
rückwärtiger —, backstop;
— **=schiene** *f*, striker bar;
— **=fläche** *f*, banking face;
— **=stift** *m*, stop pin.
anschlagen, to butt (gegen, against); to touch, to depress; to rate, to estimate; to strike; to click.
anschließen (an), to join (to), to link (with).
Anschluß *m*, an das Netz: supply; wiring; communication;
Fernsprech= —, telephone station, telephone set;

Haupt= —, main telephone station;
Neben= —, extension telephone station;
Teilnehmer= —, subscriber's station, subscriber's set: am.: substation, subset:
— **=bolzen** *m*, connecting bolt;
— **=buchse** *f* (connector) socket;
— **=dose** *f*, wall socket;
— **= —, unverwechselbare**, non-interchangeable wall socket;
— **=kasten** *m*, terminal box;
— **=klemme** *f*, binding post;
— **=schraube** *f*, connecting screw.
anschrauben, to screw (on) to, to bolt (on) to, to fasten with screws.
anschuhen, to shoe (eine Stange, a pole), *B*.
anschwellen, to bulge, to rise.
Anschwellen *n*, bulge, hump (einer Kurve, of a curve).
ansetzen, to set up (Elemente, cells).
Ansicht *f*, view, elevation;
allgemeine —, general view;
perspektivische —, perspective view;
vergrößerte —, enlarged view;
Gesamt= —, general view;
Rück= —, back view;
Seiten= —, side elevation;
Vorder= —, front view, front elevation;
— **in voller (natürlicher) Größe**, full-size view;
— **von oben**, top plan view.
anspannen, to tighten, to bend.
anspitzen, to point.
anspleißen, to joint (an, to).
ansprechen, Relais, Funkstrecke: to operate (auf, on); Magnet: to respond, to be responsive (auf, to), to pull up; Sicherung: to blow, to strike, to fuse, to melt.

Ansprechen *n*, operation, response; Sicherung: blowing, fusing, melting, striking; lineares —, linear response.

ansprechend, responsive (auf, to), operating (auf, on); langsam (schnell) —, slow (quick) operating.

Ansprech=strom *m*, (Mindest=), (minimum) operating current;

— =stromstärke *f*, Relais usw.: figure of merit; Sicherung: striking point (von .... m A, of *n* m. a.); auf eine —= von *n* m A einstellen to margin to pull up at *n* m a;

— =zeit *f*, operating time; Relais: transit time (of armature).

Anspruch *m*, claim, title (auf, to); in — nehmen, to (lay) claim to.

Anstalt *f*, office, station.

Ansteckdose *f*, wall socket.

ansteigen, to ascend, to rise.

Anstieg, *m*, rise; slope.

anstoßen, to butt (gegen, an, against); to collide: einen Stromkreis: to impulse.

anstoßend, impulsive; benachbart: adjacent.

anstreichen, to paint.

Anstrich *m*, paint, coating; feuersicherer —, fire-resisting paint.

ansuchen, to apply (bei, to; um, for).

Anteil *m*, share; component.

anteilig, component.

Antenne *f*, aerial, antenna, *pl*: antennas, antennae;
dachförmige —, roof-shaped aerial;
eingegrabene —, buried aerial;
freihängende —, trailing (wire) aerial;
geknickte —, bent aerial;
künstliche —, artificial or mute or phantom antenna;
mehrdrähtige —, multiple-wire aerial;
mehrfach abgestimmte —, multiple tuned aerial;
(stark) richtfähige —, (highly) directive aerial;
ungerichtete —, equi-radial aerial, non-directive aerial;
zweidrähtige —, two-wire aerial;
Ausgleichs= —, balancing aerial;
Beverage= —, Beverage aerial;
Dach= —, overhouse aerial;
Doppelkegel= —, double-cone aerial;
Empfangs= —, receiving aerial;
Erd= —, ground antenna, earth antenna;
Ersatz= —, phantom aerial;
Fächer= —, fan (-shaped) aerial;
Flossen= —, skid-fin aerial;
Flugzeug= —, aeroplane aerial;
Harfen= —, harp aerial;
Hoch= —, high aerial;
Käfig= —, cage aerial;
Kegel= —, **Konus=** —, cone aerial;
Kreuzrahmen= —, cross-coil aerial;
L= —, inverted L-aerial;
— = —, wagerechter Teil *m* der, top spreader;
Mehrfach= — mit abgestimmten Zweigen, multiple tuned aerial;
Peil= —, direction finder aerial;
Rahmen= —, frame aerial, coil (aerial);
— = — (mit einer Windung), loop (aerial);
— = —n, zwei gekreuzte, cross-coil aerial;

Antenne
Raumstrahl- —, radiator;
Richt- —, directional antenna. directive aerial;
Schirm- —, umbrella aerial;
Schleifen- —, loop (aerial);
Sende- —, transmitting aerial;
Spiral- —, flat spiral coil, flat coil aerial;
Spulen- —, coil aerial;
T- —, T-antenna;
— = —, **verlängerte**, extended T-aerial;
— = —, **wagerechter Teil** m **der**, top spreaders pl;
Trichter- —, funnel-shaped aerial;
Unterwasser =—, underwater antenna;
Vertikal- —, vertical wire aerial;
Vielfach- —, multiple aerial;
Zimmer- —, indoor aerial;
Zwillings- —, pair of aerials, twin aerial;

Antennen-aufhängung f, aerial suspension;
— **-effekt** m, aerial effect;
— **-einführung** f, aerial lead-in;
— **-ersatzstromkreis** m, dummy aerial, phantom aerial, mute aerial;
— **-gebilde** n, aerial structure, aerial network;
— **-gerüst** n, aerial support;
— **-verkürzungskondensator** m, aerial series condenser;
— **-kreis** n, aerial circuit;
— **-leistung** f, aerial power;
— = —, **zugeführte**, antenna input;
— **-paar** n, pair of aerials, twin aerial;
— **-selbstinduktion** f, aerial inductance;
— **-strom** m, aerial current;
— **-umschalter** m, aerial change-over switch;

— **-verkürzungskondensator** m, aerial series condenser, short-wave condenser, shortening condenser;
— **-verlängerungsspule** f, aerial loading inductance.

Antikathode f, anticathode.
Antikohärer m, anticoherer.
Antimon n, antimony (Sb);
— **-blei** n, antimonial lead, antimonious lead.

antimonhaltig, antimonious.
antreiben, to drive, to rotate, to run, to impel.
antreibend, impulsive.
antreten, den Dienst, to assume duty.

Antrieb m, drive, mover; rotation, running;
direkter —, direct drive;
elektrischer —, electric drive;
Federkraft- —, clockwork drive, clockwork train;
Fremd- —, machine drive, separate drive;
— = —, **mit**, separately driven;
Riemen- —, belt drive;
Zahnrad- —, (thoothed wheel) gear drive;

Antriebs-achse f, driving shaft;
— **-gewicht** n, driving weight;
— **-kraft** f, motive force;
— **-magnet** m, driving magnet, drive magnet;
— **-motor** m, mover, driving motor;
— **-welle** f, driver shaft.

antworten, to answer; **Teilnehmer antwortet nicht**, there is no reply, F.

anwachsen, to increase, to grow, to rise.

Anwachsen n, increase, growth, rise;
— **mit dem Quadrat von . . .**, increase in proportion with the square of . . . .

anweisen, to instruct, to direct; to appopriate.
Anweisender *m*, assignor.
Anweisung *f*, instruction, direction, specification; eines Platzes: allocation; assignation;
Dienst- —, service instruction.
anwendbar sein, to apply (auf, to).
anwenden, to apply, to employ, to use;
sparsam —, to economize.
Anwendung *f*, application (auf, to), employment;
Anwendungs-gebiet *n*, field of application.
Anzahl *f*, number;
zulässige —, allowance.
anzapfen, eine Batterie, Spule: to tap.
Anzapf-spule *f*, tapped coil;
— -stelle *f*, tapping (point), tap;
— -—, mittlere, centre tapping point;
— -transformator *m*, split transformer.
Anzapfung *f*, tap, tapping.
Spulen- —, coil tap, inductor tap.
Anzeichen *n*, sign; warning.
Anzeige *f*, indication; Bekanntgabe: notification; warning;
— -vorrichtung *f*, indicating device, indicator.
anzeigen, to indicate, to signal, to detect.
Anzeiger *m*, indicator; index; detector, responder;
Anruf- —, call indicator;
Erdschluß- —, leakage indicator;
Lautstärken- —, volume indicator;
Nullstrom- —, zero current indicator;
Schwingungs- —, oscillation detector;

Strom- —, detector, galvanoscope, current indicator;
Wellen- —, oscillation detector, wave detector.
Anzeigung *f*, detection, indication;
hörbare (sichtbare) —, audible (visual) detection.
anziehen, to attract; Relais: to pull up, to be pulled up; eine Schraube: to tighten, to screw down (a nut).
anziehend, (at)tractive.
Anziehung *f*, attraction;
Anziehungs-kraft *f*, force of attraction, attractive power.
Anzugsstellung *f*, attracted position (eines Relais, of a relay).
anzünden, to ignite, to light.
aperiodisch, Entladung: impulsive, non-oscillatory, unidirectional; Stromkreis: aperiodic(al), non-resonant;
Meßinstrument: dead-beat, aperiodic.
Aperiodizität *f*, aperiodicity.
A-Platz *m*, A-position, *F*.
Apparat *m*, instrument, apparatus;
— -aufbau *m*, assemblage (of apparatus);
— -aufsicht *f*, operator-in-charge, dirigeur, *T*;
— -beamter *m*, operator;
— -karren *m*, instrument cart, *R*;
— -klinke *f*, instrument jack;
— -körper *m*, body;
— -raum *m*, — -saal *m*, instrument galle y, instrument room; switch room *F*; auto room *A*;
— -satz *m*, set, assembly;
— -teil *m* (*n*), item, part;
— -tisch *m*, instrument table;
— -—, vierteiliger, quartette operating table, *T*;

Apparat
— **-zuleitungen** *pl*, instrument leads *pl*.
Apparatur *f*, apparatus, instrumentality.
Äquator *m*, equator.
Äquatorial- ...., equatorial.
äquidistant, equidistant.
äquipotentiell, equipotential.
Äquipotential- ...., equipotential;
— **-fläche** *f*, equipotential surface.
äquivalent, equivalent.
Äquivalent *n*, equivalent;
Widerstands- —, equivalent resistance;
— **-gewicht** *n*, equivalent weight.
Aräometer *n*, areometer, densimeter, hydrometer;
Skalen- —, gratuated hydrometer;
Arbeit *f*, labour; physikal.: work;
elektrische —, electric work;
laufende —en *pl*, routine work;
Instandhaltungs- —en, *pl*, maintenance work;
Instandsetzungs- — *pl*, repair work;
Nutz- —, useful work.
arbeiten, to work, to labour (an, at); Apparat: to operate, to perform, to function, to run.
Arbeiten *n*, performance, functioning; operation, working;
— **in beiden Richtungen**, two way working; Duplex: full duplex operation, *T*;
— — **einer Richtung, einseitiges,** one-way working;
einseitiges — in **Gegensprechschaltung**, half-duplex working *T*.
arbeitend, operative.
Arbeiter *m*, workman, labourer, hand;

Arbeits-einheit *f*, unit of work;
— **-ersparnis** *f*, labour saving;
— **-fläche** *f* **der Zähne**, working faces *pl* of teeth;
— **-geschwindigkeit** *f*, speed of operation, rapidity of action;
— **-gleichung** *f*, energy equation, power equation;
— **-kenngrößen** *pl*, performance characteristics *pl* (eines Relais, of a relay);
— **kontakt** *m*, operating contact, make contact; off-normal contat; marking contact, *T*;
— **-lage** *f*, operative position, operated position;
— **-** — **auf der Röhrenkennlinie** working point of the valve characteristic;
— **-modell** *n*, working model;
— **-platz** *m*, operator's position;
— **-schema** *n*, schedule of operation;
— **-schiene** *f*, marking stop, *T*;
— **-stellung** *f*, operated position, off-normal position; in der
— — — befindlich, off-normal:
— **-strombetrieb** *m*, open circuit operation;
— **-stromschaltung** *f*, open circuit connection;
— **-stromsystem** *n*, open circuit system;
— **-teile** *pl*, working parts *pl*;
— **-verlust** *m*, lost work;
— **-weg** *m*, path of work, *T*;
— **-weise** *f*, method of operation.
arbeitsparend, labour saving.
Argon *n*, argon (Ar).
Argument *n*, argument *M*.
Arithmetik *f*, arithmetics *pl*.
arithmetisch, arithmetic(al);
— **es Mittel**, *n*, arithmetic mean.
Arldsche Kupferröhre *f*, copper jointing sleeve.

Arm *m*, arm; mit Armen ver=
sehen, armed:
a/b= —, minus plus wiper,
line wiper, *A*;
Bürsten= —, brush arm, brush
gear, *T*;
c= —, private wiper, *A*;
Prüf= —, private wiper, *A*;
Wähler =—, wiper, *A*.
armieren, Kabel: to sheath, to
armour; Beton: to reinforce.
armiert, armoured, sheathed;
reinforced;
stahlband= —, steel tape armoured;
— er Beton *m*, reinforced concrete.
Armierung *f*, sheath(ing), ar-
mour(ing); reinforcement;
Eisen= —, iron sheathing;
geschlossene —, closed or
locked armour;
leichte —, light armour;
offene —, open armour;
schwere —, heavy armour.
arretieren, to stop; to secure.
Arretierung *f*, stop(ping), detent, catch.
Arsen *n*, arsenic (As).
arsenhaltig, arsenic.
Art *f*, type; nach — des . . ., on
the . . . . plan.
Artikulation *f*, articulation.
artikuliert, articulate.
Asbest *m*, asbestos;
— =pappe *f*, asbestos board.
A=Schrank *m*, A-(switch)board, *F*.
Asphalt *m*, asphalt(um),
asphalte.
asphaltieren, to asphalt.
Astasie *f*, astaticism.
astasieren, to astaticise.
astatisch, astatic(al);
— er Zustand *m*, astaticism;
— es Nadelpaar *n*, astatic couple,
two compound needles *pl*.
Asymmetrie *f*, asymmetry.

asymmetrisch, asymmetrical.
— e Leitfähigkeit *f*, asymmetrical
or non-linear or unilateral
conductance (der Kristalle,
of crystals).
Asymptote *f*, asymptote *M*.
asymptotisch, asymptote, asymp-
totic(al).
asynchron, asynchronous, non-
synchronous.
Asynchron=Funkenstrecke *f*, non-
synchronous rotating spark
gap;
— =motor *m*, asynchronous motor.
Äther *m*, ether;
— =welle *f*, ether wave.
Atmosphäre *f*, atmosphere;
Erd= —, earth's atmosphere;
Gas= —, gaseous atmosphere.
Atmosphärendruck *m*, atmospheric pressure.
atmosphärisch, atmospheric(al);
— e Elektrizität *f*, atmospheric
electricity;
— e Störungen *pl*, strays *pl*,
atmospherics *pl*, X.'s *pl*;
Atom *n*, atom; corpuscle;
— =gewicht *n*, atomic weight;
— =zahl *f*, atomic number.
atomistisch, atomic(al); cor-
puscular.
ätzen, to etch.
Ätz=kali *n*, caustic alkali;
— =kalk *m*, quicklime;
— =mittel *n*, etching reagent.
Audion *n*, audion, (thermionic)
valve detector, detecting
tube or valve, rectifier
triode;
Rückkopplungs= —, regenerative or retroactive audion or
(valve) detector;
Schwing= —, oscillating detector;
— =empfänger *m*, (sekundärer)
(secondary) audion receiver;

Audion
— **-verstärker** *m*, amplifying detector;

aufarbeiten, to clear, *F, T*.

Aufbau *m*, structure, design; aus mehreren Teilen: assemblage, assembly; building-up; **periodischer —, aus gleichen Gliedern bestehender** periodic (recurrent) structure.

aufbauen, to design; to assemble to build-up; ein Magnetfeld: to set up, to create.

aufbohren, to (re)bore.

aufbrauchen, to use up; Energie: to dissipate.

aufdrehen, Wähler: to step up; (sich) — to untwist.

aufdrucken, to imprint.

aufdrücken, to impress; einem Stromkreis eine Spannung —, to impress a voltage upon a circuit.

Aufdrücken *n*, impression (einer Spannung, of a voltage).

aufeinanderfolgend, consecutive, successive.

Aufeinanderschichten *n*, superposition.

auffangen, to pick up, *R*.

Auffangspule *f*, search coil.

aufflammen, to flash (Lampe, lamp).

auffrischen, to recuperate.

Auffrischung *f*, recuperation.

auffüllen, to fill up, Elemente: to top up.

Auffüllen *n*, **Auffüllung** *f*, filling, topping-up.

Aufgabe *f*, filing, handing-in, *T*; — **-zeit** *f*, code time, time of acceptance, *T*.

aufgearbeitet, clear, *F, T*.

aufgeben, to file, to hand in, (ein Telegramm, a message).

aufgehängt, suspended (an, from).

aufgekeilt, keyed (auf, to).

aufgeladen, charged (auf... *V*, to .... volts).

aufgeschraubt, threaded (auf, on).

aufgeschweißt, welded (auf, to) **elektrisch —**, electro-welded.

aufgestiftet, pinned (auf, to).

aufgezogen, Feder: wound up.

aufhalten, to check; to stop.

Aufhänge-draht *m*, suspending wire, suspension wire;
— **-öse** *f*, suspension eye, suspension loop (am Hörer, of the receiver);
— **- punkt** *m*, point of suspension;

Aufhänger *m*, hanger, suspender (für Luftkabel, for aerial cables);
Rohaut- —, raw-hide suspender.

Aufhängung *f*, suspension;
gegenfäßige —, top-bottom suspension (einer Drehspule, of a moving coil);
kardanische —, cardanic suspension;
Schneiden- —, knife-edge suspension;
Spitzen- —, pivot suspension.

aufhaspeln, to reel up, to reel in.

aufheben, to lift; eine Verbindung: to clear, to suspend, to take down a connection; to nullify, to neutralize;
einander —, to cancel, to annul each other, *M*.

Aufheben *n*, lifting; clearing (einer Verbindung, of a connection); neutralization; cancellation.

Aufhebung *f*, clearing; neutralization; cancellation.

aufhissen, to hoist.

aufhören, to finish, to discontinue, to stop.

Aufhören *n*, stopping, cessation.

aufkeilen, to key on (auf eine Welle, to a shaft).

aufklappbar, hinged.

aufkleben, to gum (auf, to).

aufladen, to charge (auf, to); **wieder=—**, to recharge.

Aufladen *n*, charging.

Aufladung *f*, charge; **Wieder= —**, recharge.

Auflage *f*, Stütze: rest, seat; Überzug: coating; **Metall= —**, metal(lic) coating; **— =fläche** *f*, seat.

Auflager *n*, abutment, seat.

auflegen, den Hörer, to clear, to restore (the telephone).

Auflegen *n*, replacement, restoring.

aufleuchten, to light (up); to flash.

Aufleuchten *n*, lighting; flashing; illumination.

auflösen, *M*: to resolve (in, into); to solve, to analyze; in Wasser usw.: to dissolve; (sich) — in seine Bestandteile: to disintegrate; eine Verbindung: to clear out, to release.

Auflösen *n*, solution, solving; dissolving; clearing-out, release.

Auflöserelais *n*, clear-out relay, *A*.

Auflösung *f*, (re)solution, analyzation; dissolution; disintegration; release, clear(ing-)out;
— **einer Gleichung nach** *n*, solution of an equation for *n*;
rückwärtige —, back release, *A*;
vorzeitige —, premature release, *A*.

aufmauern, to brick up.

Aufnahme *f*, reception; detection; von Kurven: plotting (of curves);
— **=beamter** *m*, checker *T*;
— **=fähigkeit** *f*, capacity; susceptibility; receptivity;
— **=mikrophon** *n*, pick-up transmitter *R*;
— **=raum** *m*, studio (des Rundfunksenders, of broadcast transmitter);
— **=relais** *n*, receiving relay;
— **=segment** *n*, receiving segment, *T*;
— **=vermögen** *n*, capacity, von hohem (geringem) — = —, high (low) capacity . . . . ; receptivity.

aufnehmen, to receive; to detect; Kurven: to trace, to plot; Kabel: to remove, to pick up; am Klopfer: to copy (messages); Strom: to take (current); räumlich: to accommodate (Hauptverteiler, der 500 Leitungen aufnimmt, MDF accommodating 500 wires).

Aufnehmen *n*, reception; tracing, plotting; removal; picking-up.

aufrauhen, to roughen; to get rough.

Aufrauhen *n*, roughening.

aufrechterhalten, to maintain.

aufreihen, to thread (auf, on).

aufrichten, die Klappe: to restore (the shutter).

Aufriß *m*, sketch, elevation.

aufrollen, to coil up; to reel in.

Aufsatz *m*, der Stange: finial, head piece.

aufsaugen, to soak; to absorb.

Aufschaukeln *n*, resonant rise.

aufschieben, to delay, to postpone.

aufschießen, to coil (Kabel, cable).

Aufschießen *n*, coiling.

aufschrauben, to thread on, to screw on.
Aufschrift f, address; label; **mit einer —** versehen, to address; to label.
aufschweißen, to weld (auf, to).
Aufschweißen n, welding; **elektrisches —**, electro-welding.
Aufsicht f, inspection (über, of, over), supervision; **Apparat- —**, operator-in-charge; dirigeur: **Fernamts- —**, trunk supervisor; **Störungs- —**, fault clerk:
Aufsichts-beamter m (**-beamtin** f), supervisor, chief operator; **— -platz** m, supervisor's position; **— -tisch** m, chief operator's desk, monitoring desk.
aufspalten, to split up.
Aufspalten n, splitting-up (von Jonen, of ions).
aufspeichern, to store up, to accumulate; to register, A.
Aufspeicherung f, storing, accumulation.
aufspleißen, to fan out (ein Kabel, a cable).
aufspulen, to spool, to coil.
Aufspulen n, coiling.
aufstecken, to slip on.
Aufsteck-Kappe f, slip-on cap; **— -spule** f, plug-in coil, R.
aufstellen, to mount, to fit up, to erect; to assemble; to allocate; to set (eine Stange, a pole).
Aufstellung f, mounting, fitting (-up), erection; allocation.
aufstiften, to pin (auf, to).
aufsuchen, to hunt out (einen freien Wähler, an idle selector) A; to find.
Aufsuchen n, **Aufsuchung** f, hunting-out, A; finding.

aufstreifen, to slip on, to thread on (auf, to).
auftakeln, to rig.
auftragen, Farbe: to ink; Kurven: to plot, to trace (curves).
Auftragsröllchen n, ink roller, ink wheel.
auftreffen (auf), o strike, to impinge (upon); to abut (against); to encounter.
auftreten, to occur, to appear.
Auftreten n, occurrence, appearance.
Aufwand, m, expenditure; input; **Energie- —**, expenditure of energy; **Leistungs- —**, power input.
Aufwärts-bewegung f, upward motion; **— -transformator** m, step-up transformer; **— -transformierung** f, step-up.
aufwärts transformieren, to step up.
aufwenden, to expend.
Aufwendung f, expenditure.
aufwiegen, to counterbalance.
aufwinden, to lift (up), to jack up.
aufzehren, to dissipate.
Aufzehrung f, dissipation.
aufzeichnen, to record, to register.
Aufzeichnung f, record.
Aufzeigung f, notification.
aufziehen, to wind up; den Nummernschalter, auch: to pull round (the dial), A.
Aufziehen n, winding-up.
Aufzug m, elevator.
Auge n, eye, eyelet.
Augenblicks- ..., instantaneous; **— -wert** m, instantaneous value.
ausästen, to trim (Bäume, trees).

Ausästwerkzeug *n*, tree trimmer.
Ausbau *m*, establishment;
im erſten —, on first establishment.
Ausbauchung *f*, bulge.
ausbauen, to equip.
ausbeuten, to exploit.
Ausbeutung *f*, exploitation.
ausbilden, to train.
Ausbildung *f*, training.
ausblaſen, to blow out (Funken, sparks).
ausbleien, to lead (Verbindungen, joints).
ausbreiten, (ſich), to spread, to expand; to diffuse; Wellen: to propagate, ſich ausbreitende Welle, travelling wave.
Ausbreitung *f*, spreading, expansion; diffusion; propagation; Wellen- —, wave propagation.
Ausbreitungswiderſtand *m*, diffusion resistance.
ausbrennen, to burn out.
Ausbrennen *n*, burn-out (einer Wicklung, of a coil).
ausdehnen, (ſich), to stretch, to extend, to expand.
Ausdehnung *f*, expansion, extension;
Ausdehnungskoeffizient *m*, coefficient of expansion.
ausdrehen, to (turn) hollow.
Ausdruck *m*, expression, term;
cos- —, cosine term;
Fach- —, technical term.
ausdrücken, to express.
auseinandergehen, to diverge.
Auseinandergehen *n*, divergence.
auseinandergehend, divergent.
auseinandernehmen, to disassemble, to dismantle, to strip, to take to pieces.
Ausfluß *m*, emanation.
ausformen, to lace out, to form out (ein Kabel, a cable), *B*.

Ausformen *n*, lacing-out, forming-out.
ausfranſen, to fray.
ausfräſen, to mill out.
ausführen, to carry out, to perform.
ausführlich, detailed, in detail.
Ausführung *f*, performance;
äußerlich: finish;
Tropen- —, tropical finish.
ausfüllen, to fill in (Formulare forms).
Ausfütterung *f*, lining.
ausfüttern, to line (mit Kupfer, with copper).
Ausgabe *f*, expense.
Ausgang *m*, outlet, *A*; Tür: exit;
Not- —, emergency exit.
Ausgangs-klemmen *pl*, output terminals *pl*;
— -kreis *m*, output circuit;
— - — impedanz *f*, output impedance;
— -ſtellung *f*, home position (eines Wählerarmes, of a wiper);
— -ſtromkreis *m*, output circuit; plate-filament circuit *V*;
— -übertrager *m*, output transformer.
ausgefräſt, milled out.
ausgefreſſene Kontakte *pl*, pitted contacts *pl*.
ausgeglichene Belaſtung *f*, balanced load (eines Drehſtromſyſtems, of a triphase system).
ausgehen, to originate; to emanate.
ausgehender Verkehr *m*, originating traffic.
ausgekehlt, fluted, chamfered.
ausgelichteter Raum *m*, clearance, *B*.
ausgeſchieden werden, to exude, *B*.
Ausgeſetztſein *n*, weather exposure, *B*.

ausgestrahlte Energie *f*, radiated energy.

Ausgleich *m*, balancing, balance; equalization; compensation; **Dämpfungs--**, attenuation equalization; **Kapazitäts--**, capacity balance;

— - — verfahren *n* durch Zusatzkondensatoren, condenser balancing method, *K*;

— - — - — durch Abernkreuzung, test-balancing method, *K*.

ausgleichen, to balance, to compensate (for), to equalize, to equilibrate; durch eine Gegenkraft: to bias out, to counterbalance; neu —, to rebalance.

Ausgleichen *n*, balancing, compensation, equalization, equilibration; neues —, re-balancing, rebalance.

Ausgleicher *m*, equalizer; **Dämpfungs--**, attenuation equalizer.

Ausgleichs-kapazität *f*, balancing capacity;

— **-kondensator** *m*, balancing condenser;

— **-leitung** *f*, balancing network, artificial (balancing) line or circuit, compensation circuit;

— **-luftdraht** *m*, balancing aerial;

— **-spannung** *f*, transient voltage;

— **-strom** *m*, compensating current; flüchtiger Strom: transient current;

— - — **-kreis** *m*, compensation circuit;

— **-transformator** *m*, hybrid coil or transformer, balanced differential transformer, balanced three-winding transformer, *K*;

— **-vorgang** *m*, transient;

— - — bei Stromschließung (Stromunterbrechung), make (break) transient;

— - —**-s**, Dauer *f* des, transient period;

— **-vorgänge** *pl*, transient effects *pl*;

— **-wicklung** *f*, compensation winding (des Linienrelais, of the line relay);

— **-widerstand** *m*, compensating resistance, balancing resistance.

Ausgleichung *f*, balance, balancing, equilibration, compensation; **schlechte —**, want of balance, unbalance;

Ausgleichungs-fehler *m*, balance error, unbalance;

— **-schaltung** *f*, equalizing network, *K*.

ausglühen, to anneal.

Ausglühen *n*, annealing.

aushängen, to lift (den Hörer, the telephone).

ausheben, to lift clear of (eine Klinke, a pawl).

aushöhlen, to hole, to hollow; to channel.

auskehlen, to chamfer, to flute, to channel.

Auskehlung *f*, channel, chamfer.

auskleiden, to line.

Auskleidung *f*, lining.

Ausklingeln *n* der Adern, circuit identification *B*.

ausklinken, Zahn: to unlatch.

auskoppeln, to tune out, to balance out, to neutralize.

Auskoppelung *f*, tuning-out, balancing-out, neutralization.

auskreuzen, Aderpaare: to testsplice, to cross-joint, *K*.

Auskreuzen *n*, test-splicing (method), cross-jointing, *K*.

Auskreuz-lötstelle *f*, test-splice, *K*.

auskunden, to survey (eine Leitung, for a line).

Auskundung f, surveying.

Auskunft f, information;

Auskunfts-beamter m, enquiry clerk, information operator;
— **-platz** m, information desk, F;
— = —, **Fernamts-**, toll information desk;
— **-stelle** f, enquiry (position).

auskuppeln, to uncouple, to declutch.

ausladend, projecting.

Ausladung f, projection.

Auslands-leitung f, international line.
— **-saal** m, foreign gallery;
— **-telegramm** n, foreign message;

Auslaß m, outlet, A;
— **-ventil** n, escape valve.

auslassen, to omit.

Auslassung f, omission.

auslaufen, Kabel: to end; Motor: to run down, to stop gradually; Lager: to wear out.

Auslauflänge f, length of end section, K.

Auslege-maschine f (für Seekabel), paying-out machine;
— **-trommel** f, paying-out drum.

auslegen, to interpret; Patente: to publish; Kabel: to lay (cables).

Ausleger m, cantilever; am Kran: derrick; Spreize: outrigger, B.

Auslegung f, laying, B; publication; interpretation.

auslöschen, Lampen: to darken, to extinguish; to quench; durch Blasen: to blow out.

Auslöse-daumen m, resetting cam, releasing cam;
— **-fehler** m, failure to release, A;
— **-hebel** m, trip(ping) lever; am Hughesapp.: detent lever;
— **-kontakt** m, release contact;
— **-magnet** m, Einrücken: trigger magnet, trip magnet; Rückstellen: release magnet, A;
— = —, **Druck-**, printing trip magnet, T;
— **-relais** n, tripping relay; clear-out relay, A;
— **-spindel** f, trip spindle, A;
— **-strom** m, releasing current, A;
— **-stromstoß** m, starting impulse, T;
— **-taste** f, release key.

auslösen, einrücken: to trip, to start; ausrücken: to release, to reset; langsam auslösend, slow-to-release.

Auslöser m, trigger.

Auslösung f, tripping, starting; release, releasing, resetting;
— einer Verbindung durch Einhängen des rufenden Teilnehmers, calling party release, A;
— — —, sobald einer von beiden Teilnehmern einhängt, first party release, A;
— — — durch Einschneiden vom Amt aus, telephonist release, A;

Null- —, no-load release;

Nullspannungs- —, no-volt(age) release;

Nullstrom- —, no-load release;

Rück- —, back release, A;

Vor- und Rück- —, first party release A.

Ausmaß n, dimension.

ausnützen, to utilize, to exploit.

Ausnützung f, utilization, exploitation.

auspumpen, Kabelbrunnen: to unwater; Sammler: to run down; Luft: to exhaust, to evacuate.

Auspumpen *n*, unwatering; running-down; exhaustion, evacuation.

ausradieren, to erase.

ausrichten, to straighten (Stangen, poles), to dress; nivellieren: to level; geradlinig: to align, to aline.

Ausrichten *n*, straightening, dressing; levelling; alignment, alinement.

— der Federn (eines Relais), aligning of springs (of a relay).

ausrücken, to throw out of gear, to stop.

Ausrück-hebel *m*, Ein- und, starting and stopping lever.

ausrüsten, to fit out, to equip.

Ausrüstung *f*, fitting, equipment, outfit;

Stangen- —, pole fittings *pl*;

— für erste Hilfe, first-aid outfit.

aussägen, to saw out.

ausschalten, to cut out, to switch off, to disconnect.

Ausschalter *m*, cut-out, circuit breaker;

Maximal- —, overload circuit breaker;

Null- —, no-load cut-out;

Nullspannungs- —, no-voltage circuit breaker;

Überstrom- —, overload circuit breaker.

Ausschaltstellung *f*, off-position.

ausscheiden, to exude; to separate; to eliminate.

Ausscheidung *f*, exudation; separation; elimination.

Ausschlag *m*, deflection, swing, throw;

End- —, full deflection;

Nadel- —, needle throw, deflection of the needle;

Winkel- —, angular deflection.

ausschlagen, to deflect, to swing, to throw; mit Kupfer: to line (with copper);

zu weit —, to overthrow.

Ausschlagen *n*, deflection, swinging, throwing; lining;

zu weites —, overthrowing (des Heberschreibers, of the syphon recorder).

Ausschluß *m*, disconnection:

— -feder *f* (am Hughesapparat) disconnecting spring.

Ausschnitt *m*, aperture, window.

ausschütten, to spill (Säure) acid).

ausschwingen, to decay, to die away, to die out.

Ausschwingen *n*, decay, dying-out;

Verzerrung *f* durch Ein- und —, transient distortion.

Ausschwing-Komponente *f*, dying-out (transient) component;

— -vorgang *m*, dying-out transient.

ausschwitzen, to exude (Harz, resin).

Ausschwitzung *f*, exudation.

Außen- ...., external;

— -anlage *f*, external plant;

— -dienst, Störungssucher *m* im, external faultsman;

— -durchmesser *m*, external diameter, outer diameter;

— -leiter *m*, outer main;

— -leitung *f*, external leads *pl*;

— -linie *f*, contour;

— -maß *n*, overall dimension;

— -nebenstelle *f*, external extension;

— -strom *m*, foreign current;

— -wecker *m*, extension bell;

— -wirkung *f*, external effect.

aussenden, to emit.

Aussenden *n*, **Aussendung** *f*, emission.

äußere(r), external, outer.

außereuropäischer **Vorschriften-bereich** m, Extra-European range, T'.
aussetzen, to expose; unterbrechen: to intermit; versagen: to fail.
Aussetzen n, failure.
Aussetzung f, exposure.
aussieben, to filter (out), to select, to sift.
aussondern, to separate (out), to eliminate.
Aussonderung f, separation, elimination.
ausspannen, to extend, to stretch, to spread; Drähte: to string (wires).
Aussparung f, hollow.
Aussprache f, pronunciation; deutliche —, articulation, pronunciation; undeutliche —, inarticulateness.
aussprechen, deutlich, to pronounce, to articulate.
ausstanzen, to blank out.
Ausstattung f, equipment, layout.
Ausstellung f, exhibition.
aussteuern, to modulate, R.
Aussteuerung f, modulation, R. prozentuale —, percentage modulation; vollständige —, complete modulation.
Aussteuerungsgrad m, amount or degree of modulation.
ausstrahlen, to radiate, to emit
Ausstrahlung f, radiation, emission;
— in den Raum, radiation into space.
Ausstrahlungsvermögen n, emissivity, emissive power; radiating capacity.
ausströmen, to emanate; — (lassen), to emit.
Ausströmen n, emanation.
Austausch m, interchange.

austauschbar, interchangeable.
Austauschbarkeit f, interchangeability.
austauschen, to interchange, to exchange.
austreiben, to expel (Feuchtigkeit, humidity), to drive out (Gas, gas). [out;
austrocknen, to dry (up), to dry Holz — lassen, to season wood.
Austrocknen n, Austrocknung f, drying-(up or out); seasoning.
ausüben, to practice.
Ausübung f, practice.
Auswahl f, selection.
auswählen, to select.
auswalzen, to roll out.
auswechselbar, replaceable; interchangeable.
auswechseln, to replace, to interchange, to exchange.
Auswechseln n, replacing, replacement.
auswirken, (sich), to work out.
Auswüchse bilden to fan out (Sammlerplatten, storage cell plates).
ausziehen, to ink (Zeichnungen, drawings).
Auszug m, extract.
Autodynempfang m, autodyne reception.
autoinduktiv, autoinductive.
Automatenleitung f, coin box circuit, F.
Autotransformator m, autotransformer.
automatisch, automatic(al), self-acting.
Automobil n, motor car, automobile.
axial, axial.
Axt f, ax(e).
Ayrtonscher Nebenschluß m, universal shunt box, compensating resistance.

B.

Backe f, jaw; die; bit;
 Gewinde= —, Schneid= —, screw
 die.
Backstein m, brick;
 — =schicht f, course of bricks.
Bad n, bath.
baggern, to drag.
Bahn f, path, way, Eisenbahn:
 railway, am.: railroad;
 Fahr= —, carriageway.
Bajonettverschluß m, bayonet joint.
bakelisierter Faserstoff m, phenol fibre.
Bakelit n, bakelite.
Balata f, balata.
Balken m, beam, Zug=, Streck=balken: balk, Dielenbalken: joist.
Ball m, ball;
 — =senden n, re-radiation, re-broadcasting, radio repeating;
 — =sendestelle f, — =station f, radio repeating station.
Ballastwiderstand m, ballast resistane, loading resistance.
Ballen m, pack, bale.
ballistisch, ballistic(al).
Ballon m, carboy, demijohn;
 Glas= —, Meidingerelement: bell jar.
Bambusrohr n, malacca cane.
Band n, tie, strap, binder; ribbon, tape, band, Lasche: bond;
 Gürtel: belt; Reifen: hoop; mit — bewickeln, to tape (together).
 Förder= —, band conveyer;
 Frequenz= —, frequency band, band of frequencies;
 Hitz= —, hot band;
 Isolier= —, insulating tape;
 Seiten= —, side band, R;
 Spiral= —, helical tape (Kraruptleiter, loaded conductor);
 Wellen= —, wave band, R;
 — =breite f, band width, R;
 — =eisen n, hoop iron;
 — = —=bewehrung f, hoop iron sheathing;
 — =filter n, band (pass) filter, wave band filter, acceptor circuit;
 — = — von großer Lochbreite, broad band filter;
 — =förderer m, band conveyer, belt carrier;
 — =kabel n, ribbon (-shaped) cable;
 — =kupplung f, belt coupling;
 — =post f, band conveyer, belt carrier;
 — =spule f, ribbon coil;
 — = —, hochkant (flach) gewickelte, edgewise (flatwise) wound ribbon coil;
 — =stahl m, ribbon steel;
 — =umwicklung f, taping;
 — =wickler m, taping machine.
Bandage, binding, bandage.
bandförmig, band-shaped.
Bank f, bench, bank;
 Kontakt= —, contact bank;
 Werk= —, bench;
 — =kontakt m, bank contact;
 — =schraubstock m, bench vice (am.: vise).
Bariton m, baritone.
Barium n, barium (Ba).
Barometer n, barometer;
 — =stand m, barometric pressure.
Base f, base.
Basis f, basis; base line.
basisch, basic(al).
Baß m, bass, bass voice;
 — =geige f, kleine: bass-viol, violoncello; Kontrabaß: contrabass;
 — =pfeife f, bassoon.
Batterie f, battery, pile;

Batterie
eine — anlegen, to apply a battery;
aus einer — betreiben, to run from a battery;
geerdete —, earthed or grounded battery;
gemeinsame —, common battery, central battery, universal battery;
geteilte —, split battery;
Amts- —, exchange battery;
Anoden- —, anode or plate battery, B-battery;
Bleisammler- —, lead storage battery;
Gitter- —, grid battery, C-battery;
Haupt- —, main battery;
Heiz- —, filament battery, A-battery;
Meß- —, testing battery;
Mikrophon- —, speaking or transmitter battery;
Not- —, emergency battery;
Primär- —, primary battery;
Prüf- —, testing battery;
Besetztzeichen: engaged test battery F;
Puffer- —, buffer battery, floated battery;
Sammler- —, Sekundär- —, secondary or storage battery;
Sprech- —, speaking battery;
Taschen(lampen)- —, flash lamp battery;
Trenn- —, spacing battery, T;
Trocken- —, dry (cell) battery;
Zähler- —, meter battery;
Zeichen- —, marking battery;
Zentral- —, common battery, ab: c. b.; central battery;
Zusatz- —, booster battery;
— -anruf, m, battery ringing;
— = —, Fernsprecher m. mit, battery ringing telephone;

— =galvanometer n, battery gauge;
— =gestell n, battery stand or rack or frame;
— =heber m, battery syringe;
— =kasten m, battery box, battery container;
— =klinke f, battery or power jack;
— =ladesatz m, battery charger, battery charging set;
— =prüfer m, battery tester or gauge;
— =raum m, battery room;
— =schrank m, battery cupboard;
— =wärter m, battery attendant;
— =widerstand m, battery resistance; Schutzw.: protective resistance; Erdungsw.: earthing resistance T;
— =zuführung f, battery lead.
Bau m, erection, construction;
gedrungener —: compactness;
im —, under erection, under construction;
— =arbeiter m, wireman, B;
— =art f, construction;
— =holz n, timber.
— =kolonne f, construction unit or gang;
— =lager n, store, B;
— =strecke f, field;
— =tagebuch n, log book;
— =trupp m, gang, unit, construction gang; Störungstrupp: repair gang;
— =werk n, building;
— =zeug n, line material;
— =zug m, construction unit.
Bauch m einer Schwingung, bulge; loop, antinode.
Baum m, tree; beam; Bäume ausästen, to trim trees B.
Baumwoll-draht m, cotton covered wire;
— =faden m, cotton thread;
— =garn n, cotton twine;

Baumwoll
— **=kabel** *n*, cotton-covered cable;
— **=samenöl** *n*, cotton seed oil;
— **=(seiden)kabel** *n*, (silk and) cotton covered cable.

Baumwolle *f*, cotton;
mit — **umsponnen**, cotton-covered, *ab*: c. c.;
— — —, **doppelt**, double cotton covered, *ab*: d. c. c.;
merzerisierte —, mercerised cotton.

Beamter *m*, official; operator, employe(e), executive, clerk; leitender B.: officer (-in-charge); technischer B.: engineering officer;
Abnahme= —, testing officer;
Apparat= —, operator;
Aufnahme= —, checker, *T*;
Aufsichts= —, supervisor;
Auskunfts= —, enquiry clerk;
Funk= —, wireless or radio officer;
Prüf= —, testing officer; checker *T*.

Beamtin *f*, operator, employe(e); telephonist;
ferne —, distant operator;
— — **für ankommenden (abgehenden) Verkehr**, distant in (out) operator;
freie —, idle operator;
A= (B=) —, A- (B-) telephonist or operator;
Abfrage= —, answering operator, A-operator;
Aufsichts= —, supervisor;
Auskunfts= —, information operator;
— — —, **Fernamts=**, toll information operator;
Fern(schrank)= —, toll operator, l. d. operator;
Melde= —, record (table) operator;

— = —, **Spitzenplatz=**, record transfer operator;
Nachtdienst= —, night operator;
Sende= —, transmitter operator, *T*;
Stanz= —, perforator operator, *T*.

beanspruchen, to stress, to strain; **Rechte**: to claim.
Beanspruchung *f*, strain, stress (des Kabels, on the cable);
elastische —, elastic strain;
Biege= —, bending strain;
Druck= —, compressive stress;
Zug= —, tensile stress.

beantragen, to apply (for).
beantworten, to answer (einen Ruf, a call).
bearbeiten, to dress, to finish; to tool, to machine; bearbeitete Fläche *f*, tooled, machined surface.
Bearbeiten *n*, machining, dressing.
beaufsichtigen, to supervise.
Beaufsichtigung *f*, supervision.
Bedarf *m*, demand, need, requirement, want;
Energie= —, energy requirement.
bedecken, (sich), to cover.
Bedeckung *f*, covering, capping.
bedienen, to attend (to); to work, to operate, to manipulate, to handle.
Bedienung *f*, attendance (to), operation, manipulation, handling.
Bedienungs=kosten *pl*, cost of attendance, cost of operation;
— **=vorschrift** *f*, instruction;
— **=zeit** *f*, handling time.
bedrahten, to wire (up).
Bedrahtungs=plan, wiring diagram.

Bedürfnis *n*, need, want, requirement;
Verkehrs= —, traffic requirements *pl.*
beeinflussen, to influence; to control.
Beeinflussung *f*, control; störende —: interference;
gegenseitige —, interaction (between);
induktorische —, inductive interference, inductive trouble;
Gitter= —, grid control, *V*;
— durch die Sprache, voice control, *V*;
Beeinflussungsröhre *f*, modulator valve.
Beeinträchtigung *f*, impairment.
beendigen, to finish, to end.
befestigen, to attach, to fasten, to fix, (an, to), mit Klammern: to cleat; befestigt (auf, an), solid (with).
Befestigung *f*, fastening, attachment, fixing; cleating.
Befestigungs=lappen *m*, fastening or fixing lug;
— =mittel *n*, fastener;
— =schraube *f*, tightening screw.
befördern, to transmit, to transport, to convey.
Beförderung *f*, transmission, transport(ation), conveyance;
Telegramm= —, transmission of messages.
befreien, to free, to liberate; von der Umhüllung: to strip, to unwrap.
begehen, to patrol (Leitungen, lines) *B*.
Begehen *n*, **Begehung** *f*, patrolling.
Beginn *m*, start; Ursprung: origin; initiation; commencement, beginning.
beginnen, to start, to begin; to originate, to initiate; to commence.

begrenzen, to limit; to terminate.
begrenzend, limiting, terminal.
begrenzt, limited, (de)finite.
Begrenzer *m*, limiter, limiting device;
Strom= —, current limiter.
Begrenzung *f*, limitation, termination, definition.
Begrenzungswiderstand *m*, limiting resistance.
Behälter *m*, container, receptacle, vessel, für Flüssigkeit: well, reservoir, tank;
Farb= —, ink well;
Zink= —, zinc containing vessel (des Trockenelements, of the dry cell).
behandeln, handhaben: to handle; to treat.
Behandlung *f*, handling; treatment;
Schutz= — (der Stangen), preservative treatment (of the poles).
Behandlungsvorschrift *f*, instruction.
Beharrungsvermögen *n*, inertia, moment (of inertia).
behauen, to trim, to dress.
beheizen, to heat; zu stark —, to overheat.
beheizt, heated;
schwach —, dull, *V*;
stark —, bright, *V*.
behindern, to impede.
behobeln, to plane, to chip.
Beihilfe *f*, (staatliche), subsidy.
Beil *n*, ax(e).
beilegen, einen Wert, to assign a value (to).
Bein *n*, bone.
Beißzange *f*, cutting pliers *pl*, nippers *pl*.
Beitel *m*, chisel.
Beitrag *m*, share, part.
beizen, to etch.

Bekanntmachung *f*, notification, publication.
Bekleidung *f*, sheathing, coating;
  Innen= —, lining.
Belag *m*, coating.
Belastbarkeit *f*, carrying capacity.
belasten, to load; Konto: to debit;
  mit Ableitung —, to leak-load;
  gleichförmig —, to load continuously;
  induktiv —, to load inductively;
  punktförmig —, to lump-load;
  mit Querspulen —, to leak-load;
  ungleichförmig —, to lump-load.
Belasten *n*, loading.
belastet, loaded, *K*;
  gleichförmig —, continuously loaded;
  induktiv —, inductively loaded;
  leicht —, besonders, extra light loaded, *ab*: X. L. L.;
  mittelstark —, medium heavily loaded, *ab*: M. H. L.;
  punktförmig —, lump-loaded;
  spulen= —, coil-loaded;
  stark —, heavily loaded, *ab*: H. L.;
  stetig —, continuously loaded;
  für Hochfrequenz —, loaded for carrier.
Belastung *f*, stress, strain; load; loading;
  bei — mit 200 A, on load 200 amps;
  ausgeglichene —, balanced load (eines Drehstromsystems, of a triphase system);
  elastische —, elastic strain;
  gleichförmige —, continuously or evenly distributed load;
  induktive —, inductive or inductance load;

kapazitive —, condenser load;
punktförmige —, lumped (series) load;
reaktive —, reactive load;
Biege=—, bending strain, bending load;
Bruch= —, breaking strain, breaking load;
Nacht= —, night load, *F*, *T*;
Schnee= —, snow load, *B*;
Spitzen= —, peak load;
Spulen= —, coil loading;
Strom= —, current load;
Tages= —, day load;
Verkehrs= —, traffic load;
Vierer= —, phantom loading;
— mit Ableitung, leak-load(ing);
— — Querspulen, lumped leak load(ing);
— — Reihenspulen, lumped series load(ing).
Belastungs=abschnitt *m*, loading section *K*;
— =bereich *m*, load range;
— =spule *f*, loading coil, *K*; loading inductance *R*;
— =tabelle *f*, stress table (für Drähte, for wires) *B*;
— =widerstand *m*, loading resistance.
belegen, mit Gebühren: to charge; Besetzzeichen: to mark engaged, to busy, to seize; Zeitdauer: to hold.
Belegen *n*, **Belegung** *f*, busying, seizure, marking engaged, *F*, *A*.
belegt, busy, engaged, *F*, *A*.
Belegungs=dauer *f*, holding time *F*, *A*;
— =vorrichtung *f*, make-busy arrangement, *A*;
beleuchten, to illuminate, to light.
Beleuchtung *f*, illumination, lighting.
bemerken, to observe, to perceive.

bemessen, to dimension; zeitlich: to time.

benennen, genau, to specify.

benutzen, to use, to employ; to utilize.

Benutzung *f*, use, employment; utilization.

Benzin *n*, petrol, benzine.

— **=motor** *m*, petrol motor.

Benzol *n*, benzole, benzine.

beobachten, to observe; to watch; to detect.

Beobachtung *f*, observation; detection.

Beobachtungsfehler *m*, error of observation.

beratend, advisory.

berechenbar, calculable.

berechnen, to calculate; **einfach (doppelt, dreifach)** —, to charge 1, 2, 3 fees, *F*.

Berechnung *f*, calculation.

Bereich *m*, range; band; zone; **Frequenz=** —, range of frequencies; **Wellen=** —, wave band.

bereifen, to rime.

bereift, rimy.

Bereitstellung *f*, maturation (eines Gesprächs, of a call).

Berg *m*, mount(ain); einer Kurve: peak, crest; **Wellen=** —, wave crest.

bergig, mountainous.

Bericht *m*, report, information; **Wetter=** —, weather report, weather forecast.

berichten, to report, to inform.

berichtigen, to correct.

berichtigend, corrective.

Berichtigung *f*, correction.

— **der Verzerrung**, correction of distortion.

Bernstein *m*, (yellow) amber.

bersten, to burst.

Bersten *n*, bursting, disruption.

beruhen, to be based (auf, on).

berühren, to touch; Leitungen: to be in contact.

Berührung *f*, touch(ing); contact; **zeitweise Leitungs=** —, tapping or intermittent contact; **Schleifen=** —, constant or permanent loop, short-circuit; **Wetter=** —, weather contact; — **zwischen Stöpselspitze und =schaft**, tip and sleeve contact.

Berührungs=elektrizität *f*, contact electricity;

— **=fläche** *f*, contact (sur)face.

beschädigen, to damage, to impair, to injure, to mutilate.

Beschädigung *f*, damage, injury, mutilation.

Beschaffenheit *f*, state, condition.

beschalten, to wire (up) (mit, for).

Beschaltung *f*, wiring; **vorder= (rück=) seitige** —, surface (panel) wiring (einer Schalttafel, of a switchboard).

Beschaltungsplan *m*, wiring scheme.

Beschlagen *n* (der Lampenbirnen), age-coating (of lamp bulbs).

beschleunigen, to accelerate; to speed up.

Beschleunigung *f*, acceleration; speeding-up.

Beschleunigungswicklung *f*, acceleration coil (des Gulstab= relais, of the Gulstad relay).

beschneiden, to trim.

beschränken, to confine, to restrict (auf, to), to reduce.

Beschränkung *f*, restriction, reduction.

Beschreibung *f*, characterization; specification; **Patent=** —, patent specification.

beschriften, to designate, to label.

Beschriftung *f*, designation; label; einer Zeichnung: legend.

Beschwerde *f*, complaint (bei, with; über, about);

— **-stelle** *f*, complaint section, complaint desk.

beschweren, (sich), to complain.

beseitigen, to remove, to clear (Fehler, faults).

Beseitigung *f*, removal, clearing.

besetzt, busy, engaged;

— **e Verbindungsleitung**, busy trunk;

— **er Platz** *m*, occupied position; als — kennzeichnen, to busy, to mark engaged.

Besetzt-einrichtung *f*, make-busy arrangement;

— **-erde** *f*, busy earth, *A*;

— **-lampe** *f*, engaged lamp;

— **-prüfung** *f*, busy test, engaged test, checking;

— — —, Knackgeräusch bei der, engaged click;

— **-sein** *n*, engagement; engaged condition;

— **-spannung** *f*, engaged battery, busying potential;

— — **anlegen**, to establish the busy condition, to apply busying potential;

— **-stellung** *f*, busy condition;

— **-ton** *m*, busy (back) tone;

— **-zeichen** *n*, engaged signal; busy (back) tone;

— — —, **optisches**, visual engaged signal;

— **-zustand** *m*, busy condition.

Besetzung *f*, staff, personnel.

besichtigen, to inspect, to survey.

Besichtigung *f*, inspection, surveying.

besprechen, to control, *V*.

Besprechung *f*, einer Röhre (voice) control, *V*;

Gitter- —, grid control, talking to the grid.

besprochen, speech-modulated, voice actuated.

Besselsche Funktion *f*, Bessel's function.

Bessemerstahl *m*, Bessemer steel.

beständig, steady, stable, continuous, invariable, constant.

Beständigkeit *f*, constancy, invariability, continuity, stability, steadiness.

Bestandteil *m*, constituent, ingredient, part, component.

bestehen, to exist; to be composed of.

Bestehen *n*, existence.

Bestellanstalt *f*, delivering office, *T*.

bestellen, to order; Telegramme: to deliver (up); einen Vertreter: to assign.

Besteller *m*, messenger, *T*.

Bestellung *f*, order; delivery, *T*; assignment.

bestimmen, to fix, to determine, to define; Kurven: to plot.

bestimmt, fixed, definite, precise.

Bestimmung *f*, determination, definition; finding.

Bestimmungs-gleichung *f*, defining equation;

— **-land** *n*, country of destination;

— **-ort** *m*, destination;

— **-stück** *n*, characteristic.

bestreichen, Kontakte: to brush, to sweep over; Magnete: to stroke.

betakeln, to rig; bewickeln: to whip.

betätigen, to actuate, to operate; bedienen: to manipulate.

Beton *m*, concrete;

aus — **bestehend**, concrete ...;

armierter —, reinforced concrete;

Beton
— -**bettung** *f*, concrete bed;
— -**deckplatte** *f*, concrete cover;
— -**fundament** *n*, concrete bed, concrete foundation;
— -**fußboden** *m*, concrete floor;
— -**kasten** *m*, concrete box;
— -**mast** *m*, concrete pole;
— -**mischer** *m*, concrete mixer;
— -**platte** *f*, concrete slab;
— - **rohr** *n*, concrete pipe.

Betrag *m*, amount, magnitude;
— **und Phase**, magnitude and phase.

betragen, to amount to.

betreiben, to run, to work, to operate; to exploit; nach einem selbsttätigen System —, to operate on an automatic system; aus einer Batterie —, to run from a battery; mit Arbeitsstrom —, to work on open circuit.

Betrieb *m*, working, operation, running, performance; exploitation; service.
außer —, out of gear; (vorübergehend, temporarily) out of service; off;
— — setzen, to put out of gear, out of action;
im —, in gear; in service; on;
in — setzen, to set to work, to put into service.

Arbeitsstrom- —, open-circuit operation;
Reihen- — mehrerer Ämter hintereinander, tandem operation;
Ruhestrom- —, closed-circuit operation.

Betriebs-Abteilung *f*, service section, *F*, *T*;
— -**apparat** *m*, working set, working instrument;
— -**bedingungen** *pl*, operating or working or service condition(s *pl*);
— -**fähigkeit** *f*, working order, clearness;
— -, **Prüfung** *f* **auf**, clear test;
— -**geschwindigkeit** *f*, (commercial) working speed;
— -**kapazität** *f*, mutual capacity, wire-to-wire capacity;
— - **der Viererkreise**, pair-to-pair capacity, *K*;
— -**kosten** *pl*, working cost, operating expense, cost of operation;
— -**luft** *f*, air draught;
— -**raum** *m*, — -**saal** *m*, operating room;
— -**sicherheit** *f*, constancy of performance, reliability of operation;
— -**spannung** *f*, operating voltage;
— -**störung** *f*, breakdown;
— -**strom** *m*, working current, (minimum) operating current;
— -**versuch** *m*, field test, field trial;
— -**vorschrift** *f*, service instruction, operating rule;
— -**weise** *f*, method of operation;
— -**welle(nlänge)** *f*, operating wavelength, *R*;
— -**zustand** *m*, working order.

betriebs-fähig, clear, perfect, in working order;
— -**mäßig**, workable; commercial;
— -**sicher**, reliable (in operation).

Bett *n*, **Bettung** *f*, bed(ding).

beugen, (sich), to bend; to diffract, to inflect.

Beugen *n*, **Beugung** *f*, bending; diffraction, inflection.

Beutel *m*, bag, sack;
— -**elektrode** *f*, bag electrode;
— -**element** *n*, sack cell.

Beverageantenne *f*, Beverage aerial.

Bevölkerungsdichte f, density of population.
bevollmächtigen, to assign.
Bevollmächtigter m, assignee.
Bevollmächtigung f, assignment, assignation.
bewegen,(sich),to move;to travel;
hin- und her- —, (–), to reciprocate;
zu weit —, (–), to overshoot, to overthrow.
bewegend, moving, motive.
beweglich, movable;
— es System m, moving system (eines Galvanometers, of a galvanometer).
Bewegung f, movement, motion; travelling, migration;
hin- und hergehende —, reciprocating motion;
zu weite — des Zeigers usw., overthrow(ing), overshoot(ing);
fortschreitende —, progression;
periodische —, periodical motion;
rückläufige —, retrogression;
schnelle unregelmäßige —, flutter(ing);
sinusförmige —, reine, simple harmonic motion;
Abwärts- —, downward motion;
Anker-—, armature travel;
Aufwärts- —, upward motion;
Schwing- —, vibratory movement;
Wellen- —, wave motion;
Winkel- —, angular motion;
Bewegungs-geschwindigkeit f, velocity of motion;
— -studie f, motion study.
bewehren, to armour, to sheath.
bewehrt, armoured, sheathed;
stahlband- —, steel tape armoured;

stark —, heavily armoured.
Bewehrung f, sheath(ing), armour(ing);
geschlossene —, closed or locked armour:
leichte —, light armour;
offene —, open armour;
schwere —, heavy armour;
Bandeisen- —, hoop iron sheathing;
Draht- —, wire sheathing;
Eisen- —, iron armouring;
Flachdraht-—, flat wire sheathing;
Stahlband- —, steel tape armouring.
Bewehrungsdraht m, armouring or sheathing wire.
Beweis m, proof;
strenger —, rigorous proof, M.
beweisen, to prove; to verify.
bewickeln, to wrap (up), to whip;
mit Band —, to tape.
Bewicklung f, wrapping(-up), whipping; taping;
Krarup- —, iron (tape or wire) wrapping or whipping.
bezeichnen, to denote, to designate; to mark; to label.
Bezeichnung f, designation; label; mark(ing);
Bezeichnungs-schild n, designation card;
— -streifen m, designation strip (am Klappenschrank, of the switchboard).
— -system n, notation;
— - —, dezimales, decimal notation, A.
Beziehung f, relation, relationship;
in — stehen, setzen, to relate (zu, to).
Phasen- —, phase relation.
Bezirk m, district, zone, area;
Amts- —, exchange area, F.
Bezugnahme f, reference (auf, to).

Bezugstromkreis *m*, reference circuit.

bezüglich, relative (to).

Biege-belastung *f*, bending load;
— **-festigkeit** *f*, bending strength
— **-zange** *f*, bending pliers *pl*, pipe bending tongs *pl.*

biegen, to bend, to inflect.

Biegen *n*, bending, inflecting.

biegsam, flexible, pliable.

Biegsamkeit *f*, flexibility, pliability.

Biegung *f*, bend; (point of) inflection (einer Kurve, of a curve).

Bienenwachs *n*, beeswax.

bifilar, double-wound, bifilar;
— **gewickelt**, double-wound, wound in duplicate;

Bifilar- ...., bifilar ....

Bild, *n*, picture, image, diagram;
sprechendes —, talking-motion picture;
Schalt- —, circuit diagram;
Stangen- —, pole diagram, *B*;
— **-telegraphie** *f*, picture or image telegraphy;
— **-übertragung** *f*, picture transmission.

bilden, (**sich**), to form, to set up.

Bildung *f*, formation;
Blasen- —, formation of bubbles;
Kristall- —, formation of crystals.

Bimetalldraht *m*, bimetallic wire; copper-clad steel wire, *B*.

bimetallisch, bimetallic.

binär, binary.

binaural, binaural.

Binde *f*, bandage, binding;
— **-draht** *m*, binding wire, binder, *B*.
— **-glied** *n*, link.

binden, to bind.

Binder *m*, girder.

Bindfaden *m*, packthread, twine.

Bindung *f*, binding;
— **im oberen (seitlichen) Drahtlager**, top (side) binding, *B*.

binomischer Satz *m*, binominal theorem.

Birne *f*, (glass) bulb.

Bittererde *f*, magnesia (MgO).

Bittersalz *n*, sulphate of magnesium ($MgSO_4$). epsom salt ($MgSO_4 + 7 H_2O$).

Bitumen *n*, bitumen.

bituminös, bituminous.

blank, Draht: bare; Metall: bright;
— **gewalzt**, bright rolled.

Blank *n*, blank, spacing signal, *T*;
Buchstaben- —, letter blank, unshift signal;
Zahlen- —, figure blank, shift signal;
— **-taste** *f*, blank key.

Blasbalg *m*, bellows *pl.*

Blase *f*, bubble;
Luft- —, air bubble; **eingeschlossene**: air cavity.

blasen, to blow.

Blasenbildung *f*, formation of bubbles.

Bläser *m*, ventilator.

Blaszylinder *m*, dust bellows *pl.*

Blatt *n*, leaf; Papier: sheet, page; Metall: foil; Messer: blade; Eisenkern: lamina (*pl*: laminae), lamination.
leeres —, (paper) blank;
Gold- —, gold foil;
Kurven- —, graph, curve sheet;
Säge- —, sawblade;
— **-druck** *m*, page printing, *T*.
— — **-telegraph** *m*, page printing telegraph;
— **-drucker** *m*, page printer, *T*;
— **-feder** *f*, flat spring, leaf spring, plate spring;
— **-vorschub** *m*, page-feed, paging-up, *T*.

Blätter=kern *m*, laminated core;
— = — =**spule** *f*, laminated iron core coil.
blättern, to laminate.
Blätterung *f*, lamination.
blau, blue.
Blaupause *f*, blue-print;
— n **herstellen**, to blue-print.
Blaustein *m*, vitriol ($CuSO_4$).
Blech *n*, sheet, plate; des Kerns: lamination, lamina (*pl*: laminae);
aus — **gestanzt**, blanked out from sheet;
gelochtes —, perforated sheet;
Dynamo= —, dynamo sheet iron;
Eisen=—, sheet iron, iron plate;
Gitter= —, perforated sheet;
Riffel= —, channeled plate;
Weiß= —, tin (plate), tinned sheet iron;
Well= —, corrugated sheet iron;
Zink= —, sheet zinc;
— =**gefäß** *n*, can;
— =**kern** *m*, laminated core;
— =**schere** *f*, cutting shears *pl*.
blechern, (Klang) tinny.
Blei *n*, lead;
mit— ausgeschlagen, lead-lined;
schwefelsaures —, sulphate of lead ($PbSO_4$);
Antimon=—, antimonial or antimonious lead;
Hart= —, hard lead;
Walz= —, rolled lead;
Weich= —, soft lead;
— =**gitter** *n*, lead grid;
— =**glanz** *m*, galena, lead sulphide (PbS);
— — **detektor** *m*, galena or lead sulphide detector;
— -**glätte** *f*, lead monoxide, litharge (PbO);
— =**kabel** *n*, lead(-covered) cable;

— = — = **löter** *m*, plumber joiner, *B*;
— =**kappe** *f*, lead cap, *B*;
— =**karbonat** *n*, carbonate of lead ($PbCO_3$);
— =**löter** *m*, plumber;
— =**mantel** *m*, lead sheath(ing), lead covering;
mit einem — = — **umpreßt**, lead-sheathed;
mit einem (neuen) — = — **versehen**, to (re)lead;
Abreißen *n* **des**— = —**s**, parting of the lead sheath;
— =**mennige** *f*, red lead, minium (Pb_3O_4);
— =**muffe** *f*, lead sleeve;
— = —, **Lötwulst** *m* **oder Plombe** *f* **der**, plumber's wiped joint;
— =**oxyd** *n*, lead monoxide, litharge (PbO);
— = —, **kohlensaures**, carbonate of lead ($PbCO_3$);
— =**rohr** *n*, lead tubing;
— = — = **kabel** *n*, lead-covered cable;
— = — = —, **zweiadriges (vieradriges)** lead-covred twin (four) wire cable, two- (four-) wire lead cable;
— = — = — **zwischen Überführungskasten und Freileitung**, pothead tail, tail end, *B*;
— =**sammler** *m*, lead storage cell, lead sulphuric acid cell;
— =**schwamm** *m*, spongy lead;
— =**staub** *m*, lead dust;
— = — =**sammler** *m*, lead-dust storage cell;
— =**stift** *m*, pencil;
— — =**skizze** *f*, pencil drawing;
— =**sulfat** *n*, sulphate of lead ($PbSO_4$);
— **superoxyd** *n*, lead peroxide (PbO_2);
— =**weiß** *n*, white lead ($H_2PbO_2 \cdot 2\ PbCO_3$);

bleiben, in der Leitung, to hold the line, F.
bleiern, leaden, lead.
blind, Blind=..., blind; dummy; wattless;
—**es Stanzen** n **oder Maschine-schreiben,** touch-typing, T;
—**e Stelle** f **im Funkempfang,** blind spot, R;
Blind=komponente f, wattless component, reactive component, reactance component;
— **=leitwert** m, susceptance;
— **=sicherung** f, dummy fuse;
— **=spannung** f, reactance voltage;
— **=spannungskomponente** f, wattless component of e. m. f.;
— **=stöpsel** m, dummy plug;
— **=strom** m, wattless current, reactance current;
— **= =komponente** f, wattless component of ourrent;
— **=widerstand** m, reactance. **mit** — **=** — **behafteter Stromkreis** m, reactive circuit; **induktiver** — **=** —, inductive reactance, inductance reactance, inductance;
kapazitiver — **=** —, condensive or capacity reactance, condensance.
blinken, to flash, to flicker.
Blinken n, flash(ing);
durch — **das Amt zum Eintreten veranlassen,** to flash in the exchange.
Blinklampe f, flash lamp.
Blinkzeichen n, (lamp) flashing, flash or flickering signal;
— **geben,** to flash.
Blitz m, lightning;
der — **schlägt in eine Leitung ein,** the lightning strikes a wire;
Abspringen n **des** —**es,** side flashing of lightning;

— **=ableiter** m, (lightning) protector, lightning arrester;
Hörner= — **=** —, horn-shaped lightning arrester;
Kohle= — **=** —, carbon protector;
Luftleer= — **=** —, vacuum lightning arrester;
Platten= — **=** —, plate lightning arrester;
Schneiden= — **=** —, spark(ing) gap, knife or wedge-shaped lightning arrester.
Spindel= — **=** —, reel protector;
Spitzen= — **=** —, point lightning arrester;
Stangen= — **=** —, pole lightning protector;
— **=** **=streifen** m, protector strip;
— **=entladung** f, lightning discharge;
— **=pfeil** m, danger arrow;
— **=schlag** m, lightning stroke or flash or discharge;
— **=schutz** m, protection against lightning discharge; protectors, pl;
— **=** **=spirale** f, inductance spiral;
— **=** **=vorrichtung** f, lightning protector.
Block m, **Holz:** block, log; **Eisenbahn, Flaschenzug:** block;
— **=kondensator** m, block(ing) or stopping condenser;
Gitter= — **=** —, grid blocking condenser;
Luftdraht= — **=** —, aerial blocking condenser;
— **=schrift** f, block signals pl, T;
— **=signal** n, **Eisenbahn:** block signal;
— **=strecke** f, **Eisenbahn:** block (section).
blocken, blockieren, to block.
bloß, bare;
— **legen,** to bare.

Bock *m*, pedestal, trestle; jack;
Hebe= —, lifting jack.
Boden *m*, soil; base;
— =brett *n*, baseboard;
— =satz *m*, sediment.
Bogen *m*, bow, curve, bend;
Grab= —, bow;
Kreis= —, arc;
Rohr= —, bend.
— =bildung *f*, arcing, formation of arcs;
— =lampe *f*, arc lamp;
— =lampenkohle *f*, carbon;
— =lineal *n*, bow, French curve;
bogenbildend, Metall: arcing;
nicht —, non-arcing.
bohren, to bore, to drill;
auf= —, nach= —, to rebore.
Bohrer *m*, borer, drill, auger;
Brust= —, breast drill;
Erd= —, (earth) auger;
Gewinde= —, screw tap;
Spiral= —, twist drill;
Zentrum= —, centre bit.
Bohr=futter *m*, drill chuck;
— =käfer *m*, Holz=, wood boring beetle;
— =knarre *f*, ratchet drill;
— =lehre *f*, drilled jig;
— =loch *n*, bore;
— =maschine *f*, drilling machine;
— = —, Stangenloch=, post hole drilling machine;
— =röhre *f*, boring tube;
— =spitze *f*, bit;
— =winder *m*, hand brace, boring brace, breast drill.
Bohrung *f*, bore; boring;
Anker= —, armature bore.
Boje *f*, buoy.
Bolometer *n*, bolometer.
bolometrisch, bolometric(al).
Bolzen *m*, bolt, screw;
Anker= —, tie bolt;
Anschluß= —, connecting bolt;
Ösen= —, eye bolt;
Ring= —, ring bolt, eye bolt;
Verbindungs= —, tie bolt.
bombardieren, to bombard, *V*;
Bombardement *n*, bombardment;
Elektronen= —, electron bombardment.
Bor *n*, boron (B).
Borax *m*, borax ($Na_2B_4O_7$).
Bord *m*, board;
an —, on board ship;
— =funker *m*, wireless officer;
— =funkstelle *f*, ship radio station.
Börse *f*, exchange.
Börsendrucker *m*, stock ticker.
Böschung *f*, slope.
Bote *m*, messenger.
Botenlohn *m*, porterage.
Bottich *m*, (wood) tank.
boucherisieren, to boucherize, *B*.
Boucherisierung *f*, boucherization, *B*.
B=Platz *m*, B-position, *F*;
halbautomatischer —=—, semi-B-position, *A*.
Bradfieldisolator *m*, Bradfield insulator, *R*.
Brand *m*, fire; Kollektor, Spule: burn-out.
Bratsche *f*, (alto) viol(a).
braun, brown.
Braunsche Röhre *f*, Braun tube.
bräunen, (brünieren) to burnish.
bräunlich, brownish.
Braunstein *m*, manganese dioxide, pebble manganese (MnO_2)
Brech=eisen *n*, —=stange *f*, crowbar.
brechen, to break, to crack, to fracture, reißen: to rupture, to break; Morsezeichen: to split; Strahlen: to refract.
Brechen *n*, breakage, fracture; splitting.
Brechung *f*, refraction.

breit, wide, ... in width;
Morsezeichen: lengthened.
Breite f, width, breadth; geographisch: latitude;
....° nördlicher (südlicher) —, lat ....° N (S).
Breitseite f, breadth.
Bremse f, brake.
bremsen, to brake, to check (the motion).
Brems=filz m, tow brush, T;
— =magnet m, brake magnet, T;
— =ring m, brake ring;
— =zylinder m, dash-pot;
— = — mit Ölfüllung, oil dashpot.
brennbar, combustible, inflammable.
brennen, to burn. [fire.
brennend, burning, alight; on
Brennen n, burning; der Lampen: illumination.
Brenner m, burner; Röhren: heated filament;
Bunsen= —, Bunsen burner.
Nernst= —, Nernst needle.
Brenn=punkt m, focus;
im — = — vereinigen, to focus;
— =stempel m, marking or branding iron, B;
— =stunde f, burning hour, lighting hour;
— =weite f, focal distance.
Brett n, board, panel; shelf;
Boden= —, baseboard;
Grund= —, base, baseboard;
Kabel= —, cable shelf;
Brief m, letter;
— =beutel m, mail bag;
— =einwurf m, slit;
— =kasten m, letter box, mail box;
— =marke f, post stamp;
— =post f, mail;
— =telegramm n, night telegraph letter;
— =träger m, postman;
Land= —=—, rural postman.

brodeln, to boil.
Brodeln n, boiling (noise).
Bronze f, bronze, gunmetal;
Silber= —, silver bronze;
Silizium= —, silicon bronze;
— =draht m, bronze wire.
Bruch m, breakage, fracture;
M: fraction;
muscheliger — des Porzellans, conchoidal fracture of porcelain;
speckiger —, lardaceous fracture;
— =belastung f, breaking load;
— =festigkeit f, breaking strength,
— =last f, breaking weight or load; breaking strain;
— =spannung f, breaking strain;
— =strich m, bar of fraction;
— =stück n, fraction;
— =teil m, fraction.
brüchig, brittle;
kalt= —, cold-short brittle;
warm= —, hot-short brittle.
Brücke f, bridge; des Stöpsels: contact plate;
in — liegen, to be in bridge (zu, across);
in — schalten, to bridge, to tee, (zu, across),
in — geschaltet, in leak, teed or bridged across;
einfache —, single bridge;
magnetische —, permeability bridge;
Wheatstonesche —, Wheatstone bridge;
Scheitel(punkt) m der —, bridge apex;
Verhältnisarme pl der —, ratio arms pl;
Doppel= —, double bridge;
Gleichstrom= —, d. c. bridge;
Gleitdraht= —, slide wire bridge;
Scheinwiderstands= —, impedance bridge;

Brücke
Meß- —, measuring bridge;
Schleifdraht- —, slide wire bridge;
Wechselstrom- —, a. c. bridge;
Brücken-arm m, bridge arm; duplex or bridge coils pl T;
feste — - —e pl, ratio arms pl;
— -draht m, bridge wire;
— - — mit Gleitkontakt, (differential) slide wire;
— -gegensprechsystem n, bridge duplex system, T;
— -gleichgewicht n, bridge balance; duplex balance T;
— -Isolationsmesser m, — -megger m, bridge megger;
— -messung f, bridge test, bridge measurement;
— -scheitel m, bridge apex.
brummen, to hum.
Brummen n, hum(ming).
brünieren, to burnish.
Brust-bohrer m, breast drill;
— -leier f, hand or boring brace, breast drill;
— -mikrophon n, breastplate transmitter.
B-Schrank m, B-(switch)board, F;
Buche(nholz n) f, beech.
Buchsbaumholz n, boxwood.
Buchse f, collet, sleeve, bush; bushing; Nabe: hub;
Anschluß- —, (connector) socket;
Isolier- —, insulating bush;
Lager- —, hub;
Nab- —, bushing;
Steck(er)- —, plug socket, connector socket.
Büchse f, box;
Rohrpost- —, (pneumatic dispatch) carrier.
Buchstabe m, letter;
großer —, capital;
Buchstaben pl, Gegensatz zu Zahlen und Zeichen: letters, capi-

tals, lower case characters, pl, T;
— -abstand m, letter space, letter blank, T;
— -bezeichnung f, lettering;
— -blank n, letter blank or space, T;
— -umschaltung f, — -wechsel m, letter shift (signal), unshift, T;
— -vorschub m, letter feed, T;
— -weiß n, letter blank or space, T;
— -zählvorrichtung f, letter-counting device.
buchstabieren, to spell.
Bucht f, bay (des Zwischenverteilers of IDF);
Verstärker- —, repeater bay.
Buckel m, hump, bulge (einer Kurve: of a curve).
Buffer m, pad, buffer;
— -kontakt m, buffer contact;
— -feder f, buffer spring.
Bug m, bow.
Bügel m, clamp, clip, strap, bail, bow;
durch einen — verbunden, clamped or strapped (together).
Bühne f, platform, gallery; Theater: stage.
Bund m, binding; Absatz: collar, shoulder; Bündel: bunch;
Ober- —, top binding, B;
Seiten- —, side binding, B;
Bündel n, bundle, bunch, group; Papier: file, pack;
Leitungs- —, ankommendes (abgehendes), bundle of incoming (outgoing) trunks, A;
Verbindungsleitungs- —, trunk group A;
Vierer- —, four-wire core;
— -staffelung f, group grading A.
bündeln, to bunch (together).
bündig, compact.
Bündigkeit f, compactness.

Bunsen-brenner *m*, Bunsen burner;
— -element *n*, Bunsen cell.
Bürgersteig *m*, footway.
Büro *n*, office, bureau.
Bürste *f*, brush, wiper;
 die —n verstellen, to shift the brushes;
 a/b- —n *pl*, line wipers *pl*, *A*;
 e- —, private wiper, A;
 Gaze- —, gauze brush;
 Kupfergewebe- —, copper gauze brush;
 Steuer- —, private wiper, *A*;
 Stromzuführungs- —, current supply brush.
Bürsten-abhebevorrichtung *f*, brush lifting device;
— -arm *m*, brush arm, brush gear, trailer;
— - — -spindel *f*, brush shaft, *T*; wiper shaft, *A*;
— -detektor *m*, catwhisker detector, *R*;
— -stellung *f*, brush position;
— -halter *m*, brush holder;
— -paar *n*, coupled brushes *pl*, pair of brushes;
— -rahmen *m*, Mc Berth-Wähler: brush carriage;
— -träger *m*, brush gear, brush carriage;
— -verstellung *f*, brush shifting.
Büschel *m*, bunch; brush;
 sich in —n entladen, to brush;
— -entladung *f*, brush discharge;
— -licht *n*, brush light.

C.

Ccm, cubic centimetre(s).
Ceder *f*, cedar;
 rote virginische —, red cedar.
Cello *n*, (violon)cello, bass viol.
Celsiusgrad *m*, ... degree Centigrade.
Cerussit *m*, cerusite ($PbCO_3$) *R*.
CGS-Einheit *f*, cgs-unit.
Charakteristik *f*, characteristic property, characteristic curve; Wellenwiderstand: characteristic impedance, surge impedance, reeller Teil: characteristic resistance;
 Kurzschluß- —, short-circuit impedance;
 Leerlauf- —, no-load impedance;
 Maschinen- —, speed-load characteristic;
 Polar- —, polar characteristic.
charakteristisch, characteristic(al).
Chattertonmasse *f*, Chatterton's compound.

Chaussee *f*, road.
chaussieren, to macadamise.
Chaussierung *f*, macadam; macadamisation.
Chemie *f*, chemistry;
 Elektro- —, electrochemistry.
Chemikalien *pl*, chemicals, chemical drugs, *pl*.
Chemiker *m*, chemist.
chemisch, chemical;
— rein, chemically pure.
Chiffre *f*, cipher.
Chiffern-schlüssel *m*, cipher(ing) code;
— -telegramm *n*, ciphered message.
chiffrieren, to cipher, to code, to encipher.
Chiffriermaschine *f*, ciphering machine.
Chlor *n*, chlorine (gas) (Cl);
— -ammonium *n*, ammonium chloride (NH_4Cl);
— -kalzium *n*, calcium chloride ($CaCl_2$);

Chlor
— -natrium *n*, sodium chloride (NaCl);
— -wasserstoffsäure *f*, hydrochloric acid (HCl).

Chloridsammler *m*, chloride storage cell.

Chrom *n*, chromium (Cr);
— -säure *f*, chromic acid;
— - — -element *n*, chromic acid cell, bichromate cell;
— -stahl *m*, chromium steel.

Clarkelement *n*, Clark cell.

c-Leitung *f*, third conductor, c-wire, *F*.

Cosekante *f* (*ab*: cosec), cosecant, cosec.

Cosinus *m*, (*ab*: cos), cosine, cos;
— -reihe *f*, cosine series.
hyperbolischer —, (*ab*: Cof, cosh), hyperbolic cosine, cosh.

Cotangente *f*, (*ab*: cot), cotangent, cot.

Coulomb *n*, coulomb;
— -sche Wage *f*, Coulomb's balance.

Curbsenden *n*, curbed signalling, curbing, *T*.

D.

Dach *n*, roof, shed.
dachartig, roof-like, roof-shaped.
Dach-first *m*, ridge;
— -gebälk *n*, roof timber, roof beams *pl*;
— -gestänge *n*, roof standard, overhouse structure;
— -sparren *m*, rafter, spar;
— -ständer *m*, roof pole, roof standard;
— - —, Abspann-, roof end standard.

Damm *m* der Straße, roadway, carriageway.

Dampf *m*, steam, vapour, fume;
Quecksilber- —, mercury vapour;
Säure- —, acid fume;
— -kessel *m*, boiler;
— -maschine *f*, steam engine;
— -turbine *f*, steam turbine.

dämpfen, to attenuate, to damp; gedämpft werden, to attenuate; vollständig dämpfen, to damp out.

Dämpfer *m*, deafener, silencer, anti-hum, *B*; sourdine, damper;
— -fahne *f*, vane;

— -feder, *f*, damping spring;
— -kammer *f*, damping chamber.

Dämpfung *f*, attenuation, damping; (transmission) loss, transmission equivalent, *L*;
geometrische —, geometrical attenuation;
sekundäre —, current supply loss (in 3. B.-Netzen, in c. b. systems);
spezifische —, attenuation constant;
zulässige — der Teilnehmerleitung auf der Empfangs- (Sende-)seite, local line receiving (sending) allowance;
Ableitungs- —, leakance loss *L*;
Flüssigkeits- —, liquid damping;
Gesamt- —, total cable equivalent, total transmission equivalent, total loss, attenuation length, transmission efficiency (of...m. s. c.);
— - —, zulässige, total permissible transmission equivalent;
Leitungs- —, line loss;

Dämpfung
Luft= —, air damping;
Nebensprech=—, crosstalk transmission equivalent;
Nutz= —, useful resistance;
Rest= —, net attenuation, net transmission equivalent, overall transmission equivalent, overall transmission loss;
Schein= —, apparent attenuation;
Übersprech= —, crosstalk transmission equivalent;
Verlust= —, loss damping;
Widerstands=—, resistance loss;
Dämpfungs=äquivalent n, attenuation equivalent;
— =ausgleich m, attenuation equalization, K;
— =ausgleicher m, attenuation equalizer, equalizing network, K;
— =dekrement n, decrement;
logarithmisches — = —, logarithmic decrement;
scheinbares logarithmisches —=— equivalent logarithmic decrement;
— =entzerrer m (line) attenuation equalizer;
— =entzerrung f, attenuation equalization, correction;
— =faktor m, attenuation factor, damping factor;
— =frequenzkurve f, attenuation-frequency curve;
— =konstante f, attenuation constant, damping factor or coefficient;
— =kurve f, attenuation curve;
— =maß n, attenuation equivalent, total transmission equivalent, total attenuation;
— =messer m, transmission (efficiency) measuring set, attenuation measuring device, L; decremeter, R;

— = —, Strecken=, transmission measuring set (for straight-away tests).
— =messung f, attenuation measurement, transmission test;
— = —, Rest=, overall (toll circuit) transmission test $K F$;
— = —, Strecken=, transmission efficiency test K, F;
— =verlauf m, attenuation curve, attenuation characteristics;
— =verminderung f, regeneration, R; gain, improvement, L;
— = —, erhöhte oder übertriebene, super-regeneration, R;
— =zahl f, attenuation factor;
— =ziffer f, attenuation coefficient.

Daniellelement n, Daniell cell.
darstellbar, representable.
darstellen, to represent, als Kurve: to plot, to graph, to depict, schematisch: to skeletonize;
mit starken (schwachen) Linien—, to show heavy (light);
vektoriell —, to represent vectorially;
einen Wert in Abhängigkeit von einem andern —, to plot a value against another magnitude.
Darstellung f, representation; graph; plotting (of a curve); vektorielle —, vector representation.
Daten pl, data pl.
Datum n, date.
Dauer f, duration;
Schwingungs= —, period or time of oscillation;
— =brenner m, permanent call F.
dauerhaft, durable.
Dauer=haftigkeit f, durability;
— =magnet m, permanent magnet;

Dauer
— **-magnetisierung** *f*, permanent magnetization;
— **-strom** *m*, permanent current;
— **-wert** *m*, steady (state) value;
— **-zustand** *m*, steady state.
dauern, to last.
dauernd, permanent.
Daumen *m*, cam, tappet;
 axial wirkender —, face cam;
 versetzte — *pl*, staggered cams *pl*;
 Anschlag- —, stop cam;
 Auslöse- —, releasing cam;
 Hebe- —, lifting cam;
 Korrektions- —, correcting cam *T*;
 Mantel- —, edge cam;
 Rückführ- —, resetting cam;
 Transport- —, spacing cam, (paper) feed cam;
— **-regel** *f*, thumb rule;
— **-scheibe** *f*, cam.
dechiffrieren, to decipher, to decode.
Decke *f*, ceiling; cover; mat; envelope.
Deckel *m*, cover, lid; cap; shutter;
 Mannloch- —, man hole cover;
 Glas- —, glass top, glass cover.
decken, to cover; sich teilweise —, to overlap.
Deck-glas *n*, lamp cap, *F*;
— **-licht** *n*, skylight;
— **-platte** *f*, cover (plate), top-plate;
— **-rahmen** *m*, manhole frame (des Kabelbrunnens, of cable manhole).
defekt, defective.
Defizit *n*, deficiency.
deformieren, to deform.
Deformierung *f*, deformation.
dehnbar, ductile.
Dehnbarkeit *f*, ductility.

dehnen, to rack, to stretch; **sich** —, to expand.
Dehnung *f*, elongation.
 elastische —, elastic elongation.
Dehnungsgrenze *f*, elastic limit.
Definitionsgleichung *f*, defining equation.
Dekade *f*, decade; level, *A*.
Dekaden-vielfach *n*, level multiple, *A*;
— **-widerstand** *m*, decade resistance box, decimal resistance.
Deklination *f*, declination.
Dekrement *n*, decrement;
 lineares —, linear decrement;
 logarithmisches —, logarithmic decrement;
 — —, **scheinbares**, equivalent logarithmic decrement;
 Empfänger- —, receiver decrement;
 Sender- —, transmitter decrement;
 — **Null**, zero decrement;
— **-messer** *m*, decremeter,
Dekremeter *n*, decremeter.
Demodulation *f*, demodulation.
Demodulator *m*, demodulator, translating circuit.
demodulieren, to demodulate.
demontieren, to disassemble, to dismantle; to break down.
Depolarisation *f*, depolarization.
Depolarisator *m*, depolarizer.
depolarisieren, to depolarize.
Destillation *f*, distillation.
destillieren, to distill.
destilliertes Wasser *n*, distilled water.
Detektor *m*, detector, responder;
 elektrolytischer —, electrolytic detector or responder;
 Bleiglanz- —, lead sulphide detector, galena detector;
 Bürsten- —, brush detector, cat(s)whisker detector;

Detektor
Elektrolyt= —, electrolytic detector,
Kristall= —, crystal detector;
Magnet= —, magnetic detector;
Perikon= —, Perikon detector or rectifier;
Pinsel= —, cat(s)whisker detector;
Röhren= —, (thermionic) valve detector;
Thermo= —, thermo-electric detector;
— =empfänger *m*, detector receiver;
— =Fernhörerkreis *m*, detector-phone circuit;
— =gerät *n*, detector receiving set;
— =röhre *f*, audion, detecting or detector valve or tube;
— =wirkung *f*, detector or detecting action.

Determinante *f*, determinant.

deutlich, pronounced, articulate(d);
— e Aussprache *f*, (good) articulation, pronunciation.

Deutlichkeit *f*, clearness; definition; Sprache: articulation.

Dezentralisation *f*, decentralisation.

dezentralisieren, to decentralise.

Dezimal=bruch *m*, decimal fraction;
— =stelle *f*, decimal place.

Dezimeter *n*, (*ab*: dm), decimetre (= 3.937 ins.);
Kubik= —, (*ab*: cdm, dm³), cubic decimetre (= 61.026 cub. ins.);
Quadrat= —, (*ab*: qdm, dm²), square decimetre (= 15.501 squ. ins.).

diagonal, diagonal.

Diagonale *f*, diagonal.

Diagonalstrebe *f*, diagonal strut, *B*;

Diagramm *n*, diagram:
Polar —, polar diagram;
Vektor —, vector diagram.

Diamagnetismus *m*, diamagnetism.

diamagnetisch, diamagnetical.

diametral, diametrical;
— **gegenüber(liegend)**, diametrically opposite.

Diaphragma *n*, diaphragm.

dicht, dense; consistent.

Dichte *f*, density; consistence, consistency;
mittlere —, average density;
spezifische —, specific density;
Fluß= —, flux density;
Ladungs= —, density of charge.

Dichtigkeit *f*, density; consistency.

dick, thick, heavy;
— =drähtig, heavy gauge wire ..;
— =wandig, thick-walled.

Dicke *f*, thickness.

Dielektrikum *n*, dielectric;
schlechtes —, poor dielectric.

dielektrisch, dielectric(al);
— e Festigkeit *f*, dielectric strength;
— e Verluste *pl*, dielectric losses
— er Verlustwinkel *m*, phase angle difference (of a condenser);
— e Widerstandsfähigkeit *f*, elastance.

Dielektrizitätskonstante *f*, specific inductive capacity, *ab*: s. i. c., dielectric constant, electric inductivity (of medium), permittivity (selten).

Dienst *m*, service; duty;
im —, Beamter: on duty;
den — antreten, to assume duty;
— =anruf *m*, official call, service call;

Dienst
— =anweisung f, service instruction;
— =gespräch n, service call;
— =leitung f, order wire, o. w.; speaker wire, call wire; service circuit; zwischen 2 Plätzen eines Amtes: transfer circuit; Verbindungsleitung: order wire junction;
— = —, unmittelbare, straight order wire;
— = —, Sammel=, split order wire;
— =leitungs=betrieb m, order wire operation, o. w. trunking;
— = — =taste f, order wire key, assignment key;
— =telegramm n, service message;
— =vorschrift f, service rules pl;
— =wähler m, service connector A;
— =überwachung f, service observation;
— =überwachungsplatz m, observation desk;
— =zeichengeber m, auto-control, T.
Dieselhorst=Martin=Kabel n, m(ultiple t(win) cable;
— = — =Verseilung f, m(ultiple) t(win) formation.
differential, differential;
— gewickelt, differentially wound
Differential n, differential, increment, M;
— =galvanometer n, differential galvanometer;
— =Gegensprechsystem n, differential duplex system, T;
— =gleichung f, differential equation;
— =kondensator m, differential twin condenser;
— =rechnung f, differential calculus;

— =relais n, differential relay;
— =transformator m, — =übertrager m, differential transformer, differential repeating coil;
— = —, dreispuliger, hybrid coil V, F;
— =vorschub m des Morselochers, differential (tape) feed (of the Wheatstone perforator);
— =wicklung f, differential winding;
— =wirkung f, differential action.
Differentialtät f, differentiality.
Differentiation f, differentiation.
Differenz f, difference;
halbe —, semi-difference;
differenzieren, to differentiate (nach x, with respect to x).
diffundieren, to diffuse.
Diffusion f, diffusion.
Dimension f, dimension.
dimensionieren, to dimension.
Dimensionsgleichung f, dimensional equation.
Diplex=betrieb m, diplex operation;
— =system, n, diplex system;
— =telegraph m, diplex telegraph.
Dipol m, dipole.
direkt, direct; Kopplung: conductive.
Direktor m, director, A;
— =system n, director or controller or translator system A.
Dirigeur m, dirigeur, T.
disruptiv, disruptive.
Dissoziation f, dissociation.
dissoziieren, to dissociate.
Distanz f, distance;
— =stück n, distance piece.
divergent, diverging, divergent.
Divergenz f, divergence.
divergieren, to diverge (from).
Dividendus m, dividend.

dividieren, to divide (durch, by).

Divisor m, divisor.

D. M.-Kabel n, multiple twin cable, ab: m. t. cable;

D. M.-Verseilung f, multiple twin formation, ab: m. t. formation.

Docht m, wick;

— -öler m, wick lubricator;

— -ölung f, — -schmierung f, wick lubrication.

Donner m, thunder.

Doppel n, duplicate.

Doppel- ...., doppel- ...., double, twin, duplicate;

— -ader f, twin wire, twin leader, pair;

gekreuzte — - —, crossed pair;
verdrallte — - —, twisted pair;
Treusen- — - —, worming pair;

— -adrig, two-wire..., pair..., twin..., bifilar;

— - -es Kabel n, twin cable, bifilar cable;

— -anker m, V-stay, B;

— -arbeitskontakt m, double make contact;

— -blockkondensatoren pl (Seekabel), double block condensers pl;

— -brücke f, double bridge;

— -daumen m, shuttle cam (für Hin- und Rückgang, for reciprocating motion);

— -drähtig, double-wire ...., two-wire ....;

— - e Leitung f, metallic circuit;

— -empfang m, dual reception, double reception;

— -gegensprechen n, quadruplex system T;

— - — für Gabelverkehr, split quadruplex;

— - — — Staffelverkehr, extended quadruplex;

— -gegensprechsystem n, quadruplex system, T;

— -gestänge n, H-pole, B;

— -gitterröhre f, double grid valve;

— -glocke f, double shed, double petticoat;

— -glockenisolator m, double shed or petticoat insulator;

— -hebelschalter m, double lever switch;

— -kapselmikrophon n, double button transmitter;

— -kegelantenne f, double cone aerial;

— -klemme f, double terminal;

— -kondensator m, twin condenser;

— -kontakt m, double or collateral contact, A;

— -kopfhörer m, headphones pl;

— -kurbelumschalter m, double lever switch;

— -leitung f, metallic (return) circuit, two wire circuit, loop-(ed) circuit, double conductor(s pl);

Widerstand je Meile — - —, resistance per loop mile;

gekreuzte — - —, transposed pair;

verdrallte — - —, twisted pair or loop;

Fernsprech- — - —, metallic telephone circuit;

Teilnehmer- — - —, subcriber's loop;

— -leitungs-betrieb m, two-wire operation;

— - — -widerstand m, loop resistance;

— -metalldraht m, bimetallic wire, B;

— -mikrophon n, double button transmitter, pushpull transmitter;

Doppel=...., doppel=....,
— =nabeltelegraph m, double needle telegraph;
— =polig, bipolar, double pole, double polar;
—=-e Umschaltung f, double commutation;
— =rohrverstärker m, two-valve repeater;
Zweidraht= — =—, two-valve two-wire (intermediate) repeater K;
— =schließkontakt m, double-make contact;
— =schließrelais n, double make relays;
— =seitig, to both sides;
— =senden n, double transmission, R;
— =sperrklinke f, double dog, pair of pawls, double pawls pl, double detent, A;
— =sprechen n, phantom telephony;
— =sprech=betrieb m, phantom telephone operation;
— =— =ringübertrager m, phantom repeating coil;
— =—=schaltung f, phantom telephone connection;
— =stecker m, — =stöpsel m, biplug, double plug, two-pin plug;
— =strom m, double current, T;
— =— =gegensprechen n (auf Doppelleitungen), (metallic) polar duplex;
— =— =gegensprechsystem n, polar duplex system;
— =— =taste f, double current key, T;
— =T=Anker m, H-armature, shuttle armature;
— =trennkontakt m, double break contact;
— =unterbrechungs=klinke f, double break jack;

— =— =relais n, double break relay;
— =verbindung f, double connection F;
— =wandig, double-walled, hollow-walled;
— =zellenschalter m, double cell switch;
— =zwilling m, two pair core, K.
doppelt, double, duplicate;
— =gerichtet, both-way. two-way F;
— =—=e Verbindungsleitung f, both-way junction, two-way trunk circuit;
— =wirkend, double acting.
Dorn m, mandrel, mandril, spike, drift.
Dose f, box;
Abzweig= —, connector box; Anschluß= —, Ansteck= —, wall socket;
Dosen=fernhörer m, watch receiver, watch-case telephone;
— =relais n, box relay;
— =wecker m, circular bell; Gleichstrom= — =—, circular trembler.
Drachen m, kite;
— =schnur f, kite string.
Draht m, wire; leader, conductor; filament;
einen — führen, verlegen, to run a wire;
einen — ziehen, to string a wire, B;
blanker —, bare wire, bare conductor;
dünner —, fine wire, small gauge wire;
geflizter —, stranded wire;
(hart)gezogener —, (hard) drawn wire;
isolierter —, insulated or covered wire;
starker —, heavy gauge wire;
umklöppelter —, braided wire;

Draht
umsponnener —, einfach (doppelt, dreifach), single (double, triple) covered wire;
— —, mit Baumwolle, cotton covered wire;
unterteilter —, composite wire, stranded wire;
verkupferter —, copper clad or coppered wire;
verzinkter —, galvanized wire;
verzinnter —, tinned wire;
a- (b) —, a- (b-) wire, a- (b-) limb or leg;
Abspann- —, span wire B;
Anker- —, stay wire B;
Aufhänge- —, suspension wire, suspending wire;
Baumwoll- —, cotton-covered wire;
Bewehrungs- —, armouring or sheathing wire;
Bimetall- —, bimetallic wire;
Binde- —, binding wire, binder, tie wire, B;
Bronze- —, bronze wire;
— - —, Phosphor-, phosphor bronze wire;
— - —, Silizium-, silicon bronze wire;
Brücken- —, bridge wire, (differential) slide wire;
c- —, local lead, c-wire F, A;
Doppelmetall- —, bimetallic wire;
Einführungs- —, lead-in wire;
Einzieh- —, draw wire, B;
Eisen- —, iron wire;
— - —, verzinkter, galvanized iron wire, g. i. wire;
Emaille- —, enamel(led) or enamel insulated wire;
Erd- —, earth or ground(ed) wire;
Fahr- —, trolley or contact wire;
Flach- —, flat wire;

Gleit- — der Meßbrücke, differential slide wire of the measuring bridge;
Haar- —, Wollaston wire;
Hartkupfer- —, hard drawn copper wire;
Heiz- —, heated filament, V;
Kupfer- —, copper wire;
Leitungs- —, line wire;
Litzen- —, stranded or composite wire; litzendraht, R;
Pol- —, connection wire;
Prüf- —, pilot wire, testing wire;
Rund- —, round wire;
Schalt- —, jumper (wire), cross-connecting wire;
Schleif- —, slide wire; Meßbrücke: differential slide wire;
Schmelz- —, fuse wire;
Schutz- —, armouring or sheathing wire;
— - —, geerdeter, earthed or grounded guard wire;
Sicherungs- —, fuse wire;
Spann- —, span wire;
Stachel- —, barbed wire;
Trag- —, suspending or suspension wire;
Überbrückungs- —, jumper (wire);
Verbindungs- —, connecting wire;
Wachs- —, waxed wire;
Walz- —, rolled wire;
Wickel- —, taping wire, B;
Widerstands- —, resistance wire;
Wollaston- —, Wollaston wire;
Zuleitungs- —, lead-in wire;
— -anker m, wire stay, guy wire;
— -bewehrung f, wire sheathing;
— -bürste f, wire brush;
— -fernsprechen n, wire telephony;
drahtförmig, filamentary;

Draht=funk *m*, line radio, wired wireless;
— = — =Endschaltung *f*, carrier terminal circuit;
— = — =leitung *f*, carrier line, carrier circuit;
— = — =system *n*, wire carrier system;
— = — =technik *f*, wire carrier art;
— =gaze *f*, wire gauze;
— =gewebe *n*, wire cloth;
— =kern *m*, wire core;
— = — =spule *f*, wire core coil;
— =kluppe *f*, pliers *pl*, *B*;
— =kreuzung *f*, cross(ing), wire crossing;
— =lager *n*, oberes (seitliches), top (neck) groove, *B*;
— =lehre *f*, wire gauge;
— = — —, amerikanische, American Wire Gage, *ab*: A. W. G.;
— = — —, britische, British Standard Gauge, *ab*: B. S. G.;
— = — —, Birmingham, Birmingham wire gauge, *ab*: B. W. G.;
— = — —, Brown & Sharpe, B. & S. Gauge;
— =litze *f*, strand, stranded wire; litzendraht;
drahtlos, wireless, radio (*v*. Funk=...);
Draht=netz *n*, wire netting, *B*;
— =rundspruch *m*, electrophone (*engl.*), program transmission over wires (*am.*), *F*;
— = — =anlage *f*, electrophone system, *F*;
— =schleife *f*, wire loop;
— =schutzkappe *f*, wire cage;
— =seil *n*, wire rope, stranded wire;
— = — =anker *m*, wire rope stay, stranded wire stay, *B*;
— =spanner *m*, wire stretcher, *B*;
— =spirale *f*, wire spiral or helix;
— =stift *m*, wire nail;
— =stück *n*, filament;
— =telegraphie *f*, wire telegraphy;
— =telephonie *f*, wire telephony;
— =trage *f*, drum barrow, *B*;
— =verbindungen *pl*, wiring;
— =wellentelegraphie *f*, wired wave telegraphy;
— =zieh=bank *f*, drawing bench;
— = — =strumpf, wire grip *B*;
— =zug *m*, pull of wire, *B*;
— = — =tabelle *f*, stress table, *B*.
drahten, to wire, to telegraph;
Drahtung *f*, message, telegram.
drainieren, to drain (off).
Drainieren *n*, draining, drainage.
Drall *m*, twist;
Lagen= —, twist of the wire layers;
Links= —, left-hand twist, left-handed lay;
Rechts= —, right-hand twist, right-handed lay;
— =länge *f*, length of lay, length of twist.
Draufsicht *f*, (top) plan view, plan.
Dregganker *m*, drag, grapnel.
dreggen, to grapple, to drag (for).
Dreggen *n*, grappling.
Drehbank *f*, (turner's or turning) lathe;
Revolver= —, capstan lathe.
drehbar, rotatable, rotating, rotary, revolving;
— lagern, to pivot (in, on);
Dreh=beanspruchung *f*, torsion, torsional strain;
— =bewegung *f*, rotary motion;
drehen, (sich), to turn, to revolve, to rotate, to gyrate, to whirl; Drehbank: to turn;
Dreh=feld, *n*, rotating field;
— =festigkeit *f*, torsional strength;

Dreh
— -klinke *f*, turning pawl;
— -kondensator *m*, rotating plate condenser;
— -ling *m*, lantern;
— -magnet *m*, rotary magnet, *A*;
— -moment *n*, torque;
Anlauf- — - —, starting torque;
— -plattenkondensator *m*, rotating plate condenser;
— -punkt *m*, pivot, fulcrum; centre of rotation;
— -rahmen *m*, rotatable coil, moving frame, *R*;
— -richtung *f*, (direction of) rotation;
— -schalter *m*, rotary or revolving switch, spindle switch;
— -schritt *m*, rotary step, *A*;
— -sinn *m*, (direction of) rotation;
— -späne *pl*, turnings *pl*;
— -spiegel *m*, revolving mirror;
— -spule *f*, moving or rotating coil, rotor;
— -spul-galvanometer *n*, moving coil galvanometer;
— - — -relais *n*, moving coil relay;
— - — -strommesser *m*, moving coil ammeter;
— -stahl *m*, turning knife;
— -strom *m*, triphase or three-phase current;
— - — -motor *m*, triphase motor;
Drehung *f*, rotation (um...°, by...°), turn, angular motion, gyration; twist, torsion, Links- —, counter-clockwise rotation;
Rechts- —, clockwise rotation.
Dreh-variometer *n*, rotating coil variometer;

— -vorwähler *m*, rotary preselector or line switch, *A*;
— -wähler *m*, rotary selector or switch, spindle switch, *A*;
Heb- — - —, vertical and rotary selector, *A*;
— - — -system *n*, rotary system *A*;
— -zahl *f*, number of revolutions;
— -zapfen *m*, pivot, trunnion.
dreiadrig, triple core..., triplet, three-wire....
Dreieck *n*, triangle;
ähnliche —e, *pl*, similar triangles *pl*;
gleichseitiges —, equilateral triangle;
kongruente —e *pl*, congruent triangles *pl*;
rechtwinkliges —, right-angled triangle;
— -glied *n*, delta circuit, π-mesh;
— -schaltung *f*, mesh connection, delta (triphase) connection;
in — - —, mesh-connected, delta-connected;
— -spannung *f*, delta voltage.
Dreielektrodenröhre *f*, triode, three-electrode valve, audion.
Dreieralphabet *n*, three-unit or ternary code, *T*.
dreifach, Dreifach- ...., triple, treble.
Dreifachstecker *m*, three-pin plug, triplug.
Dreifingerregel *f*, hand rule.
Dreifuß *m*, tripod.
dreilamellig, three-bar....
Dreileiter-dynamo *f*, three-wire generator;
— -kabel *n*, triple core cable;
— -netz *n*, three-wire supply system.
dreiphasig, triphase, three-phase.
dreipolig, three-polar, three-pole....

Dreirad *n*, tricycle.
dreischenklig, three-legged.
dreistellig, three-figure....
Dreiwegeschalter *m*, three-way switch.
dreizählig, ternary, ternary.
Drell *m*, drill.
dringend, urgent, express;
— es **Gespräch** *n*, express call;
— es **Telegramm** *n*, urgent message.
Drillich *m*, drill.
Drillingstecker *m*, triplug, three-pin plug.
Drossel *f*, choke, choker;
eisenfreie —, air core choke;
Abflachungs- —, smoothing choke;
Eisen- —, iron core(d) choke or inductance;
Hochfrequenz- —, **Hf.**- —, high-frequency choke, h. f. choke;
Lösch- —, quenching choke;
Luft- —, air core choke;
Niederfrequenz- —, **Nf.**-—, low-frequency choke, l. f. choke;
Schutz- —, protecting or protective choke (coil).
— -**kreis** *m*, multiple or branched or parallel resonant circuit, rejector or stopper circuit;
— -**satz** *m*, rejector circuit; noise killer;
— -**spule** *f*, choke coil, choking coil, impedance coil, inductance (coil), reactive or reaction coil, retard(ation) coil, reactor;
veränderliche — - —, reactance regulator;
— -**wirkung** *f*, choking effect, throttle effect (auf, on).
drosseln, to choke, to iron out, to throttle.
Druck *m*, pressure, stress, thrust, compression; print(ing);

Blatt-—, **Seiten**-—, page printing, *T*;
Quecksilber- —, pressure in (terms of) mm of mercury;
Streifen- —, tape printing, *T*;
— **in mm Quecksilbersäule**, pressure in terms of mm mercury;
— -**achse** *f*, printing shaft, *T*;
— -**apparat** *m*, printer, *T*;
— -**auslösemagnet** *m*, printing trip magnet;
— -**beanspruchung** *f*, compressive or crushing stress;
— -**buchstaben** *pl*, printed characters *pl*;
— -**daumen** *m*, printing cam, *T*;
— -**festigkeit** *f*, compressive strength;
— -**hammer** *m*, printing hammer; *T*;
— -**hebel** *m*, printing lever, *T*;
— -**knopf** *m*, push (key), push or press button;
— -**luft** *f*, compressed air, **Rohr**-**post**: pressure;
— - — -**kondensator** *m*, compressed air condenser;
— - — -**strom** *m*, forced draught;
— -**magnet** *m*, printing or printer magnet, *T*;
— -**messer** *m*, (pressure) gauge;
— -**platte** *f*, platen, *T*;
— -**relais** *n*, printing or printer relay, *T*;
— -**rolle** *f*, platen, impression roller, *T*;
— -**schmierung** *f*, forced oil feed;
— -**schrift** *f*, printed characters *pl*;
— -**schwankung** *f*, fluctuation of pressure;
— -**stab** *m*, stay crutch, spur (einer **Stange**, of a pole) *B*;
— -**telegraph** *m*, (type) printing telegraph;
Blatt- — -—, **Seiten**- — -—, page printing telegraph;

Druck
Streifen= — = —, tape printing telegraph;
— =vorgang m, act of printing, T;
— =zeile f, line of print;
— =Zug=Mikrophon n, double button transmitter, push-pull transmitter;
— = —verstärker m, push-pull amplifier.
drucken, to print.
Drucken n, print(ing).
drücken, to press, to thrust, to bear (gegen on); Taste: to depress (a key).
Drücken n, pressing, thrusting, bearing; depression.
Drucker m, printer, printing apparatus;
Blatt= —, page printer;
Börsen= —, Fern= —, stock ticker;
Seiten= —, page printer;
Streifen= —, tape printer.
Druse f, nodule, druse.
Dübel m, peg, (wall) plug, trenail, socket.
duktil, ductile.
Duktilität f, ductility.
dumpf, Sprache: drummy, heavy.
Dumpfheit f, heaviness, drumminess.
dunkel, dark, obscure.
Dunkelraum m, dark space;
Kathoden= —, dark space round the cathode.
dunkelrot, dull red, dim red.
Dunkelrotglut f, dim or dull red heat.
dünn, thin, fine; Luft: rare;
— =drähtig, small gauge wire ...
— =wandig, thin-walled.
Dunst m, fume.
duplex betreiben, to duplex, T.
Duplex= . . . ., duplex;

— =abgleich(ung f) m, duplex balance;
— =betrieb m, duplex operation T;
— =—, einseitiger, half duplex operation T;
— =leitung f, duplex circuit.
duplizieren, to duplex, T.
Duraluminium n, duralumin.
durchbrennen, Sicherung: to fuse, to blow, to melt, to strike; Spulen: to burn out.
Durchbrennen n, fusing, blowing, melting, striking; Spulen: burn-out.
Durchbruch m, breakdown, puncture.
durchdrehen, to rotate over, A.
durchdringbar, permeable.
Durchdringbarkeit f, permeability.
durchdringen, to permeate, to penetrate.
Durchdringung f, penetration.
durchfließen, to pass, to flow, to traverse.
durchflochten, interwoven (Litze, strand).
durchführen, to carry out; Drähte: to lead through, to pass through.
Durchführungs=isolator m, wall tube insulator, leading-in insulator;
— =rohr n, wall tube.
Durchgang m, passage, transit; zwischen Tischen: gangway, aisle.
Durchgangs=fernleitung f, through toll line, am.: l. d. thru circuit;
— =fernschrank m, through switchboard;
— =gebühr f, transit charge;
— =leitung f, through circuit, transit circuit, am.: thru line;

Durchgangs
— -platz m, through-position;
— -verkehr m, transit or through traffic.
durchgebrannt, Sicherung: blown, fused; Wicklung: burnt out.
durchgehen, Motor: to race, to run away.
durchgehende Leitung f, through line, transit line.
durchgeschaltet, (cut) through (zum I. GW, to first group selector).
Durchgriff m, reciprocal of the amplification or magnification factor, throughgrip (selten), V;
Durchhang m, dip, sag, B;
— -tabelle f, table of sags, B.
durchhängen, to dip, B.
durchlassen, to pass, to let through, to transmit,.
durchlässig, permeable; Isolator: leaky;
— er Kreis m, acceptor circuit.
Durchlässigkeitsbereich m, range of free transmission, transmission range.
durchlochen, to perforate.
Durchlochung f, perforation.
Durchmesser m, diameter;
Außen- —, outer or external diameter;
Gesamt- —, overall diameter;
Innen- —, inner or internal diameter;
Kreis- —, diameter.
durchpausen, to trace.
durchprüfen, to test, to overhaul.
durchrufen, to ring through.
Durchrufen n, through-ringing;
— in Schleifenschaltung, loop ringing;
— — Simultanschaltung, composite (through) ringing;
— — — mit Erdrückleitung, differential earth ringing.
Durchruf-relais n,(through-) ringing relay, signalling relay;
Rufen n mit —-—, relayed ringing;
— -schaltung f, ringing-through scheme.
durchschalten, to cut through, to put through, to connect through, to extend to;
eine Teilnehmerleitung zum I. GW —, to extend a subscriber's circuit to the first group selector.
Durchschaltung f, cutting-through, connecting-through, extension (of ... to ...).
durchscheinend, transparent.
durchscheuern, to chafe (through)
Durchscheuern n, chafing (von Seekabeln, of submarine cables).
Durchschlag m, Werkzeug: punch, piercer; Schreibmaschine: carbon copy; Isolation: rupture, breakdown, puncture.
durchschlagen (werden), to puncture, to disrupt, to break down, to fuse.
Durchschlagen n, puncturing, breaking-down, breakdown, disruption.
Durchschlagsfestigkeit f, disruptive strength, rupturing strength.
Durchschleudern n, overthrow.
durchschmelzen, to melt, to fuse.
Durchschmelzen n, melting, fusing.
Durchschnitt m, section, profile, plan; average;
den — bilden oder nehmen, to average.
durchschnittlich, mean, average.
Durchschnitts-kosten pl, average cost pl. [ed value.
— -wert m, average value, equat-

durchsetzen, to mix, to intersperse; der Fluß durchsetzt die Windungen, the flux threads with the turns.
Durchsicht *f*, inspection.
durchsichtig, transparent, translucent.
Durchsichtigkeit *f*, transparency.
Durchsprechstellung *f*, through position.
durchverbinden, to cut or connect through, to extend through (einen Anruf, a call).
durchwählen, to dial through, *A*.
Durchwählen *n*, through-dialling, *A*.
Dyn *n*, dyne.
Dynamo *f*, dynamo;
(halb)gekapselte —, (semi-) enclosed dynamo;
Compound- —, compound (-wound) dynamo;
Dreileiter- —, three-wire generator;
Gegenverbund- —, differential compound wound dynamo;
Hauptschluß- —, series(-wound) dynamo;
Innenpol- —, internal pole dynamo;
Lade- —, (battery-)charging generator;
Nebenschluß- —, shunt (-wound) dynamo;
Puffer- —, buffer dynamo;
Reihenschluß- —, series (-wound) dynamo;
Rufstrom- —, ringing dynamo, ringer;
Tret- —, pedal dynamo;
Unipolar- —, homopolar or unipolar dynamo;
Verbund- —, compound (-wound) dynamo;
— - —, Gegen-, differentially wound dynamo;
Zusatz- —, booster dynamo;
— -blech *n*, dynamo sheet iron.
dynamoelektrisch, dynamo-electric(al).
Dynamo-karren *m*, supply cart, *R*;
— -meter *n*, dynamometer;
Elektro- -—, electrodynamometer.
dynamometrisch, dynamometric(al).
Dynatron *n*, dynatron.
Dyne *f*, dyne;
Kilo- —, kilodyne;
Mega- —, megadyne.

E.

Eben, plane, plain; wagerecht: level;
— e Fläche *f*, plane.
Ebene *f*, plane, plain;
in einer — mit, flush with;
schiefe —, inclined plane;
Windungs- —, winding plane.
Ebenholz *n*, ebony.
ebnen, to plane, to level.
Ebonit *n*, ebonite;
— -gehäuse *n*, ebonite case;
— -kapsel *f*, ebonite box, ebonite case.
Echo *n*, echo;
— -lotung *f*, echo-sounding;
— -sperrer, *m*, echo suppressor, echo killer *K*;
— -ströme *pl*, echo currents *pl*, *K*;
— -weg *m*, echo path, *K*;
— -wirkung *f*, echo effect, *K*.
Ecke *f*, corner.
Edel-gas *n*, rare gas.
— - - glühkathodengleichrichter *m*, tungar rectifier.
— -metall *n*, nobler metal.
— -stein *m*, jewel.

Edisonsammler *m*, Edison storage cell.
Effekt *m*, effect;
— **Nutz**- —, useful effect.
Effektiv- ..., effektiv, effective;
— **-spannung** *f*, r. m. s. voltage;
— **-strom** *m*, r. m. s. current;
— **-wert** *m*, virtual value, r. m. s. value, effective value.
Eiche(nholz *n*) *f*, oak.
eichen, to calibrate, to gauge (*am*: gage).
Eich-gerät *n*, — **-instrument** *n*, calibration instrument;
— **-kondensator** *m*, calibration condenser;
— **-kurve** *f*, calibration curve.
Eichung *f*, calibration, gauging; **Nach**- —, check calibration.
Eigen-frequenz *f*, natural frequency;
— **-kapazität** *f*, self-capacity (von Spulen, of coils);
— **-periode** *f*, natural period (of oscillation);
— **-schaft** *f*, property, feature, characteristic;
— **-schwingung** *f*, natural motion, natural vibration or oscillation; natural period;
eigenschwingungsfrei, aperiodic(al).
Eigen-schwingungszahl *f*, natural frequency;
— **-verluste** *pl*, internal losses *pl*;
— **-welle(nlänge)** *f*, natural wavelength;
— — — **eines Luftleiters** (ohne zusätzliche Schaltmittel) unloaded wavelength of an aerial.
eigenes Amt *n*, home station.
Eignung *f*, suitability.
Eil-, express, urgent;
— **-bote** *m*, express (messenger);
— **-brief** *m*, express letter, dispatch.
Eimer *m*, bucket.

einadrig, single-core(d), single-wire;
— **es Kabel** *n*, single-core(d) cable.
Einankerumformer *m*, rotary (converter).
ein-armig, one-armed.
— **-bauen**, to mount, to build in;
— **-betten**, to embed.
Einbettung *f*, embedding.
Einbruch-melder *m*, burglar alarm;
— **-wecker** *m*, burglar alarm bell.
Einbuchtung *f*, niche, bay.
ein-drähtig, unifilar.
— **-drehen**, to indent, to recess.
Ein-drehung *f*, indent, recess.
— **-dringen** *n*, ingress, leakage (von Luft in ..., of air into ...) penetration (in, into).
eindringen, to ingress, to penetrate, to leak, (in, into).
Eindringtiefe *f*, penetration depth.
einebnen, to level.
Einer *pl*, units *pl*, units digit *A*;
— **-wahl** *f*, dialling of units *A*;
— **wählen**, to dial units digit *A*;
— **-stufe** *f*, units digit *A*;
einfach, plain; simplex *T*; simple;
— **e Funktion** *f*, simple function *M*;
Einfach-betrieb *m*, simplex operation or working;
— **-heit** *f*, simplicity;
— **-leitung** *f*, simplex circuit *T*;
— **-strom** *m*, single current *T*;
— **-telegraph** *m*, single-channel telegraph; simplex (operated) telegraph.
ein-fädig, unifilar.
— **-fallen**, Riegel: to snap (in, into); Wellen: to incide, to come in, to arrive.

Einfallen *n*, snapping-in; incidence, arrival;
einfallend, incoming, oncoming;
Einfalls=ebene *f*, plane of incidence;
— =richtung *f*, direction of incidence;
— = —, schwankende, wandering direction of arrival *R*;
— =winkel *m*, angle of incidence.
einfärben, to ink *T*.
Ein=färben *n*, Einfärbung *f*, inking *T*;
— =fluß *m*, influence.
einfügen, to interpose.
Einfügen *n*, Einfügung *f*, interposition.
einführen, Stöpsel: to insert, Kabel, Leitung: to lead in, eine EMK: to introduce.
Einführen *n*, Einführung *f*, insertion; inlet, lead(ing)-in, entrance; introduction.
Einführungs=draht *m*, lead(ing)-in wire, drop wire;
— =isolator *m*, leading-in insulator;
— = — mit Vergußkammer, pothead insulator *B*;
— =kabel *n*, Spulenkasten: stub cable *K*, Luftleitungen: terminal cable, leading-in cable;
— =öffnung *f*, inlet;
— =pfeife *f*, inlet funnel *B*;
— =stelle *f*, entrance.
Eingang *m*, entrance, inlet;
— verboten, no admittance.
Eingangs=klemmen *pl*, input terminals *pl*;
— =kreis *m*, input circuit; grid-(to-) filament circuit *V*;
— = — =impedanz *f*, input impedance;
— =öffnung *f*, entrance;
— =übertrager *m*, input transformer.

eingehen, to enter (in eine For= mel, into a formula).
eingelötet, soldered; Kristall: solder-mounted *R*.
eingeschlossen, enclosed, housed; Gas: occluded.
eingeschwungener Zustand *m*, steady state.
eingießen, to pour in.
Eingitterröhre *f*, single grid valve.
eingleisig, single track....
eingraben, to bury, to sink.
eingreifen, to gear, to mesh, to engage, (in, with).
Eingriff *m*, engagement;
außer —, clear;
in — bringen, to force into engagement;
in — stehen, to gear, to mesh, (mit, with).
eingrenzen, to locate, to localize.
Eingrenzen *n*, Eingrenzung *f*, location, localization;
Fehler= — *f*, fault-locating, fault localization;
Eingrenzungs=messung *f*, localization test;
— =verfahren *n*, location method.
Eingußöffnung *f*, pouring-in hole.
einhängen, to replace, to restore (den Hörer the receiver), to clear.
Einhängen *n*, replacement, restoring; clearing.
Einheit *f*, unit; Wert 1: unity;
absolute —, absolute unit;
abgeleitete —, derived unit;
CGS= —, cgs unit;
elektromagnetische —, electromagnetic unit, em. unit, e. m. u.;
elektrostatische —, electrostatic unit, es unit, e. s. u.;
imaginäre —, imaginary unit;
praktische —, practical unit;

Einheit
Ersatz- —, spare unit;
Gesprächs- —, conversation unit *F*;
Längen- —, unit length;
— *—, für die, per unit length;
Strom- —, unit current.
einheitlich, uniform.
Einheitlichkeit *f*, uniformity.
Einheits-frequenz *f*, standard frequency;
- **-gestell** *n*, unit rack;
— **-ladung** *f*, unit charge;
— **-pol** *m*, magnetischer, unit magnetic pole;
— **-relais** *n*, universal relay;
— **-verstärker** *m*, basic repeater unit.
Einhüllende *f*, envelope *M*;
einkerben, to indent.
Einkerbung *f*, indent.
Einklang *m*, syntony.
Einkreis-empfang *m*, single circuit reception, primary reception;
— **-empfänger** *m*, single circuit receiver, primary receiver.
Einkünfte *pl*, revenue.
einkuppeln, to entrain, to throw in gear, to clutch.
Einlage *f*, insertion, insert; Konto: deposit.
einlagige Spule *f*, single layer coil.
Einlaßöffnung *f*, inlet.
einleiten, to initiate, to originate (ein Gespräch, a call).
Einleitung *f*, initiation, originating.
einlöten, to solder in.
Einmanntelegraph *m*, single operator telegraph (instrument).
einmauern, to brick.
Einnadeltelegraph *m*, single needle (telegraph).

einordnen, to step in (Gesprächszettel, tickets) *F*;
einpassen, to adjust.
Einphasen-bahn *f*, single phase electric railway;
— *— -speiseleitung *f*, single phase electric railway power circuit;
— **-strom** *m*, monophase current, single-phase current.
einphasig, monophase, single phase.
einpolig, single polar, single pole, unipolar;
— **er Fernhörer** *m*, single pole receiver;
— **e Umschaltung**, *f*, single commutation.
einprägen, to impress (eine Spannung, an emf).
Einquellempfänger *m*, solodyne receiver;
einreichen, to file (eine Patentanmeldung, a patent application).
einreißen, to range.
einreihig, in one row, single row ....
einrichten, to establish, to arrange; to fit.
Einrichtung *f*, device, contrivance, equipment; installation; establishment, arrangement;
Amts- —, office equipment, exchange apparatus.
Einrohr(zwischen)verstärker *m*, single valve (intermediate) repeater, single relay (intermediate) repeater, *F*; single valve amplifier *R*.
einrücken, to throw in gear, to engage (with), to trip.
Einrück-hebel *m*, trip(ping) lever, engaging lever;
— **-magnet** *m*, start or trip magnet, trigger magnet;

Einsatz *m*, inset, insert;
  auswechselbarer —, replaceable inset;
  Metall- —, metal insert;
  —**härtung** *f*, case hardening;
  —**kapsel** *f*, transmitter inset, transmitter button.
einsaugen, to soak.
einschaliger Wecker *m*, single dome bell.
einschalten, to cut in, to throw into circuit, to switch in, to insert, to interpolate (in), to connect in, to switch in; mit Stöpseln: to plug in, durch Ziehen von Stöpseln: to unplug.
Einschaltrelais, tripping relay; cut-in relay *F*, *A*;
  —**stellung** *f*, on position.
Einschaltung *f*, switching-in, interpolation.
einschenklig, single bar..., single leg....
Einschiebegestänge *n*, sweep's rods *pl*, *B*.
einschlagen, Blitz: to strike (in eine Leitung, a wire).
Einschlagwecker *m*, single stroke bell.
einschleifen, to loop in (ein Amt, an office).
einschließen, to enclose, to encase, to house; Gas: to occlude.
Einschließung *f*, enclosure, housing; occlusion.
einschmelzen, to melt; in Glas: to seal (*n*, into).
Einschmelzstelle *f*, seal;
  — *n pl* **des Glühfadens**, filament seals *pl*.
einschnappen, to snap (in, into).
einschneiden, to cut (in eine Verbindung, into a connection).
Einschnitt *m*, cut; notch, groove, kerf, slot;

mit —en versehen, to notch, to groove, to slot.
Einschnurfystem *n*, single cord system *F*.
Einschnürung *f*, recess, constriction, nick.
einschränken, to restrict (auf, to), to modify; to reduce.
Einschränkung *f*, restriction, modification, reduction.
einschrauben, to thread into.
einschwingen, to build up.
Einschwingen *n*,' building-up.
Einschwing-strom *m*, building-up current;
  —**vorgang** *m*, building-up transient.
Einseitenbandmodulation *f*, single sidle band modulation *R*;
einseitig, to one side;
  — **eingestellt**, bias(ed);
  — **gerichtet**, unidirectional;
  — **vorgespannt**, biased;
  — **wirkend**, unilateral uni-directional, bias;
  — **einstellen,—vorspannen**, to bias;
  — **es Arbeiten** *n*, operation to one side, b. Gegensprechen: half duplex operation, *T*;
  — e **Wirkung** *f*, bias.
einsetzen, Stöpsel: to insert; einen Wert: to substitute, *M*; beginnen: to start.
Einsetzen *n*, insertion; substitution; start.
einsinken, to sink.
einspannen, to clamp (den Rand der Membran, the edge of the diaphragm).
einspritzen, to inject *B*.
Einspritzung *f*, injection *B*.
einspulig, single coil....
Einsteigöffnung *f*, manhole *B*.
einstellen, to set, to adjust, to position; anhalten: to stop; **neutral** —, to set neutral, to adjust neutrally;

einstellen
 normal —, to set at normal;
 auf Null —, to set to zero;
 auf einen Unterschied —, to margin;
 auf eine Ansprechstromstärke von n mA —, to margin to pull up at n MA.

Einstell=glied n, setting member A;
— =hebel m, adjusting lever; Hughes: unison lever, zero adjusting lever;
— =magnet m, setting magnet;
— =schieber m, adjusting slide;
— =schraube f, set or adjusting screw; Höhe: levelling screw;
— =weg m, selecting path, (common) dialling trunk, impulse circuit, A.

Einstellung f, setting, adjustment;
 Fein= —, fine adjustment;
 Grob= —, coarse adjustment.

Einstrahl=funkstelle f, beam station, R;
— =sender m, beam transmitter, uni-directional transmitter, R
— =system n, beam system.

eintauchen, to plunge, to submerge, to immerse, to dip; Spulen: to telescope (in, with).

Eintauchen n, plunging, submerging, immersion, dipping.

einteilen, to divide, to graduate; zeitlich: to time; to space.

eintragen, to enter (in Formulare, into forms).

Eintragung f, entry.

eintreten, to enter; in eine Leitung —, to come in on a circuit; Gas: to ingress.

Eintritt m, entering; ingress (von Feuchtigkeit, of moisture).

Ein= und Ausrückhebel m, starting and stopping lever.

— =— Ausschwingen n, make and break transients pl;
— =——, Verzerrung durch, transient distortion.

Einwege= ...., single channel ...., simplex;
— =verstärker m, one way repeater, simplex repeater.

einwegig one way, simplex, single channel;
— er Telegraph m, single channel telegraph.

einwellig, sinusoidal, simple harmonic.

einwertig, single valued, M.

einwirken, to act, to react (auf, on); to interfere (auf, with); aufeinander—, to interact.

Einwirkung f, action, reaction; interference (with); gegenseitige —, interaction.

Einwurf m, slit, slot;
 Geld= —, coin slot.

Einzel=anrufer m, selector (calling apparatus), telegraph selector, T;
— =gebühr f, message rate, measured rate, F;
— =heit f, detail;
— =leiter m, single conductor;
— =leitung f, grounded line;
— =strom m, single current T;

einzementieren, to cement, to float in cement.

Einzieh=draht m, draw wire B;
— =strumpf m, cable grip B.

einziehen, to draw in, to haul, to pull in (Kabel, cables) B.

Einziehen n, pulling-in, drawing-in.

einzügig, single (duct), single way;
— es Formstück n, single tile, single duct concrete block B;
— er Kanal m, single way duct B.

Eis *n*, ice;
— **-belag** *m*, ice coating;
— **-last** *f*, ice load;
— **-überzug** *m*, ice coating.

Eisen *n*, iron (Fe);
weiches —, soft iron, magnetic iron;
Band= —, hoop iron;
Elektrolyt= —, electrolytic iron;
Flach= —, flat (bar) iron;
Form= —, profile iron;
Guß= —, cast iron;
Holzkohlen= —, charcoal iron;
Profil= —, profile iron;
Schmiede= —, wrought iron;
Silizium =—, silicon iron;
T= —, tee iron, T-iron;
T= —, Doppel=, double T-iron, I-iron;
U= —, channel iron, U-iron;
Walz= —, rolled iron;
Weich= —, soft iron, magnetic iron;
Winkel= —, angle iron;
Z= —, Z-iron;
— **-bahn** *f*, railway, railroad;
— = —, ein= (zwei)gleisige, single (double) track railway;
— = —, elektrische, electric railway;
— = —, elektrifizierte, electrified railway;
— = —, Einphasen=, single phase electric railway;
— = —, Normalspur=, standard gauge railway;
— = —, Schmalspur=, narrow gauge railway;
— **-band** *n*, iron hoop, Ring: ferrule;
— = — = umspinnung *f*, iron taping, iron tape winding;
— **-beton** *m*, ferro-concrete;
eisenbewehrt, ironclad, iron-sheathed;
Eisen-bewehrung *f*, iron armouring, iron sheathing;

— **-blatt** *n*, iron lamina(tion);
— **-blätterkern** *m*, laminated iron core;
— = — **-spule** *f*, laminated iron core coil;
— **-blech** *n*, sheet iron, iron plate; iron lamina;
— = —, lackiertes Weich=, ferrotype;
— = —, verzinktes, zinced sheet iron;
— = — **-kern** *m*, laminated iron core;
— = — **-membran** *f*, ferrotype diaphragm, sheet iron diaphragm;
— **-bisulfid** *n*, iron disulphide, iron pyrite, (FeS$_2$);
— **-draht** *m*, iron wire;
— = —, galvanisierter oder verzinkter, galvanized iron wire, ab: g. i. wire;
— = — **-umspinnung** *f*, iron wire winding, iron whipping *K*;
— **-drossel** *f*, iron cored choke coil, iron cored inductance;
— **-feilspäne** *pl*, iron filings *pl*;
eisenfrei, ironless;
— e Drossel *f*, air core coil;
Eisen-gefäß *n*, iron tank, iron tray;
— **-gehäuse** *n*, iron case;
eisengeschlossen, iron cored, ferric, closed core...;
Eisen-gittermast *m*, steel lattice mast;
— **-gummi** *m*, iron rubber;
eisenhaltig, ferric;
Eisen-kern *m*, iron core;
— = —, geblätterter, laminated iron core;
— = —, geschlossener, closed iron core;
— **-kies** *m*, iron disulphide, iron pyrite (FeS$_2$);
— **-kreis** *m*, ferric magnetic circuit, iron circuit;

Eisen
— **-kreis, geschlossener (offener),** closed (open) magnetic or iron circuit;
— **-kupferkies** m, chalcopyrite, iron copper sulphide ($Cu_2S + Fe_2S_3$);
— **-legierung** f, iron composition; **Nickel- —,** nickel-iron composition;

eisenlos, ironless, coreless; air core....;

Eisen-mantel m, iron jacket; **mit einem — - — versehen,** iron jacketed;
— **-mast** m, iron pole;
— **-monoxyd** n, — **-oxydul** n, protoxide of iron (FeO);
— **-pulver** n, iron powder, iron dust;
— — — **-kern,** iron powder core;
— — — —, **gepreßter,** compressed iron powder core;
— **-pyrit** m, iron pyrite, iron disulphide (FeS_2);
— **-ring** m, iron ring; iron torus, ferrule;
— **-rinne** f, iron troughing B;
— **-rohr** n, iron tube, iron tubing, iron pipe;
— — — **-strang** m, iron pipe conduit B;
— **-staub** m, iron dust;
— — — **-kern** m, iron dust core;
— **-sulfat** n, green vitriol ($FeSO_4$);
— **-transformator** m, iron core transformer.

eisenumgeben, ironclad;
Eisenumspinnung f, iron whipping, iron winding, iron taping, iron wrapping, K;
eisenumsponnen, iron whipped;
Eisen-verluste pl, iron losses pl, core losses pl;
— **-vitriol** n, green vitriol ($FeSO_4$);
— **-weg** m, iron path, magnetic or iron circuit;
— **-wicklung** f, iron winding, iron whipping;
— **-widerstand** m, iron resistance, **Lampe:** iron filament ballast lamp.

eisern, iron.

elastisch, elastical, springy;
— **e Dehnung** f, elastic elongation;
Elastizität f, elasticity.
Elastizitäts-grenze f, elastic limit;
— **-modul** m, modulus of elasticity.

elektrifizieren, to electrify.
Elektrifizierung f, electrification.
Elektriker m, electrician.
elektrisch, electric(al); electrified;
gleichnamig—, similarly electrified;
ungleichnamig —, oppositely electrified.
elektrisieren, to electrify.
Elektrisierung f, electrification.
Elektrizität f, electricity;
atmosphärische —, atmospheric electricy;
Berührungs- —, contact electricity;
Glas- —, vitreous electricity;
Harz- —, resinous electricity;
Luft- —, atmospheric electricity;
Piezo- —, piezo electricity;
Reibungs- —, frictional electricity;
Thermo- —, thermo-electricity;
Elektrizitäts-erzeugung f, generation of electricity;
— **-menge** f, quantity of electricity.
Elektrochemie f, electro-chemistry.
elektrochemisch, electro-chemical.

Elektrode *f*, electrode;
heiße —, hot electrode;
kalte —, cold electrode;
negative —, negative electrode; cathode;
positive —, positive electrode; anode;
Entnahme= —, output electrode;
Steuer= —, control electrode.

Elektroden=abstand *m*, electrode spacing, electrode separation; Funkenstrecke: gap separation;
— =fläche *f*, electrode surface, electrode area.
— =kapazitäten *pl*, inter-electrode capacities *pl*, V.

Elektrodynamik *f*, electro-dynamics *pl*.

elektrodynamisch, electrodynamic(al).

Elektrodynamometer *n*, electrodynamometer.

elektro=elektrisch, electro-electric(al).

Elektroinduktion *f*, electro-induction.

elektroinduktiv, electro-inductive.

elektrokalorisch, thermoelectric(al).

Elektrolyse *f*, electrolysis.

Elektrolyt *m*, electrolyte;
— =detektor *m*, electrolytic detector;
— =eisen *n*, electrolytic iron;
— =kupfer *n*, electrolytic copper;
— =unterbrecher *m*, electrolytic interrupter.

elektrolytisch, electrolytic(al).

Elektromagnet *m*, electromagnet (*cf.* Magnet);
— =anker *m*, electromagnet armature;
— =kern *m*, electromagnet core;
— =spule *f*, electromagnet coil;
— =wicklung *f*, electromagnet winding.

elektromagnetisch, electromagnetic(al).

Elektromagnetismus *m*, electromagnetism.

Elektromechanik *f*, electro-mechanics *pl*.

elektromechanisch, electro-mechanic(al).

Elektrometer *n*, electrometer;
Faden= —, thread electrometer;
Quadranten= —, quadrant electrometer;
Saiten= —, string electrometer;
Scheiben= —, disc electrometer.

Elektromotor *m*, electromotor (*v.* Motor).

elektromotorisch, electromotive;
— e Kraft *f*, electromotive force, e. m. f. (*v.* E. M. K);
gegen= —e —, counter-electromotive force, back or counter or opposing e. m. f., c. e. m. f.

Elektron *n*, electron;
freie —en *pl*, free electrons *pl*.

Elektronen=bombardement *n*, electron bombardment;
— =emission *f*, electron(ic) emission;
— = —, sekundäre, secondary emission of electrons;
— =entladung *f*, electron discharge;
— = —, reine, pure electron discharge;
— = —, Glüh=, thermionic discharge;
— =fluß *m*, flow of electrons, electronic flow;
— =relais *n*, electron relay, valve relay, thermionic relay, discharge relay;
— =röhre *f*, electron(ic) tube (*am.*) or valve (*engl.*), ionic valve, audion;

Elektronen
— **-strom** m, ionic current, electron current, stream of electrons.
elektronegativ, electro-negative.
elektronisch, electronic(al).
Elektrophysik f, electro-physics pl.
elektrophysikalisch, electro-physical.
elektropositiv, electro-positive.
Elektroskop n, electroscope;
  Goldblatt- —, gold-foil electroscope.
Elektrostatik f, electro-statics pl.
elektrostatisch, electrostatical;
— **e Kapazität** f, electrostatic capacity.
Elektrotechnik f, electrical engineering.
Elektrotechniker m, electrician, electrical engineer.
elektrotechnisch, electro-technical.
Element n, element, cell; **chemisch**: element;
  ein — ansetzen, to set up a cell;
  galvanisches —, galvanic cell;
  nasses —, wet cell, hydroelectric cell;
  Beutel- —, sack cell;
  Bunsen- —, Bunsen cell;
  Chromsäure- —, chromic acid or bichromate cell;
  Clark- —, Clark cell;
  Daniell- —, Daniell cell;
  Grove- —, Grove cell;
  Krüger- —, Krueger cell;
  Kupfer- —, copper-zinc cell;
  Kupron- —, copper oxide cell, cupron cell;
  Leclanché- —, Leclanché cell;
  Normal- —, standard cell;
  Primär- —, primary cell, voltaic cell;
  Salmiak- —, Leclanché cell;
  Sekundär- —, secondary cell, storage cell;
  Thermo- —, thermo(-electric) couple;
  Trog- —, tray cell;
— **mit einer Flüssigkeit**, single fluid cell;
— — **zwei Flüssigkeiten**, double fluid cell;
— **-glas** n, battery or element jar, element glass;
— **-schlamm** m, battery mud.
Elementar- . . . ., elementary;
— **-welle** f, elementary wavelet R.
Elfenbein n, ivory.
Elimination f, elimination.
eliminieren, to eliminate.
Ellipse f, ellipse.
elliptisch, elliptic(al).
Email(le f) n, enamel;
  mit — überziehen, to enamel;
— **-draht** m, enamel-covered wire, enamel insulated wire, enamelled wire;
— **-lack** m, enamel lac;
— **-litze** f, enamelled strand;
— **-schild** n, enamelled plate.
emaillieren, to enamel.
emailliert, enamelled, enamel-coated.
Emanation f, emanation.
Emission f, emission;
  Elektronen- —, electronic emission, emission of electrons;
Emissions-fläche f, emitting area or surface;
— **-geschwindigkeit** f, velocity of emission;
— **-strom** m, current of emission, emission current;
— **-vermögen** n, emissive power, emissivity.
E. M. K. electromotive force, e. m. f.;
  Gegen- —, opposing or back or counter e. m. f., c. e. m. f.;
  periodische —, periodic e. m. f.;
  rein sinusförmige —, pure sine e. m. f.;

E.M.K.
zusammengesetzte —, composite e. m. f.;
— von rechteckiger Kurvenform, square-topped e. m. f.;
— der Selbstinduktion, e. m. f. of self-induction.

Empfang *m*, reception (*cf.* Empfänger);
Stellung / auf —, stand-by position;
Autodyn= —, autodyne reception;
Blattdruck=—, page printing T;
Detektor= —, detector reception;
Doppel= —, double reception;
Einkreis= —, primary reception, single circuit reception;
Einquell=—, solodyne reception
Funk= —, radio reception;
Heterodyn= —, heterodyne reception;
Homodyn= —, homodyne reception, zero beat reception;
Hör= —, audible reception;
Interferenz= —, beat reception, heterodyne reception;
Rahmen= —, loop or frame (aerial) reception;
Reflex= —, reflex reception, dual reception;
Richt= —, directional or directive reception;
Röhren= —, valve reception;
Rückkopplungs= —, regenerative or retroactive reception;
— = — mit Hilfsfrequenz, superregenerative reception;
Rundfunk= —, broadcast reception;
Schreib= —, visual reception;
Schwebungs= —, beat or heterodyne reception;
— = — mit besonderem Überlagerer, separate heteryodne reception;

Schwingaudion= —, Selbstüberlagerungs= —, autodyne or self-heterodyne reception;
Seitendruck= —, page printing T;
Sekundär= —, double circuit reception, secondary reception;
Solodyn= —, solodyne reception;
Streifendruck= —, tape printing T;
Superregenerativ= —, superregenerative reception;
Transponierungs= —, transposition reception;
Überlagerungs= —, beat reception, heterodyne reception;
— = — mit Überhörfrequenz, supertonic heterodyne reception;
Zweikreis= —, secondary reception, double circuit reception;
Zwischenfrequenz= —, superheterodyne reception.

empfangen, to receive;
mittels Überlagerung —, to heterodyne.

Empfänger *m*, receiver (*cf.* Empfang), receiving set, receiving instrument; eines Telegramms: destinator, addressee;
photographischer —, photographical recorder;
Audion= —, audion receiver;
Blattdruck= —, page printer T;
Druck= —, printer T;
Funk= —, radio or wireless receiver;
Kontroll= —, control receiver T;
Kristalldetektor= —, crystal receiver;
Lochstreifen= —, reperforator, receiving perforator, T;

Empfänger
  Maschinen- —, automatic receiver T;
  Mitlese- —, leak receiver T;
  Peil- —, wireless direction finding receiver;
  Schall-, sound receiver;
  — - —, Unterwasser-, subaqueous sound receiver;
  Seitendruck- —, page printer T;
  Streifendruck- —, tape printer T;
  Stromstoß- —, impulse receiver, impulse storing device, register, A;
  Wellen- —, wave receiver;
  — -dekrement n, receiver decrement;
  — -kreis m, receiver circuit;
  — -schaltung f, receiver circuit; receiver connections pl;
  — für gedämpfte (ungedämpfte) Wellen, spark (c. w. or continuous wave) receiver.
Empfangs-antenne f, receiving aerial;
  — -beamter m, receiving operator T;
  — -einrichtung f, receiving device;
  — -ende f, receiving end (einer Leitung, of a line);
  — -gerät n, receiving set;
  — -gleichrichterröhre f, audion;
  — -kurve f, arrival curve;
  — -leistung f, received or incoming power;
  — -loch n, (Stelle, wo kein Funkempfang), blind spot, radio shadow, radio pocket;
  — -locher m, receiving perforator, reperforator, T;
  — -lochung f, reperforation T;
  — -messung f, receiving or reception measurement R;
  — -rahmen m, receiving loop;
  — -relais n, receiving relay;
  — -ring m, receiving ring T;
  — -schaltung f, receiving circuit;
  — -satz m, receiving set;
  — -segment n, receiving segment T;
  — -seite f, receiving end;
  — -stelle f, receiving station;
  — -stellung f, receiving condition; stand-by position;
  — -versuch m, reception test;
  — -verteiler m, receiving distributor T;
  — -welle f, received wave.
empfindlich, sensible, sehr —: sensitive; responsive (für, gegen, to).
Empfindlichkeit f, sensibility, sensitiveness, hohe —: sensitivity;
  Frequenz- —, frequency sensitivity;
  Leistungs- — power sensitivity;
  Licht- —, photo-sensitivity;
  Spannungs- —, voltage sensitivity;
  Strom- —, current sensitivity.
Empfindlichkeits-grad m, sensitiveness, degree of sensitivity;
  — -kurve f, response characteristic;
  — - —, Frequenz-, frequency response characteristic;
  — -prüfung f, sensitivity test.
Empfindung f, sensation.
empirisch, empirical.
End- .... terminal;
  — -amt n, — -anstalt f, terminal office or station;
  — -ansicht f, end view;
  — -apparat m, end or terminal apparatus;
  — -ausschlag m, full deflection (der Nadel, of the needle);
  — -geschwindigkeit f, final velocity;

End
— -impedanz f, load or terminating or end impedance;
— -kabel n, terminal cable; Fernleitungs- -—, toll entrance cable;
— -kunstschaltung f, terminal circuit, terminal network;
— -satz m, terminal repeater (am.) T;
— -schaltung f, terminal circuit;
— -schaltsatz m, terminal repeater (am.) T;
Drahtfunk- -—, carrier terminal circuit;
— -spannung f, final voltage;
— -verluste pl, terminal loss;
— -verschluß m, (box) terminal, box head, cable head, B;
— -verstärker m, terminal repeater;
— -verzweiger m, terminal block B;
— -widerstand m, terminal resistance;
— -zustand m, final state.

Ende n, end, terminal, termination;
äußerstes —, extremity;
freies —, free end;
hinteres —, further end;
unteres —, bottom; der Stange: butt (end) B;
zubereitetes —, treated butt B.

end(ig)en, to end, to terminate; to finish, to cease;
an Klinken—, to be terminated on jacks.

endlich, finite M.
Endosmose f, endosmose.
endosmotisch, endosmotical.
Energie f, energy, seltener: current;
kinetische —, kinetic energy;
potentielle —, potential energy;

Sprach- —, voice power;
— -komponente f, energy component;
— -niveaulinie f, transmission level L;
— -transport m, transport of energy;
— -übertragung f, transport of energy; zwischen zwei Kreisen· transfer of energy;
— -verbrauch m, energy consumption; energy dissipation;
— — des Heizfadens, filament wattage;
— -verlauf m auf der Leitung, transmission level L;
— -verlust m, loss of energy;
— -verteilung f, energy distribution;
— -verzehrung f, energy dissipation.
— -bedarf m, energy requirement;
— -diagramm n, power level L;
— -fluß m, energy flow;
— -höhenlinie f, transmission level L;

eng, narrow, close.
Engländer m, monkey wrench, coach wrench.
engmaschig, close-meshed, fine.
entblocken, to clear.
entdämpfen, to improve.
Entdämpfung f, (repeater) gain, transmission gain, improvement.
Entdämpfungs-Frequenzkurve f, gain-frequency curve;
— -messer m, gain measuring device.
entdecken, to detect, to discover.
Entdeckung f, detection, discovery.
enteignen, to expropriate.
Enteignung f, expropriation.
entfalten, to develop.

entfernen, to remove.
entfernt, distant.
Entfernung *f*, distance; Weg=nahme: removal.
entflammbar, inflammable; nicht —, non-flammable.
entflammen, to set alight; to fire up.
entfritten, to decohere.
Entfritter *m*, decoherer, tapper.
Entfrittung *f*, decoherence.
entgasen, to outgas.
Entgasung *f*, outgassing.
entgegengesetzt, opposite, inverse;
— **gerichtet**, opposed, in opposition;
— **e Pole** *pl*, opposite poles *pl*;
— **e Ströme** *pl*, opposed currents *pl*.
entgegensetzen, to oppose; Widerstand —, to offer resistance.
Entgegensetzung *f*, opposition.
entgegenwirken, to oppose, to counteract.
Entgelt *n*, remuneration.
Entharzung *f*, extraction of resin.
entionisieren, to deionize, to scavenge.
Entionisierung *f*, de-ionization, scavenging.
entkohlen, to decarbonize.
Entkohlung *f*, decarbonization.
entkoppeln, to tune out, to neutralize, to balance out.
Entkopplung *f*, tuning-out, neutralization, balancing-out, *R*;
— **von Störern**, balancing-out of jamming.
entkuppeln, to uncouple, to declutch.
Entkupplung *f*, uncoupling, declutching.
entladen, (sich), to discharge; in Büscheln: to brush; über- —, to run down (Sammler, storage cells).

Entladen *n*, discharging, discharge.
Entlader *m*, discharger.
Entladung *f*, discharge;
aperiodische —, aperiodic or dead-beat or impulsive or non-oscillatory or unidirectional discharge;
atmosphärische —, atmospheric pulse or discharge;
disruptive —, disruptive discharge;
kathodische —, cathodic discharge;
oszillierende —, oscillating or oscillatory discharge;
Blitz= —, lightning discharge;
Büschel= —, brush discharge;
Elektronen= —, (reine), (pure) electron discharge;
Funken= —, disruptive discharge;
Glimm= —, corona discharge, glow discharge;
Lichtbogen= —, arc discharge;
Oberflächen= —, surface discharge;
Rück= —, back discharge; back kick *T*;
Schwing= —, oscillatory discharge;
Selbst= —, self-discharge.
Entladungs=frequenz *f*, discharge frequency;
— **=funken** *m*, disruption spark;
— **=gefäß** *n*, (electron) discharge tube, discharge vessel;
— **=kreis** *m*, discharge circuit;
— **=potential** *n*, discharge potential;
— **=relais** *n*, (gas) discharge relay;
— **=röhre** *f*, discharge or discharging tube;
— **=spannung** *f*, discharge potential;
— **=strecke** *f*, discharge gap;

Entladungs
— -ſtromkreis *m*, discharging circuit.
entlangſtreifen, to brush, to wipe (an, over).
entlegen, remote.
entlüften, to ventilate.
Entlüftung *f*, ventilation.
entmagnetiſieren, to demagnetize.
Entmagnetiſierung *f*, demagnetization.
Entnahme *f*, extraction;
— -elektrode *f*, output electrode;
— -kreis *m*, output circuit, load circuit;
— -punkt *m*, output terminals *pl*;
— -ſeite *f*, receiver end;
— -ſtromkreis *m*, receiver circuit.
entnehmen, to extract; Strom — to take current.
entnommene Leiſtung *f*, output.
entriegeln, to unlock, to unlatch;
Entriegelung *f*, unlocking.
Entriegelungsmagnet *m*, unlocking magnet.
entrinden, to bark *B*.
Entſchwefelung *f*, desulphurization.
entſpannen, to slacken, to relax.
Entſpannung *f*, slackening, relaxing (von Federn, of springs)
entſperren, to unlatch.
entſprechen, einer Gleichung, to fit an equation.
entſtehen, to form.
entſtellen, to mutilate, to alter, (Telegramme, messages).
Entſtellung *f*, mutilation, alteration.
entwäſſern, to drain (off).
Entwäſſerung *f*, draining (-off), drainage.
Entwäſſerungshahn *m*, drain cock.
entweichen, to escape.

entwerfen, to design, to contrive, to outline, to plan; to trace, to sketch.
Entwerfen *n*, designation, planning.
entwerten, to depreciate; Briefmarken: to cancel.
Entwertung *f*, depreciation.
entwickeln, to develop; Gas —, to evolve gas.
Entwickler *m*, developing bath.
Entwicklung *f*, development; Gas= —, formation or evolution of gas
Entwurf *m*, design, project, scheme, lay-out.
entwurzeln, to uproot.
entzerren, to correct, to eliminate distortion, to rectify, to regenerate; to equalize.
Entzerrer *m*, correcting device, anti-distortion device; equalizer, *KV*;
Querimpedanz= —, shunt-admittance type equalizer *K*;
Reihenimpedanz= —, series-impedance type equalizer *K*;
— -anordnung *f*, — -einrichtung *f*, distortion correcting device, anti-distortion device;
— -kette *f*, correcting or corrective network;
— -ſchaltung *f*, compensating or correcting circuit;
Entzerrung *f*, correcting (of distortion), correction;
Amplituden= —, correction of amplitudes;
Dämpfungs= —, attenuation equalization *K*;
Phaſen= —, correction of phase.
Entzerrungs-droſſel *f*, anti-resonant coil;
— -einrichtung *f*, distortion-corecting device;
— -filter *n*, — -kette *f*, filter-type equalizer, equalizing network, *K*.

entziehen, to extract.
entziffern, to decipher, to decode.
Entzifferung f, decoding, deciphering.
entzündbar, inflammable.
entzünden, to light, to ignite.
epizyklisch, epicyclic(al).
Epizykloide f, epicycloid.
erbauen, to erect, to design.
Erbauer m, designer.
Erd- ..., terrestrial; earth....;
— **-ableiter** m, earth arrester;
— **-alkalimetall** n, alkaline earth metal;
— **-antenne** f, earth antenna, ground antenna;
— **-anziehung** f, gravitation;
— **-anziehungskraft** f, gravitational force;
— **-arbeiten** pl, earth work B;
— **-atmosphäre** f, earth's atmosphere;
— **-bohrer** m, earth borer, auger;
— - — mit **Stoßbewegung**, thrust borer;
— **-draht** m, earth or ground wire; mit einem — - — versehen, to earth-wire;
— - — **-netz** n, ground mat R;
— **-fehler** m, earth fault;
— - — **-Schleifenmessung** f, loop test;
— -—- nach **Varley**, Varley loop test;
— **-geräusch** n, earth noise;
— **-harz** n, bitumen;
— **-kabel** n, buried cable;
— **-kapazität** f, earth capacity; wire-to-earth capacity (einer Leitung, of a line);
— **-klemme** f, ground or earth(y) terminal;
— **-krümmung** f, curvature of the earth;
— **-kruste** f, earth's crust;
— **-leiter** m, ground wire;
— **-leitung** f, earth connection;

— **-magnetfeld** n, earth's magnetic field;
— **-magnetismus** m, terrestrial magnetism;
— **-oberfläche** f, earth's surface;
— **-öl** n, mineral oil; [men;
— **-pech** n, mineral pitch; bitu-
— **-platte** f, earth plate;
— **-potential** n, earth potential;
— **-rohr** n, soil pipe B;
— **-rückleitung** f, earth or ground return, ground circuit;
— - —, **Stromkreis** m mit, earth return or grounded circuit;
— **-rutsch** m, landslide;
— **-schelle** f, earth clip, ground clamp;
— **-schleife** f, earth circuit;
— **-schluß** m, ground(ing), earth(ing);
vollständiger — - —, dead grounding;
zeitweiser — - —, intermittent or grounded earth;
— - — **-prüfer** m, leakage indicator, ground detector;
— **-schraube** f, earth screw B;
— **-störung** f, earth disturbance;
— **-strom** m, earth current;
— **-system** n, earth return automatic telephone system;
— **-verbindung** f, earth (connection);
— **-wachs** n, ozokerite.
— **-zone** f, ground line (section) (der **Stangen**, of poles) B.
Erde f, earth, ground;
an — legen, to (put to) earth, to ground;
an — liegend, grounded, earthed;
— — — de **Ader** f, positive wire A;
in die — verlegen, to bury B;
mit — verbinden, to connect to earth or ground;
Besetzt- —, busy earth, F.

erden, to earth, to ground; geerdet, earthed, grounded, earthy.

Erden n, earthing, grounding.

erdmagnetisch, earthmagnetic.

Erdung f, earthing, grounding; Schutz= —, protective ground.

Erdungs=schalter m, earthing switch;

— =widerstand m, earthing or ground resistance.

Erfahrung f, experience, observation.

Erfahrungswert m, empirical value.

erfinden, to invent, to contrive, to devise.

Erfinder m, inventor.

erfinderisch, inventive.

Erfindung f, invention, contrivance, device.

erforderlich, required, requisite.

erfordern, to require, to demand.

Erfordernis n, requirement, demand.

erforschen, to explore, to investigate.

Erforschung f, exploration, investigation.

erfüllen, eine Beziehung, to satisfy or fit a relation.

Erg n, erg.

ergänzen, to complete, to fill up, to restore.

Ergänzung f, complement; completion.

ergeben, (sich), to result.

Ergebnis n, result.

ergreifen, to seize.

erhaben, embossed, raised.

erhalten, to maintain, to keep (up); Schutz: to preserve.

erhaltend, preservative.

Erhaltung f, maintenance; preservation, conservation;

— der Masse, conservation of the mass;

— =zustand m, guter, high maintenance standard.

erhärten, to harden.

erheben, to raise; Gebühren: to levy.

erhitzen, to heat; sich —, to get hot, to heat (up);

Erhitzung f, heating, gettinghot.

erhöhen, to raise, to increase; to intensify.

Erhöhung f, raising, increase; intensification; Kurve: bulge.

erholen, sich, to recover, to recuperate.

Erholung f, recovery, recuperation.

Ericsson= Z.B.=System n, bridged impedance c. b. system.

erklären, to interpret; to declare; to explain.

Erklärung f, interpretation; declaration; explanation.

Erlaubnis f, allowance; licence; Versuchs= —, experimenter's license R;

— =inhaber m, licensee.

erledigen, to handle, to settle.

Erledigung f, handling, settling.

erlöschen, to darken, to extinguish, to blow out; Patent,. Recht: to expire.

Erlöschen n, darkening, extinguishing; expiration; zum — bringen, to darken, to extinguish.

Ermäßigung f, decrease; Preis: rebate.

ermüden, to fatigue.

Ermüdung f, fatigue.

ernennen, to assign.

Ernennung f, assignation.

erneuern, to renew, to restore; to regenerate.

Erneuerung f, renewal.

Erneuerungsgebühr f, renewal fee.

errechnen, to calculate, to compute.
Errechnung f, calculation, computation.
erregen, to excite; Relais: to energize; durch Stoß: to impulse.
Erreger m, exciter; excitant;
— -flüssigkeit f, exciting fluid;
— -funkenstrecke f, exciting spark gap;
— -kreis m, exciting circuit;
— - — -tastung f, control of excitation R;
— -masse f, excitant (der Trockenelemente, of dry cells);
— -paste f, white or exciting paste;
— -röhre f, exciter tube;
— -spannung f, exciting voltage;
— -strom m, exciting current;
— -wicklung f, exciting winding.
erregt werden, Relais to pull up, to energize.
Erregung f, excitation, energization;
Nebenschluß- —, shunt excitation;
Selbst- —, self-excitation;
Stoß- —, impact or shock excitation, (im)pulse excitation;
Verbund- —, compound excitation.
erreichen, to attain; to reach.
errichten, to erect, to construct, to build up, to mount, to instal.
Errichtung f, erection, construction, instalment.
errufen, to call, to gain the attention.
Ersatz m, renewal; reserve;
— -anker m, spare armature;
— -antenne f, phantom or mute aerial; reserve aerial;
— -apparat m, spare instrument;
— -ausrüstung f, spare equipment;
— -einheit f, reserve or spare unit;
— -leitung f, spare circuit;
— -schaltung f, equivalent circuit, equivalent network;
— -platte f, spare plate;
— -teile pl, spare parts pl.
Erscheinung f, phenomenon.
erschöpfen, to exhaust, Sammler: to run down.
Erschöpfung f, exhaustion; running-down.
erschüttern, to shake, to vibrate.
Erschütterung f, shake, shaking, vibration.
erschütterungsfrei, resilient.
ersetzbar, replaceable.
ersetzen, to renew, to replace, to supplant.
eine fehlerhafte Leitung —, to make good a faulty circuit.
Ersetzung f, renewal, replacement.
ersparen, to save, to economize.
Ersparnis f, saving (an, in).
Erst- ...., primary;
— -kreis m, primary circuit;
— -wicklung f, primary winding.
erstarren, to solidify.
erstarrt, solid(ified).
Erstarrung f, solidification.
Erstarrungspunkt m, solidifying point.
erstrecken, sich, to range, to bear, to stretch, to extend.
erwärmen, (sich) to heat.
Erwärmung f, heating.
Erwärmungsverlust m, Joulean loss.
erweitern, to expand, to multiply through (mit, by), M; to enlarge, to extend, to widen.
erzeugen, to produce, to create, (ein Magnetfeld, a magnetic field), to set up (Schwingungen, oscillations), to generate.

Erzeuger m, generator;
  Schwingungs= —, oscillation generator;
  Wellen= —, wave generator;
  — =seite f, generator end.
Erzeugnis n, product;
  deutsches —, German factory product.
Erzeugung f, production, generation;
  Strom= —, current generation.
erzwingen, to force.
erzwungene Schwingungen pl, forced vibrations pl.
Eschappement n, escapement.
Esche(nholz n) f, ash.
Esse f, smoke stack.
Etalonapparat m, reference instrument, standard instrument.
evakuieren, to evacuate.
Evakuierung f, evacuation.
Evolute f, evolute.
Evolvente f, evolvent.
Exkavator m, excavator.
Expedient m, dispatcher.
expedieren, to dispatch.
Experiment n, experiment.
experimentell, experimental.
experimentieren, to experiment.
Experimentierender m, experimenter.
Experimentierlizenz f, experimenter's license.
Explosivkonsonant m, explosive sound.
Exponent m, exponent.
Exponential=ausdruck m, exponential expression;
  — =funktion f, exponential function;
  — =gesetz n, exponential law;
  — = —, nach einem, exponentially;
  — =kurve f, exponential curve.
exponentiell, exponential.
exponieren, to expose.
Extrapolation f, extrapolation.
extrapolieren, to extrapolate.
exzentrisch, eccentric(al), non-concentric(al).
Exzentrizität f, eccentricity.

F.

Fabrik f, factory.
Fabrikation f, manufacture.
Fabrikationslänge f, manufacturing length, drum length, (eines Kabels, of a cable).
Fabrikmessung f, factory test.
fabrizieren, to manufacture, to make.
....fach, ...fold;
  40= =e Verstärkung f, 40fold amplification.
Fach n, shelf, partition, bay; art;
  — =ausdruck m, technical term;
  — =mann m, expert;
  — =presse f, technical press;
  — =welt f, technical world.
Fächer m, fan, blower;
  elektrischer —, motor fan, electric fan;
  — =antenne f, fan(-shaped) aerial;
  — =gestell n, rack shelving.
Faden m, thread, fibre, filament; fathom (= 6 feet = 1,82878 m);
  thorhaltiger —, thoriated filament;
  Glüh= —, Heiz= —, incandescent filament, heated filament;
  Seiden= —, silk fibre;
  Strom= —, current path;
  Wolfram= —, tungsten filament;

Faden
— -aufhängung f, fibre suspension;
— -elektrometer n, thread electrometer;
fadenförmig, filamentary, fibre...., thread....
Faden-galvanometer n, thread galvanometer;
— -kathode f, filamentary cathode;
— -kreuz n, crossed threads pl, spider lines pl;
— -spannung f, filament voltage;
— -strom m, filament current;
— -umschnürung f, serving of thread (der Papierkabel, of paper-insulated cables);
— -widerstand m, filament resistance.
Fading n, fading R;
— -effekt m, fading effect R.
Fagott n, bassoon.
Fahne f, lug, vane;
Dämpfer- —, (damping) vane.
Fahr-bahn f, — -damm m, roadway, carriageway;
— -draht m, trolley wire, contact wire;
— - — -aufhängung f, trolley wire suspension; trolley span wire;
— -plan m, time table, schedule.
— -rad n, bicycle;
— -straße f, road;
— -stuhl m, lift.
Faktor m, factor, coefficient; einer Zahl: sub-multiple; gemeinsamer —, common factor.
Falle f, latch.
fallen, to fall (off), to drop.
fällen, chemisch: to precipitate.
Fall-klappe f, (drop) shutter, drop indicator;
— - — mit elektrischer (mechanischer) Rückstellung, electrical

(mechanical) replacement drop indicator.
— -rohr n, gravity tube;
— -scheibe f, drop shutter;
— - —, Wecker m mit, indicator bell;
— -scheibenkasten m, indicator board;
— -tür f, trap door.
falsch, wrong;
— e Verbindung f, wrong connection.
Falte f, fold.
falten, to fold, to plait.
fangen, to intercept mischievous calls A.
Fangvorrichtung f, arrangement for the interception of mischievous calls A.
Farad n, farad, ab: f, F;
Mikro- —, microfarad, ab; mf(d), μF.
Farbe f, colour; ink;
Morse- —, ink;
— auftragen, to ink.
Farb-auftragung f, inking;
— -band n, ink ribbon;
— - — -vorschub m, ink ribbon feed;
— - — -wechsel m, ink ribbon reversal;
— -gefäß n, — -kasten m, ink well;
— -rad n, — -rädchen n, ink(ing) wheel, ink disc, inking roller;
— -rolle f, — -röllchen n, ink(ing) roller;
— -schreiber m, inker, inkwriter, writer, seltener: printer;
— - —, unmittelbar in die Leitung geschalteter, direct inker;
— - — mit vorgeschaltetem Relais, local inker.
Farbenfolge f, colour scheme B.
farbig, coloured.
Faser f, fibre.
faserig, fibrous; grained.

Faser=stärke *f*, fibre strength;
— =stoff *m*, fibre;
— = =kabel *n*, fibre-covered cable.
fassen, to seize, to grip.
Fassen *n*, seizure, grip.
Fasson *n*, shape(d);
— =eisen *n*, profile iron.
Fassung *f*, holder, socket, fitting;
Lampen= —, lamp holder;
Röhren= —, valve socket;
Steck= —, socket, plug socket; lamp jack *F*.
Fassungsvermögen *n*, (carrying) capacity.
Fäule *f*, rot.
faulen, to rot.
Faulen *n*, rotting.
Fäulnis *f*, rot(ting).
fäulnis=hindernd, — =widrig, antirot.

Feder *f*, spring; pen; feather;
eine — anspannen, to bend, to tighten a spring;
— — entspannen, to relax a spring;
an einer — angebracht, spring-supported;
durch eine — festgehalten, spring-clamped;
a= —, a-spring, short spring, (der Klinke, of the jack) *F*;
b= —, b-spring, long spring *F*;
feine —, hairspring;
mit einem Gewicht versehene —, weighted spring;
Ausrichten von —n, aligning of springs;
Abreiß= —, retracting or antagonistic spring;
Ankerumlege= —, biasing spring (des Wechselstromweckers, of magneto bell);
Ausschluß= —, disconnecting spring (Hughes);
Blatt= —, leaf or plate spring, flat spring;

Buffer= —, buffer spring;
Dämpfer= —, damping spring;
Flach= —, flat spring, plate spring;
— = —spirale *f*, flat spiral spring;
Gegen= —, reacting or opposing spring;
Haupt= —, main spring, master spring;
Klang= —, coiled wire gong *F*;
Klinken= —, jack spring;
Kontakt= —, contact spring;
Leitungs= —, line spring (der Klinke, of the jack);
Polwechsler= —, pole-changing spring;
Rückführ= —, restoring or retracting spring, control(ling) spring;
Rückzug= —, retracting spring;
Schalter= —, switch spring;
Schleif= —, wiper;
Spiral= —, Schrauben= —, helical spring, coiled spring, spiral spring;
Schreib= —, (recording) pen;
Stromzuführungs= —, current supply brush;
Torsions= —, torsion spring;
Trieb= —, driving spring, main spring;
Zentrier= —, centring spring;
— =druck *m*, spring pressure;
— =haus *n*, spring drum, spring barrel, spring box;
— =keil *m*, feather key;
— =klammer *f*, spring clip, clamping spring;
— =klemme *f*, (spring) clip;
— =klinke *f*, spring catch;
— =kontakt *m*, switch jack, spring contact;
— =kraft *f*, spring tension;
— = — =antrieb *m*, clockwork train;

Feder
— -motor m, spring (-wound) motor;
— -schnepper m, spring trigger;
— -spannschraube f, tension screw, spring tensioning screw;
— -spannung f, spring tension;
— -stift m, spring pin;
— -trommel f, spring drum, spring barrel;
— -zug m, spring pull; clockwork train.

Federn-bündel n, set of springs;
— -paket n, bank, set of springs;
— -satz m, spring assembly, spring bank.

federnd, springy, elastic(al), flexible;
— er Kontakt m, flexible contact;
— eingespannt, spring clamped.

Fehlanruf m, lost call F.
fehlen, to fail.
Fehlen n, failure.
Fehler m, error; defect, fault, failure; deficiency;
— beseitigen, to remove faults;
— eingrenzen, to locate faults;
— suchen, to trace faults;
Auslöse- —, failure to release A.
Beobachtungs- —, error of observation;
Erd- —, earth fault;
Isolations- —, insulation failure;
Telegraphier- —, operator's error;
— -beseitigung f, removal of faults;
— -eingrenzung f, fault-locating, fault localisation;
fehlerfrei, correct;
Fehlergrenze f, limit of error;
fehlerhaft, faulty, defective;
Fehler-ortsmessung f, fault location test, localisation test;
— -strecke f, faulty section;
— -suchen n, faultng;
— -widerstand m, fault resistance.

Feile f, file;
Kontakt- —, contact file;
Sägen- —, saw file.
feilen, to file.
Feilicht n, filings pl.
Feil-kloben m, handvice;
— -späne pl, filings pl;
Eisen- — - —, iron filings pl.
fein, fine.
Fein-einsteller m, vernier:
— -einstellung f, fine adjustment;
— -steller m, vernier;
— -stellvorrichtung f, vernier.
feinkörnig, fine grained.
Fein-meßschraube f, micrometer screw;
— -sicherung f, heat coil, ab: h. c.;
— -sicherungs-einsatz m, — - — -patrone f, heat coil;
— -stellschraube f, micrometer screw.

Feld n, field; Teil: division, partition, section; panel;
das — wandert, the field travels;
äußeres —, extraneous field;
bewegliches —, moving field;
elektrostatisches —, (electro-)static field;
feststehendes —, fixed field;
homogenes —, homogeneous field;
magnetisches —, magnetic field;
ruhendes —, static field;
schwingendes —, oscillating or oscillatory field;
überlagertes —, super(im)posed field;
umlaufendes —, rotating field;
veränderliches —, variable field;

Feld
wanderndes —, travelling field;
zusammengesetztes —, composite field;
— —, aus zwei senkrechten Komponenten, quadrature field;
Dreh= —, rotating field;
Gleich= —, steady (magnetic) field;
Induktions= —, induction field;
Längs= —, longitudinal field;
Magnet= —, magnetic field;
Quer= —, cross field, transverse field;
Spulen= —, loading section K;
Strahlungs= —, radiation field;
Streu= —, stray or leakage field, extraneous field;
Tafel= —, panel;
Verstärker= —, repeater section;
Wander= —, moving or travelling field;
Wechsel= —, alternating field;
schnelles — = —, oscillatory or oscillating field;
— =dichte f, field density;
— =intensität f, field intensity;
— =komponente f, component field;
— =länge f, length of section; span lenght B;
— =magnet m, field magnet;
— =messer m, surveyor;
— =regler m, field rheostat, field regulator;
— =spule f, field coil, magnetizing coil;
— =stärke f, field strength, field intensity;
— =system n, field system n;
— =verteilung f, field distribution;
— =verzerrung f, field distortion, field deformation;
— =wicklung f, field coil, field winding;
— =widerstand m, field rheostat, field regulator.

Fels, Felsen m, rock.
felsig, rocky.
Fenster n, window;
— =umschlag m, window envelope.
fern, distant, remote;
— es Amt n, distant station.
Fern= ..., toll, l. d. (= long-distance);
— =amt n, trunk exchange, toll exchange, toll office;
Haupt= — = —, main zone centre
Neben= — = —, sub-zone centre;
— =amts=aufsicht f, trunk supervisor;
— = — =auskunft(sstelle) f, trunk enquiry, toll information desk; trunk directory enquiry;
— = — =beamtin f, trunk operator, long-distance operator;
— = — =einrichtung f, toll office equipment;
— = — =meldeleitung f, (trunk) record circuit;
— = — =meldestelle f, trunk record section;
— = — =trennung f, breaking of local calls for toll calls, FA;
— =anruf m, trunk call; trunk signalling;
— =beamtin f, toll operator, l. d. operator; operator;
fernbesetzt, toll busy, engaged on trunk call;
Fern=besetztsein n, toll busy condition;
— =drucker m, stock ticker, teletyper; Siemens printer;
— =gespräch n, toll call, l. d. or toll line conversation;
— = — zu bestimmter Zeit, fixed time call;

Fern=, fern=
— =getaftet, remote controlled;
— =hörer m, telephone, (telephone) receiver, phone, talker;
doppelpoliger — = —, double pole or bipolar receiver;
einpoliger — = —, single pole receiver;
verzerrungsfreier — = —, pantelephone;
zweiter — = —, second receiver F;
Dofen= — = —, watch-case telephone, watch receiver;
Kopf= — = —, h. g. receiver head(gear) receiver;
— = — = —, Doppel=, head phones;
Normal= — = —, standard receiver;
O. B. — = — = l. b. or local battery receiver;
Z. B. — = —, c. b. or common battery receiver;
— = —, Kiffen n, phone cushion;
— = — =fchnur f, phone cord;
— = — =normal n, receiver standard;
— =kabel n, trunk telephone cable, long-distance telephone cable, toll cable;
Normal= — = — —, standard long distance cable;
Überland = — = —, overland l. d. cable;
— = — =leitung (mit Sprechstromverftärkern), (repeatered) toll cable circuit;
— = — =netz n, toll cable system, long-distance cable system;
— =leitung f, toll line, long-distance circuit;
Durchgangs= — = —, through toll line;
— = — ohne Verftärker, non-repeatered toll circuit;

— =leitungs=bezirkskabel n, trunk zone cable;
— = — =endkabel n, toll entrance cable;
— = — =gefellfchaft f, longlines company;
— = — =klinke f, toll line jack;
— = — =netz n, toll plant, trunk system;
— = — =fchnur f, trunk cord;
— = — =zwifchenkabel n, toll intermediate cable;
— = — =wähler m, l. d. connector A;
— = — =ftörung f, toll line fault;
— =linie f, toll line, trunk route;
— =melde=leitung f, communication circuit;
— = — =technik f, communication art;
— =photographie f, telephotography;
fernphotographifch, telephotographical;
Fern=platz m, trunk position;
— =prüffchrank m, trunk or toll test board;
— =rohr n, telescope;
Ablefe= — = —, reading telescope;
— =fchalter m, remote control switch, remotely controlled switch, teleswitch;
— =fchaltung f, distant or remote control;
— =fchrank m, long-distance or trunk (telephone) switchboard, toll switchboard;
Durchgangs= — = —, l. d. through switchboard;
— = —=beamtin f, l. d. operator;
— =fehen n, television;
— =feher m, television apparatus;
Fernfprech= ...., telephonic, telephone ....;
— =amt n, exchange, central office, (cf. Amt);

Fernsprechamt
automatisches — = —, automatic exchange;
fliegendes — = —, temporary exchange;
halbautomatisches oder halbselbsttätiges — = —, semiautomatic exchange;
kleineres — = —, minor exchange;
öffentliches — = —, public exchange;
unbedientes — = —, unattended exchange A;
Hand- — = —, manual exchange;
— = — = — mit **selbsttätiger Schlußzeichengebung**, c. b. s. (= central battery signalling) exchange;
Not- — = —, temporary exchange;
O. B. — = —, local battery exchange;
— — = — mit **Induktoranruf**, magneto exchange;
Orts- — = —, local exchange;
— = — = —, **vereinigtes Fern- und**, combined trunk and local exchange;
Selbstanschluß- — = —, automatic exchange, mechanical telephone office;
— =**anlage** f, telephone plant;
Privat- — = —, private telephone plant; house telephone plant;
— = — mit **mehreren Vermittlungsstellen** multi-office exchange;
— =**anschluß** m, telephone station;
— =**apparat** m, telephone instrument;
— =**automat** m, coin collector telephone station, pay station;
— =**betrieb** m, telephone service;
— =**buch** n, telephone directory;
— =**dienst** m, telephone service;
— =**doppelleitung** f, metallic telephone circuit, two wire telephone circuit, telephone loop;

fernsprechen, to telephone, to phone;
Fernsprechen n, telephony;
drahtloses —, wireless or radio telephony;
Draht- —, wire telephony;
Funk- —, radio or wireless telephony.
Mehrfach- —, multiple telephony;
— = — mit **hochfrequenten Trägerströmen**, h. f. multiple telephony, carrier wave telephony;
Fernsprecher m, telephone; telephone station;
— **für 10 Linien**, 10-way telephone station;
— =**für Batterie-(Induktor-)Anruf**, battery ringing (magneto) telephone station;
— **mit Schutzschaltung gegen Mikrophongeräusch**, anti-side tone telephone set;
tragbarer —, portable telephone set;
Hand- —, hand telephone;
Münz- —, coin box call office, pay station, coin collector telephone station;
— = — **für Fernbetrieb für mehrere Geldsorten**, multi-coin box call station;
Säulen- —, desk stand telephone station;
Ständer- —, pedestal desk telephone;
Strecken- —, portable telephone set;
Tisch- —, table telephone station;
Wand- —, wall telephone station;

Fernsprech-frequenzen, *pl* telephone or telephonic frequencies *pl*;
— -hauptanschluß *m*, main set, (suberiber's) main station;
— -kabel *n*, telephone cable;
— -klappenschrank *m*, telephone switchboard;
— -leitung *f*, telephone circuit, telephone line;
(un)gekreuzte — - —, (non-) transposed telephone line;
verdrallte — - —, twisted telephone circuit;
— -Meßtechnik *f*, telephonometry;
— -Nebenstelle *f*, extension set;
— -netz *n*, telephone network, (outside) telephone plant;
— - — mit mehreren Vermittlungsämtern, multi-exchange system (*engl.*), multi-office exchange (*am.*);
— -reihenanlage *f*, intercommunication or house telephone plant;
— -relais *n*, telephone relay, telephonic relay;
— -Ringübertrager *m*, toroidal repeating coil;
— -Schnellverkehr *m*, no-delay telephone service;
— -stelle *f*, telephone station; öffentliche — - —, public call office;
— -störfaktor *m*, telephone interference factor;
— -system *n*, telephone system;
— -Teilnehmerverzeichnis *n*, telephone directory;
— -Telegrammaufnahme *f*, phonogram section;
— -übertrager *m*, telephone transformer, repeating coil;
— -Vergleichsstromkreis *m*, standard reference telephone circuit;
— -verkehr *m*, telephone traffic;

Orts- — - —, local telephone traffic;
— -verkehrsschreiber *m*, telephone traffic recorder;
— -vermittlungsstelle *f*, exchange, central office (*am.*);
— -verstärker *m*, telephone repeater, telephonic repeater;
— - - -amt *n*, telephone repeater station;
— - — -betrieb *m*, telephonic repeater operation;
— - — -röhre *f*, telephone amplifying valve, telephone repeater tube;
— -weitverkehr *m*, long-distance telephone service;
— -wesen *n*, telephony;
— -zelle *f*, telephone cabin(et); schalldichte — - —, silence cabinet;
— -zone *f*, telephone zone;
— -zonenhauptpunkt *m*, telephone zone centre;
— -zwischenverstärker *m*, telephone intermediate repeater.
Fern-steuerung *f*, remote or distant control;
— -tastung *f*, remote control;
— -thermometer *n*, distance thermometer;
— -und Ortsamt, vereinigtes, combined trunk and local exchange;
— -verkehr *m*, toll or long-distance traffic;
Nah- — - —, short haul toll traffic (*am.*);
— -verkehrszone *f*, telephone trunk zone;
— -vermittlungsleitung *f*, trunk junction circuit, toll switching trunk (*am.*);
— -wahl *f*, l. d. selection, toll switching, *A*;
— -wirkung *f*, distant effect; radiation effect.

ferromagnetisch, ferromagnetic(al).
Ferromagnetismus *m*, ferromagnetism.
Fertigkeit *f*, skill.
fertigen, to make, to manufacture.
fertigstellen, to finish.
Fertigstellung *f*, finish.
Fertigung *f*, manufacture; wirtschaftliche —, economical manufacture.
fest, rigid, solid (auf, with), straff: tight, stabil: stable; feststehend: fixed;
— gekoppelt, tightly coupled;
— er Körper *m*, solid;
— e Spule *f*, fixed coil, stator;
— =halten, to retain, to hold;
— =haltend, retentive.
Festigkeit *f*, strength, tightness, stability; Bündigkeit: compactness; Dichte: consistency;
dielektrische —, dielectric strength, elastance;
mechanische —, mechanical strength;
Biege- —, bending strength;
Bruch- —, breaking strength;
Dreh- —, torsional strength;
Druck- —, compressive strength;
Durchschlags- —, disruptive or rupturing strength;
Scher- —, shearing strength;
Torsions- —, torsional strength
Zug- —, tensile strength.
fest-keilen, to wedge;
— =klammern, to clamp, to grip in;
— =klemmen, to clamp; sich —, to wedge, to jam (tight);
— =laufen, to jam tight;
— =legen, to fix; to fasten, to clamp, to secure in position, to anchor;

Festlohn *m*, fixed wages *pl*;
festmachen, to lash, to fasten;
Festpunkt *m*, section point (für Leitungskreuzungen,ʊ for line crossings; für Spulenabstände, for loading coil spacing);
fest-schrauben, to fasten with screws, to screw on;
— =setzen, to fix; Gebühren: to assess charges (upon);
Festsetzung *f*, fixing;
fest-sitzen, to be jammed tight;
— =stampfen, to tamp, to pun;
Feststampfen *n*, tamping, punning;
fest-stehend, fixed, stationary;
— =stellen, Fehler: to state, to locate; sichern: to stop, to secure (in position); prüfen: to verify; Zeit: to time; Übereinstimmung: to identify.
Feststellung *f*, statement; location; stopping, securing; verification; timing; identification.
fest-ziehen, to screw down (eine Mutter, a nut).
Fett *n*, grease, lubricant;
konsistentes —, consistent lubricant;
— =schmierung *f*, grease lubrication.
fetten, to grease.
feucht, wet, damp, moist, humid;
— e Räume *pl*, damp rooms *pl*.
Feuchtigkeit *f*, dampness, humidity, moisture;
Eindringen von —, ingress of moisture;
— austreiben, to expel humidity.
feuchtigkeitsdicht, damp-proof.
Feuchtigkeitsmesser *m*, hygrometer.
feuchtigkeitssicher, damp-proof
Feuer *n*, fire;

Feuer
— -löschdecke f, fire-exstinguishing cover;
— -löscher m, fire exstinguisher;
— -löschgerät n, fire extinguishing appliances pl;
— -melder m, fire alarm (signal box);
— - —leitung f, fire alarm circuit.
feuern, to spark, to flash.
Feuern n, flashing, sparking.
Feuerschlauch m, fire hose;
Feuersgefahr f, fire risk;
feuersicher, flame-proof, fire-resisting;
— machen, to flame-proof;
— er Anstrich m, fire-resisting paint;
Feuerversicherung f, fire insurance;
feuerverzinnt, fire-tinned;
Feuer-wache f, fire station;
— -wehrtelegraph m, fire alarm telegraph.
Fiber f, fibre, am.: fiber;
Phenol- —, phenol fibre;
Vulkan- —, vulcanized fibre;
— -hülse f, fibre sleeve;
— -rohrstrang m, fibre duct, fibre [conduit.
Fichte f, pine;
Fichten-harz n, pine resin;
— -holz n, pine.
fieren, to lower, to veer.
Figur f, figure;
Figurenwechsel m, shift (signal), inversion, T;
fiktiv, ficticious.
Filter n, filter, sifter;
Band- —, band (pass) filter, wave band filter;
— - — mit großer Lochbreite, broad band filter;
Sperr- —, rejector or stopping circuit, suppression filter (circuit);
Übertragungs- —, transmission filter (circuit);

Wellen- —, wave filter;
— -glied n, filter section;
— -kreis m, filter(ing) circuit.
Filz m, felt;
mit — aus- oder unterlegt, felted;
Brems- —, tow brush (am Regler, of the governor);
— -unterlage f, felt underlayer.
Finder m, finder.
Finger m, finger;
Kontakt- —, contact finger;
— -anschlag m, finger stop A;
— -öffnung f, finger hole A;
— -scheibe f, finger disc or wheel A.
Firnis m, varnish.
firnissen, to varnish.
First m, ridge B.
Fischplatte f, fishplate.
Fixierbad n, fixing bath.
fixieren, to fix.
FL-Wähler m, l. d. selector. l. d. connector, A.
flach, flat, plane, squat;
— nebeneinander, on the flat.
Flach-draht m, flat wire;
— - — -bewehrung f, flat wire sheathing;
— -eisen n, flat (bar) iron;
— -feder f, flat spring;
— - — -spirale f, flat spiral spring;
— -gewinde n, square thread;
— -kabel n, ribbon (-shaped) cable, flat cable;
— -kern m, flat core;
— -kopf m, flat head; flach-köpfig, flat-headed;
— -spirale f, flat spiral;
— -spule f, flat coil;
— - —, quadratische, flat square coil, square plane coil, pancake coil R;
— -zange f, flat nose pliers pl.
Fläche f, Ebene: plane; Größe: area; Oberfläche: surface; face;

Fläche
  bearbeitete —, machined or tooled surface;
  Äquipotential= —, equipotential surface;
  Typen= —, type surface;
  Windungs= —, turn area;
  — konstanten Potentials, equipotential surface;
Flächen=einheit f, unit area;
  — =inhalt m, area;
  — =maß n, superficial measure.
Flachs m, jute;
  mit Tannin getränkter —, tanned jute;
  — =garn n, jute yarn.
flackern, to flicker, to flash.
Flackern n, flickering, flash(ing);
Flackerzeichen n, flickering or flashing signal;
  durch — zum Eintreten veranlassen, to flash in(to circuit);
  — geben, to flash.
Flamme f, flame;
Flammenmikrophon n, flame transmitter.
Flanell m, flannel.
Flanke f, flank (des Zahnes, of the tooth).
Flankenstreuung f, side leakage.
Flansch m, flange, socket; Rippe: vane;
  Kühl= —, cooling vane;
  Rohr= —, pipe flange;
  Spulen= —, spool flange.
Flasche f, bottle;
  Leydener —, Leyden jar;
Flaschenzug m, tackle, luff.
flattern, to flutter.
Flatterwirkung f, flutter(ing) [effect.
flechten, to plait.
Fleck m, spot;
  Sonnen= —, sun-spot.
fliegende Anlage f, temporary plant.
Fliehkraft f, centrifugal force;
  — =regler m, centrifugal governor.

Fliese f, tile, flag;
  mit — n belegen, to flag.
fließen, to flow, to circulate, to stream, to pass (in, to, into).
Fließen n, flow, flowing, circulation.
Fließgrenze f, yield point.
Flinse f, flaw.
Flintglas n, flint glass.
Floß n, float.
Flosse f, fin, skid;
Flossenantenne f, skid-fin aerial (der Flugzeuge, of airplanes).
Flöte f, flute.
Flotte f, fleet.
Fluchtentafel f, straight-line chart, self-computing chart.
flüchtig, Körper: volatile; Vorgang: transient;
  — e Spannung f, transient voltage;
  — er Stromstoß m, transient impulse;
  — er Vorgang m, transient.
Flüchtigkeit f, transientness.
Flügel m, wing, blade, vane;
  — =anker m, vane armature;
  — = — =relais n, vane armature relay;
  — =gebläse n, fan;
  — =mutter f, wing(ed) nut, butterfly nut;
  — =schraube f, wing screw, thumb screw.
Flugzeug n, aeroplane, airplane;
  — =antenne f, aeroplane aerial;
  — =station f, aircraft station R.
Fluidum n, fluid.
Fluktuation f, fluctuation.
fluktuieren, to fluctuate.
Fluktuieren n, fluctuation.
Fluß m, flow, flux, circulation;
  magnetischer —, magnetic flux;
  wirksamer —, net flux;
  Elektronen= —, flow of electrons;

Fluß
Energie- —, energy flow;
Gesamt- —, total flux;
Induktions- —, induction flux;
Kreis- —, circular flux;
Magnet- —, magnetic flux;
Streu- —, stray or leakage flux, magnetic crossflux;
Wechsel- —, alternating flux;
— -bett n, river bed;
— -dichte f, flux density;
— -kabel n, river cable, subfluvial cable;
— -kreuzung f, river crossing;
— -mittel n (zum Löten), (soldering) flux;

flüssig, liquid.

Flüssigkeit f, fluid, liquid;
Erreger- —, exciting fluid;
Kühl- —, cooling fluid;

Flüssigkeits-anlasser m, liquid starter;
— -dämpfung f, liquid damping;
— -stand m, liquid level;
— - -anzeiger m, liquid level indicator;
— -strahl m, liquid jet;
— - -mikrophon n, liquid jet transmitter or microphone;
— - -relais n, jet relay (für Seekabel, for submarine cables);
— -widerstand m, water resistance.

Fluttereffekt m, flutter effect.

Flux m, flux.

Fokus m, focus.

Folge f, sequence, succession;
— -kontakt m, make-before-break contact, continuity-preserving contact;
— -pol m, consequent pole;
— -schalter m, sequence switch A;
— - — und Relaissatz für L W (= Leitungswähler), final sequence switch and relay set, A.

Folie f, foil;
Gold- —, gold foil;
Zinn- —, tin foil.

Förder-anlage f, conveying plant;
— -band n, band conveyer, conveying belt, belt carrier;
— -wagen m, lorry, truck.

Förderer m, conveyer, conveyor;
Band- —, m, belt carrier, band conveyer.

Form f, shape, form, figure;
Gieß- —, mould;
— -eisen n, profile iron, profiled iron;
— -faktor m, form factor;
— -stück n, moulded body;
— - —, Isolier-, moulded insulation;
— - —, Ton-, tile, clay conduit, earthenware block, B;
— - —, Zement-, concrete block B;
— - — -kanal m, block conduit (einzügiger, single; mehrzügiger, multiple).

Formaldehyd n, formaldehyde.

Formation f, formation.

Formel f, formula.

formen, to shape, to mould.

formieren, to form (Sammlerplatten, storage cell plates).

formierte Platte f, formed plate.

Formierung f, forming, formation.

Formular n, form, paper blank;
Telegramm- —, message blank, telegraph blank.
— -rolle f, (paper) web.

forschen, to search (nach, for); to study, to research.

Forschung f, study, research.

Forschungsarbeit f, research work.

fortleiten, to conduct.

Fortleitung f, conduction.

fortpflanzen, to propagate.

Fortpflanzung *f*, propagation, communication; convection; **auf — beruhend**, convective; **Wellen-—**, wave propagation;

Fortpflanzungs-geschwindigkeit *f*, velocity of propagation;

— **-größe** *f*, propagation constant, (hyperbolical) line angle, complex attenuation constant, $(\gamma = \beta + ja), L$;

— **-konstante** *f*, propagation constant or coefficient L;

— **-maß** *n*, (hyperbolical) line angle L;

— **-richtung** *f*, direction of propagation.

Fortsatz *m*, extension, prolongation.

fortschaffen, to transport.

Fortschaffung *f*, transportation.

fortschalten, to step on, to step up.

Fortschaltwerk *n*, stepping mechanism.

Fortschellwecker *m*, continuous(ly) ringing bell.

fortschreiten, to progress, to advance.

fortschreitend, progressive, advancing;

— **e Abnahme** *f*, progressive diminution;

— **e Welle** *f*, advancing wave.

Fortschreiten *n*, progression, advance.

Fortschritt *m*, progress.

Foucaultströme *pl*, Foucault or eddy currents *pl*.

Fouriersche Reihe *f*, Fourier's series;

— **r Satz** *m*, Fourier's theorem.

Frankesche Maschine *f*, Franke machine.

frankieren, to frank.

Frankiermaschine *f*, franking machine.

fräsen, to mill.

Fräsmaschine *f*, milling machine.

frei, free, clear; **Leitung usw.**: disengaged; **portofrei**: frank; **räumlich**: clear (von, of, from); **Wähler usw.**: idle;

— **er Raum** *m*, clearance;

— **e Wahl** *f*, hunting (operation) *A*.

Freigabe *f*, liberation.

freigeben, to liberate, to free, **Blockstrecke**: to clear.

freihängend, trailing.

Freileitung *f*, open line.

frei machen, to free; **Telegramme**: to frank.

freistehend, free standing, **ohne Stütze**: self-supporting.

frei suchen, to hunt, to find, *A*.

freitragend, self-supporting.

Freiwahl *f*, hunting (action) *A*.

frei wählen, to hunt, to find, *A*.

Freizeichen *n*, ringing tone, audible ringing signal, *A*;

Amts- — -, dialling tone *A*.

Fremdantrieb *m*, separate drive; **mit** —, separately driven.

fremderregt, separately excited.

Fremd-erregung *f*, separate excitation.

— **-strom** *m*, foreign current.

— **-überlagerer** *m*, separate heterodyne (local oscillator).

Frequenz *f*, frequency, periodicity;

von gleicher —, equifrequent;

mittlere —, mean frequency; medium frequency;

Eigen- —, natural frequency;

Einzel- —, component frequency;

Entladungs- —, discharge frequency (eines Kondensators, of a condenser);

Fernsprech- —, telephone or telephonic frequency;

Funk- —, radio frequency, *ab r. f.*;

Funkfrequenz
— ⸗ — ⸗bereich *m*, radio frequency range;
— ⸗ —, Über⸗, ultra-radio frequency;
Grenz⸗ —, obere (untere), upper (lower) limiting or cut-off frequency;
Grund⸗ —, fundamental or base frequency;
Hoch⸗ —, high frequency; radio frequency;
Hör⸗ —, acoustic or audio or audible frequency;
— ⸗ —, Über⸗ (Unter⸗), ultra- (sub-) audio frequency;
— ⸗ — ⸗verstärker *m*, note magnifier or amplifier;
Impuls⸗ —, impulse frequency *A*; [cy;
Interferenz⸗ —, beat frequen-
Kombinations⸗ —, combination frequency;
Kommutierungs⸗ —, ripple frequency;
Kreis⸗ —, angular velocity, frequency in radians;
— ⸗ —, Grenz⸗, cut-off angular velocity;
Ladungs⸗ —, charge frequency;
Modulations⸗ —, modulating frequency;
Nieder⸗ —, low frequency;
Nutenwellen⸗—, slot ripple frequency;
Radio⸗ —, radio frequency;
Resonanz⸗ —, resonance or resonant frequency;
Resonanz=Grund⸗ —, first resonating frequency;
Schwebungs⸗ —, beat frequency, combination frequency;
— ⸗ — Null, zero beat frequency;
Seitenband⸗ —, obere (untere), upper (lower) side frequency;

Sprach⸗ —, Sprech⸗ —, voice frequency, telephonic frequency;
— ⸗ —, mittlere, mean frequency of speech;
Stör⸗ —, interfering frequeny;
Telegraphier⸗ —, signalling frequency, telegraph(ic) frequency;
Telegraphier=Grund⸗ —, dot frequency;
Ton⸗ —, acoustic frequency, audible or audio frequency;
— ⸗ — ⸗telegraphie *f*, voice-frequency telegraphy;
Träger⸗ —, carrier frequency;
Überlagerer⸗ —, local oscillation frequency;
Wellenzug⸗ —, group or wave train frequency (gedämpfter Sender, of damped senders);
Welligkeits⸗ —, ripple frequency;
Zeichen⸗ —, signal frequency;
Frequenz null, zero frequency;
frequenzabhängig, variable with frequency, dependent on frequency, resonant;
Frequenz=abhängigkeit *f*, variability with frequency, dependency on frequency;
— ⸗band *n*, frequency band, band or range of frequencies;
— ⸗ —, Sprech⸗, speech band;
— ⸗bereich *m*, range of frequencies;
— ⸗ —, durchgelassener, transmitted band of frequencies (eines Siebgebildes, of a filter);
— ⸗ —, Nachbildungs⸗, frequency range of simultation;
— ⸗ —, Sprech⸗, speech frequency range;
— ⸗empfindlichkeit *f*, frequency sensitivity;
— ⸗Empfindlichkeits=Kennlinie *f*, frequency response characteristic;

Frequenz
- -komponente *f*, component frequency;
- -kurve *f*, frequency characteristic;
- - —, Dämpfungs-, attenuation frequency curve;
- - —, Entdämpfungs-, gain-frequency curve;
- -messer *m*, frequency meter;
- - —, Resonanz-, resonance frequency meter;
- -schwankung *f*, variation of frequency;
- -sieb *n*, frequency sifter;
- -spektrum *n*, frequency spectrum;
- -umformer *m*, frequency transformer;
- -umformung *f*, frequency transformation;

frequenzunabhängig, independent of frequency;

Frequenz-unabhängigkeit *f*, independence of frequency, invariability with frequency;
- -verdopplung *f*, doubling of frequency;
- -verzerrung *f*, frequency distortion;
- -wandler *m*, frequency changer, frequency multiplier;
- - —, ruhender, static frequency changer;
- -wandlung *f*, frequency transformation. [ings).

fressen, to seize (Lager, bear-

Friktion *f*, friction.

Friktions-antrieb *m*, friction drive;
- - —, nachgiebiger, yielding or slipping drive;
- -rad *n*, friction wheel;
- -scheibe *f*, friction disc.

fritten, to cohere.

Fritter *m*, coherer;
  Körner- —, granular coherer;

Pulver- —, powder coherer;
- -wirkung *f* coherer action.

Frittung *f*, coherence.

Front *f*, front;
  Wellen- —, wave front.

Frosch-klemme *f*, — -zug *m*, Dutch tongs *pl*, draw vice, eccentric grip.

Frost *m*, frost.

Frühmessung *f*, (regelmäßige) morning (routine) test

Fuchsschwanz *m*, hand saw.

Fuge *f*, joint.

führen, to lead, to conduct, to guide, to pilot; Strom: to carry a current, Leitungen: to run wires.

Führung *f*, guide, guiding, lead(ing), conduction.

Führungs-loch *n*, guiding hole; feed hole, (des Streifens, of the slip) *T*;
- - — in der Mitte des Stanzstreifens, central feed hole of the perforated tape;
- - — -abstand *m*, feed hole space;
- - — -reihe *f*, row of feed holes;
- -rolle *f*, guiding pulley, guide roller;
- -stange *f*, guide rod;
- -stift *m*, guide pin;
- -stück *n*, guide piece.

füllen, to fill.

Fülleinlage *f*, filler.

Füllen *n*, filling.

Füll-masse *f*, — -paste *f*, filling paste;
- -säure *f*, accumulator acid.

Füllung *f*, filling,
  Gas- —, gas filling.

Fundament *n*, foundation;
  Beton- —, concrete foundation, concrete bed;
  Mast- —, pole or mast foundation;

fundamental, fundamental, basic.
Fundament-schraube f, foundation bolt;
— -zeichnung f, foundation sketch.
Fünfelektrodenröhre f, pentode.
Fünferalphabet n, five-unit code T.
fünffach, quintuple.
Fünfströmealphabet n, five-unit code T.
Fünftastengeber m, five-key transmitter T.
Funk-...., radio, wireless;
Draht- — m, line radio, wired wireless;
— - -system n, wire carrier or line radio system;
— - —-technik f, wire carrier art;
Rund- — m, broadcasting;
— - -anlage f, broadcasting plant;
— - -empfang m, broadcast reception;
— - -empfänger m, broadcast receiver;
— - -gerät n, broadcast receiving set;
— - -sender m, broadcast transmitter; [plant;
— -anlage f, wireless or radio
— -apparat m, wireless apparatus, wireless set;
— -beamter m, radio or wireless officer;
— -einrichtung f, wireless equipment;
— -empfang m, wireless or radio reception;
— -empfänger m, radio receiver, wireless receiving set;
Peil- — — —, wireless direction finder;
funken, to (send by) wireless; Kollektor: to spark, to flash;

Funken n, sparking; wireless transmission.
Funke(n) m, spark; ... Funken in der Sekunde, ... sparks per second;
tönender —, musical spark;
Entlade- —, disruptive spark;
Öffnungs- —, spark at break;
Schließungs- —, spark at make;
Unterbrechungs- —, spark at break;
Funken-ausbläser m, spark blow-out;
— -entladung f, disruptive discharge;
funkenfrei, non-arcing (Metalle, metals); sparkless;
— e Unterbrechung f, clean break;
Funken-induktor m, spark or induction coil, Ruhmkorff coil;
— — mit Hammerunterbrecher, hammer break spark coil;
— -länge f, spark length, sparking distance;
— -löscher m, spark extinguisher, spark blow-out;
— -löschkondensator m, spark quenching condenser;
— -mikrometer n, micrometric spark discharger, spark micrometer;
— -potential n, spark potential;
— -schutz m, spark killer;
— -sender m, spark transmitter;
tönender — -, musical spark transmitter;
Lösch- — -, quenched spark transmitter;
Ton- — -, musical spark transmitter;
— -spannung f, spark potential;
— -strecke f, (spark) gap, discharger;
feste — -, plain spark discharger;

Funken-strecke
feststehende – = –, static or stationary spark gap;
rotierende oder **umlaufende** – = –, disc discharger, rotary spark gap;
– – = – **für die Erzeugung ungedämpfter Wellen**, timed spark discharger;
unterteilte – = –, multiple spark gap;
Ansprechen n der – = –, operation of gap;
Läufer m der – = –, spark gap rotor;
Asynchron- – = –, non-synchronous rotating spark gap;
Erreger- – = –, exciting spark gap;
Kugel- – = –, sphere gap;
Lösch- – = –, quench(ed) or quenching spark gap;
Luftleer- – = –, vacuum spark gap;
Mehrfach- – = –, multiple spark gap;
Reihen- – = –, multiple spark gap;
Scheiben- – = –, (asynchrone, synchrone), (asynchronous, synchronous) disc discharger;
– = – = –, **glatte**, smooth disc discharger;
Sicherheits- – = –, safety (spark) gap;
Synchron- – = –, synchronous rotating spark gap;
Überspannungs- – = –, surge arrester;
Vielfach- – = –, multiple spark gap;
– = – = –, **rotierende**, timed spark discharger;
Zahnscheiben- – = –, studded disc discharger;
– = – **n-elektrode** f, spark gap face, spark knob, spark gap terminal;
– **-ton** m, spark note;
– **-überschlag** m, spark gap breakdown;
– **-zahl** f, spark rate, spark frequency;
– **-ziehen** n, spark drawing.

Funk-frequenz f, radio frequency;
– = – **-bereich** m, radio frequency range;
– **-freund** m, radio amateur;
– **-kompaß** m, radio or wireless compass, radio-goniometer;
– **-leitsender** m, radio beacon (für die Navigation, for navigation);
– **-mast** m, radio mast m, radio tower;
– = –, **freitragender**, self-supporting radio tower;
– **-netz** n, radio system, radio network;
– **-offizier** m, radio or wireless officer;
– **-peileinrichtung** f, radio direction finder;
– **-schatten** m, dead spot, radio shadow, radio pocket;
– **-senden** n, radio transmission;
– **-sender** m, radio transmitter;
Peil- – = –, radio beacon;
– **-sprech-** ...., radiophonic(al);
– **-sprecher** m, radiophone;
– **-spruch** m, radiophone message; wireless message, radiogram;
– **-station** f, radio or wireless station;
fahrbare – = –, cart type radio station;
Karren- – = –, wagon radio set;
– **-stelle** f, radio station;
Bord- – = –, ship radio station;
Einstrahl- – = –, beam station;

Funk-stelle
 Groß- — - —, long-distance radio station;
 Küsten- — - —, coastal radio station;
 Schiffs- — - —, ship radio station;
 — -technik f, radio art, radio engineering;
 — -telegraph m, wireless telegraph;
 — -telegraphenanlage f, radio telegraph plant;
 — -telegraphie f, radio telegraphy;
 abgestimmte — - —, syntonic wireless telegraphy;
 gerichtete — - —, directional wireless telegraphy;
 — - — -sender m, radiotelegraphic transmitter;

funktelegraphisch, radiotelegraphic(al);

Funk-telephon, radiophone, wireless telephone; [phony-;
 — -telephonie f, radio tele-
 — - — -sender m, radiotelephonic transmitter;

funktelephonisch, radio(tele)phonic(al);

Funk-turm m, radio tower;
 — -verbindung f, radio communication.

Funktion f, function;
 einfache —, simple function;
 einwertige —, single valued function;
 gerade —, even function;
 hyperbolische —, hyperbolic function;
 periodische —, periodic function;
 ungerade —, odd function;
 Exponential- —, exponential function;
 Sinus- —, sine function.

funktionieren, to function.

Fuß m, bottom, foot; pedestal, shoe, support, footing; Maß: foot, ab: ft. = 12 inches = 30,4797 cm;
 Kubik- —, cubic foot, ab: cub. ft. = 28,316 dm³;
 Quadrat- —, square foot, ab: squ. ft. = 9,29 dm²;
 Mast- —, pole pedestal, footing of a pole;
 Stangen- —, pole pedestal, pole footing;
 — -boden m, floor;
 Zement- — - —, concrete floor;
 — -brett n, foot rest;
 — -lager n, footstep bearing. vertical bearing;
 — -leiste f, skirting; foot rest;
 — -pfund n, foot-pound;
 — -schalter m, foot switch;
 — -steig m, footway;
 — -tritt m, foot pedal, foot treadle; an der Stange: foot step B.

Futter n, Drehbank: chuck; zum Montieren: fixture, jig; Auskleidung: lining;
 Bohr- —, drill chuck.

G.

Gabel f, fork; des Tischfernsprechers: cradle; [fork;
 zweizinkige —, double-pronged

gabelförmig, fork-shaped;

Gabel-schaltung f, (Gegensprech-), split (duplex) connection T;

Gabel-schaltung
— = —, Mehrfachtelegraph in, forked multiplex telegraph;
— =übertragung f, forked repeater T;
— =umschalter m, cradle switch F;
— =zinken pl, fork tines pl.
gabeln, sich, to fork, to bifurcate.
Gabelung f, bifurcation.
Galerie f, gallery.
galvanisch, galvanic(al);
— niederschlagen, to electrodeposit;
— es Element n, galvanic cell;
— e Kopplung f, galvanic coupling;
— er Niederschlag m, electro-deposit.
galvanisieren, to galvanise.
Galvanisierung f, galvanisation, electro-deposition.
Galvanokohle f, coppered carbon.
Galvanometer n, galvanometer;
ballistisches —, ballistic galvanometer;
Lichtzeiger m des —s, galvanometer spot;
Batterie- —, battery gauge, battery tester;
Drehspul- —, moving coil galvanometer;
Faden- —, thread galvanometer;
Kugelpanzer- —, ball shield galvanometer;
Saiten- —, string galvanometer;
Spiegel- —, reflecting or mirror galvanometer;
Thermo- —, thermo-galvanometer;
Vibrations- —, vibration galvanometer;
Zeiger- —, pointer galvanometer;

— =konstante f, galvanometer constant.
Galvanoskop n, galvanoscope, detector.
Gang m, corridor, way, aisle, gangway; einer Maschine: running; Gewinde: thread (10 Gänge auf 1 cm, 10 threads per centimetre); in — setzen, to start, to entrain, to throw in gear;
gleich schneller —, isochronism;
toter —, backlash, lost motion;
Haupt- —, main aisle (zwischen den Verstärkergestellen, between the repeater racks);
— =höhe f der Windungen, pitch of turns;
— =maß n, tempo;
— =unterschied m, phase difference, path difference (der Wellen, of waves).
ganz, whole, entire, complete, total;
— es Vielfaches n, integral multiple;
— e Zahl f whole number, integer, integral number;
— =zahlig, integral.
Garantie f, guarantee, warrant.
garantieren, to guarantee, to warrant.
Garantiewert m, guaranteed value.
Garn n, yarn, twine;
Flachs- —, jute yarn;
Glanz- —, glazed or glace cotton, glaced yarn;
Jute- —, jute yarn.
— = — =umklöppelung f, glazed cotton braiding;
— =trense f, yarn worming.
Gas n, gas;
okkludierte — e pl, occluded gases pl;
Leitung f der — e, conduction through gases;

Gas
  Edel- —, rare gas;
  — entwickeln, to evolve gas;
  — -atmosphäre f, gaseous atmosphere;
  — -austritt m, gas leakage (in Kabelbrunnen, in cable manholes);
gasen, to gas, to evolve gas;
Gasen n, gassing;
Gas-entladungsrelais n, (gas) discharge relay;
  — -entwicklung f, formation of gas, gassing;
gas-förmig, gaseous;
  — -frei, free of gases;
Gas-freiheit f, absence of gases;
  — -füllung f, gas filling, gas content;
gasgefüllt, gas-filled;
  — e Röhre f, gas-content tube;
Gas-gehalt m, — -inhalt m, gas content;
  — -motor m, gas engine;
  — -rückstand m, residual gas;
  — -spuren pl, trace(s pl) of gas;
  — -strecke f, gas(eous) path;
  — -teer m, gas tar.
Gasolin n, gasoline.
Gauß n, gauss.
Gaze f, gauze;
  Draht- —, wire gauze;
  Kupfer- —, copper gauze;
  — -bürste f, gauze brush.
Gebäude n, building.
geben, to transmit to send, T;
Geber m, sender, transmitter;
  Impuls- —, impulse sender A;
  Maschinen- —, auto(matic) transmitter T;
  Speicher- —, storage transmitter T;
  Streifen- —, (perforated) tape transmitter;
  Tasten- —, keyboard transmitter;
  — -=—, Fünf-, five key transmitter;
  — -amt n, transmitting station;
  — -ende n, — -seite f, sending end (einer Leitung, of a line).
Gebiet n, area, district;
  Stadt- —, city area;
  Vorort- —, suburban area.
Gebläse n, blower, blast;
  Flügel- —, fan;
  Knallgas- —, oxyhydrogen blow pipe;
  Sandstrahl- —, sand blast;
  Turbinen- —, turbine blower.
geblättert laminated.
geblockt, blocked.
gebogen, curved, bent, buckled.
Gebrauch m, application, employment, utilization.
gebrauchen, to use, to apply, to employ, to utilize, to handle; nötig haben: to want.
Gebühr f, fee, rate, charge;
  mit einer — belegen, to charge;
  eine — erheben, to levy a charge;
  — -festsetzen, to assess charges;
  ermäßigte —, deferred rate;
  — n— en, Telegramm zu, deferred telegram;
  Abonnements- —, subscription rate;
  Durchgangs- —, transit charge;
  Einzel- —, measured rate, message rate;
  Erneuerungs- —, renewal fee;
  Fern- —, toll rate;
  Gesprächs- —, message rate, message fee;
  Jahres- —, annual subscription (rate);
  Nacht- —, night rate;
  Pausch- —, flat rate;
  Staffel- —, graduated rate;
  Tages- —, day rate;
  Teilnehmer- —, subscription rate;

Gebühr
  Verlängerungs- —, renewal fee;
  Zeit- —, measured rate;
  Zonen- —, zone rate;
Gebühren-einheit f, tariff unit;
  — -minuten pl, (perfect) paid time F;
gebührenpflichtig, chargeable;
Gebühren-tarif m, tariff;
  Fern- — - —, toll tariff;
  Gesprächs- —-—, message rate tariff;
  Pausch- —-—, flat rate tariff;
  — -zeit f, perfect paid time F;
gedämpft, damped, attenuated;
  schwach —, slightly or weakly damped;
  stark —, strongly or highly damped;
  — werden, to attenuate, to be damped, to die down;
  —e Schwingungen pl, damped oscillations pl;
  — er Sender m, spark transmitter;
  — e Wellen pl, damped waves; discontinuous waves, type B waves, pl.
gedrängt, gedrungen, compact.
geeignet, suitable, suited.
geerdet, earthed, grounded, earth-connected;
  a-Zweig —, A-leg earthed;
  — er Nullpunkt m, grounded neutral point;
  — es Schutznetz n, earthed cradling B.
Gefahr f, danger, risk.
gefährden, to risk, to endanger.
gefährdet, endangered, danger....
gefährlich, dangerous.
gefahrlos, safe (from).
Gefahrpunkt m, danger point.
gefärbt, coloured.
gefasert, grained;
  längs —, straight-grained.

Gefäß n, container, jar, pot, well, vessel, box, tank, tray;
  poröses —, porous pot;
  Eisen- —, iron tray;
  Entladungs- —, electron discharge vessel;
  Farb- —, ink well T:
  Kühl- —, cooling tank;
  Rippen- —, ribbed tank;
  Sammler- —, accumulator box or jar;
  Tränk- —, impregnating tank or vessel.
Geflecht n, mesh;
  feines (grobes) —, fine (coarse) mesh;
  Kupferdraht- —, copper mesh.
gefräst, milled.
gefrieren, to freeze.
Gefüge n, structure;
  Kristall- —, crystal structure.
gegabelt, forked.
Gegend f, region, district, zone.
  Geschäfts- —, city district, commercial district;
  Wohn- —, residential district.
gegeneinander geschaltet, opposing;
  — in Reihe, series-opposed.
Gegenfeder f, opposing or reacting spring.
Gegenfritter m, anticoherer.
gegengeschaltet, counter.....;
  —e Sammlerzelle f, counter-cell.
Gegengewicht n, counterpoise, counterweight, balancing capacity, counterbalance;
  ungeerdetes —, earth screen R;
  Anker m mit —, gravity-controlled armature.
Gegeninduktivität f, mutual inductance, mutual induction;
  — s-Koeffizient m, coefficient of mutual inductance.
Gegenkraft f, opposing force, bias;

Gegenkraft
durch eine — ausgleichen, to bias out.

Gegenlage f, stiffening piece.

Gegenmutter f, clamping screw, lock nut.

Gegennebensprechen n, far-end crosstalk K.

gegenphasig, in phase opposition.

Gegenphasigkeit f, phase opposition.

Gegensatz m, opposition.

gegenseitig, mutual.

gegensinnig, opposing;
— in Reihe geschaltet, in series-opposing.

Gegenspannung f, counter voltage, counter-e. m. f.

Gegensprech- ..., duplex;
— -betrieb m, (full) duplex operation T;
— - —, einseitiger, half duplex operation, T.

gegensprechen, to duplex.

Gegensprechen n, duplex(ing), two-way working;
Doppel- —, quadruplex(ing);
Doppelstrom- —, polar duplex;
— - — auf Doppelleitungen, metallic polar duplex;
— in Staffelschaltung, echelon duplex.

Gegensprech-satz m, duplex set, terminal duplex repeater set (am);
— -system n, duplex system;
Brücken- — - —, bridge duplex system;
Differential- — - —, differential duplex system;
Doppel- — - —, quadruplex system;
— -leitung f, duplex circuit;
— -schaltung f, duplex connection;
— - —, einseitiger Betrieb m, in, half duplex operation;
— -telegraphie f, duplex telegraphy;
— -übertragung f, duplex repeater.

Gegenstrom m, reverse(d) current;
Senden von — nach jedem Stromschritt, curbing (Seekabeltelegraphie, submarine cable telegraphy);
— -relais n, reverse current relay.

Gegenverbund-dynamo f, differential compound wound dynamo;
— -motor m, differential compound wound motor.

Gegenwindung f, opposing winding (des Gußstahlrelais, of the vibration relay).

Gegenwinkel m, opposite angle M.

gegenwirken, to counteract, to react.

gegenwirkend, reactive, antagonistic.

Gegenwirkung f, counteraction, reaction.

gegerbt, tanned.

geglüht, annealed.

gegossen, cast.

Gehalt m, content, percentage;
Kohle- —, carbon content (des Stahls, of steel).

gehärtet, hardened;
im Einsatz —, case-hardened.

Gehäuse n, case, housing, chamber, cabinet, rundes: shell;
mit einem — versehen, to engehäusen,

geheim, secret;   [case.
Telegramm n in — er Sprache, secret language telegram.

Geheimhaltung f, secrecy.

Geheimnis n, secrecy.

Geheimschlüssel m, cipher code.

Geheimschrift f, cryptography.

Gehör n, hearing; — - ..., auditory;

Gehör
nach dem —, by ear;
— -maskierung f, auditory masking, clouding.
Geh-Steh-Telegraph m, start-stop telegraph;
— - -Verteiler m, start-stop distributor;
— - -Welle f, start - stop spindle.
Gei f, guy wire.
Geige f, violin;
Baß- —, kleine: bass-viol, violoncello; Kontrabaß: contra- [bass.
gekapselt, enclosed.
gekäfteltes Papier n, squared paper.
gekennzeichnet, characterized (durch, by).
geklöppelt, braided.
gekoppelt, coupled;
durch gemeinsame Kapazität —, auto-capacity coupled;
autoinduktiv —, auto-inductively coupled;
direkt —, directly coupled;
fest —, tightly coupled;
induktiv —, inductively coupled;
kapazitiv —, capacity coupled;
lose —, loosely coupled;
rück- —, back coupled;
widerstands- —, resistance coupled;
— e Kreise pl, coupled circuits pl;
— en Kreisen, Gebilde n aus, coupled circuit chain.
gekordelt, milled.
gekreuzt, crossed;
a- und b-Zweig —, a and b legs crossed.
— e Doppelader f, crossed pair, Induktionsschutz: transposed pair;
— en Leitungen, Linie f mit, transposition line.

gekröpft, cranked.
gekrümmt, curved, curvilinear, buckled.
gekuppelt, coupled;
direkt —, directly coupled.
geladen, charged;
— er Zustand m, charged condition (der Sammler, of storage cells).
gelagert, supported;
drehbar —, fulcrumed (at), pivoted (on).
gelb, yellow.
Gelbguß m, brass.
Geld n, money;
— -einwurf m, coin slot;
— -stück n, coin.
Gelenk n, joint, link, Scharnier: hinge (joint);
Kardan- —, Hooke's joint;
Knie- —, knuckle joint;
Kugel- —, ball joint;
— -kette f, link chain.
gelöschter Kalk m, slaked quicklime.
gelitzt, stranded.
gelten, to be valid, to hold M.
Geltung f, validity.
gemauert, brick ....
gemeinsam, common, joint;
—e Verbindung f, common connection;
— er Widerstand m, joint resistance.
Gemeinschaftsleitung f, YQ-circuit (engl.), omnibus circuit, way circuit (am.), T.
gemessen, measured.
Gemisch n, composition, compound.
gemischt, mixed;
— geschaltet, connected in series-multiple;
— schalten, to connect in multiple arc;
— e Schaltung f, series-multiple or multiple-arc connection, parallel series connection;

Sattelberg, Wörterbuch: Deutsch-Englisch. 7

gemischt
— **er Verkehr** *m*, mixed service (von Nebenstellenzentralen, of p. b. x.).

genau, accurate, precise, exact, Nachbildung: close, Wiedergabe: faithful;

Genauigkeit *f*, exactitude, precision, closeness, faithfulness; **Meß- —,** precision of test or of measurement.

Genauigkeitsgrad *m*, degree of accuracy.

genehmigen, to allow.

Genehmigung *f*, allowance, license, concession.

geneigt, inclined, sloping, tilted;
— **e Wellenstirn** *f*, tilted wave front.

General-nenner *m*, common denominator;
— **=stabskarte** *f*, ordnance map.

Generator *m*, generator;
 Hochfrequenz- —, high frequency generator; radio alternator;
 Induktor- —, inductor alternator;
 Innenpol- —, external armature generator;
 Sinuswellen- —, harmonic generator;
 Wechselstrom- —, alternating-current generator, alternator;
 Wellen- —, wave generator;
— **=anlage** *f*, generator plant;
— **für zwei Spannungen**, double voltage generator;
— **mit gleichförmigem Luftspalt**, non-salient pole generator;
— **mit an den Polkanten erweitertem Luftspalt**, salient pole generator.

genutet, slotted.

geöffnet, open; open-circuited, standing on open-circuit.

geometrisch, geometric(al);
— **es Mittel** *n*, geometric mean;
— **e Progression** *f*, geometrical progression.

gepanzert, shielded.

gepolt, poled, polar, polarized.

gepreßt, pressed. [ed.

gepuffert (werden), (to be) float-

gerade, straight, Zahl: even.
— **Linie** *f*, straight line;
 in eine — — bringen, to align with;
 eine — — bilden, to align with;
 in —r —, in alignment (zu, mit, with);
— **richten**, to straighten;
— **s Verhältnis** *n*, direct ratio;
— **s Vielfaches** *n*, even multiple.

geradlinig, rectilinear, straight.

geradzahlig, even numbered.

gerändelt, milled.

Gerät *n*, appliance, apparatus, instruments *pl*, gear, Werkzeug: tools *pl*;
 Empfangs- —, receiving set.

Geräusch (*n*)
 kratzendes —, scratchy noise;
 Erd- —, earth noise;
 Induktions- —, induced noise;
 Kollektor- —, commutator noise;
 Leitungs- —, line noise;
 Maschinen- —, generator hum;
 Mikrophon- —, side tone;
 Saal- —, crowd noise;
 Simultan- —, thump;
 Stör- —, interfering noise;
 Telegraphier- —, (Morse-) thump;

geräuschlos, noiseless, silent;
— **=analysator** *m*, noise analyzer *F*;
— **=messer** *m*, noise meter, noise measuring set;
— **=normal** *n*, noise standard;
— **=vernichter** *m*, noise killer.

geräuschvoll, noisy.

gereifelt, serrated, grooved, milled.
gerichtet, directive, directional; poled;
einseitig —, unidirectional;
— er Widerstand m, reactive resistance.
geriffelt, grooved, milled, serrated.
gerillt, grooved, fluted.
gerinnen, to coagulate.
Gerinnen n, coagulation.
gerippt, corrugated; ribbed.
Geröll n, pebble.
Gerüst n, rack, structure, trestle.
gesamt, total; general.
Gesamt-ansicht f, general view;
— -dämpfung f, total loss, attenuation length (in Meilen Standardkabel, in miles of standard cable), transmission efficiency, total transmission equivalent, total attenuation, total cable equivalent;
zulässige — = —, total permissible transmission equivalent;
— -durchmesser m, overall diameter;
— -fluß m, total flux.
— -schaltbild n, full connections pl;
— -strom m, total current;
— -summe f, (grand) total;
— -übertragungsmaß n, transmission efficiency;
— -verlust m, total loss;
— -verzerrung f, total distortion (einer Leitung, of a line);
— -widerstand m, total resistance;
— -wirkungsgrad m, total or overall efficiency, commercial efficiency;
— -zahl f, total.
gesättigt, saturated;
hoch- —, highly saturated.

Geschäftsgegend f, city or commercial district.
Geschiebe n, pebble (Meeresgrund, sea bottom);
geschirmt, shielded.
geschlitzt, slotted, split;
— er Kern m, split core;
— er Stöpsel m, split plug.
geschlossen, close(d), enclosed;
halb- —, semi-enclosed;
— werden, to close;
— e Dynamo f, totally enclosed dynamo;
— er Eisenweg m, closed core.
geschmeidig, flexible, pliable, ductile.
Geschmeidigkeit f, flexibility, pliability.
geschmolzen, molten.
Geschwindigkeit f, velocity, speed, rapidity;
mit einer — von, at a rate of;
geringe —, slow speed;
gleichförmige —, constant or uniform speed;
hohe —, high speed;
niedrige —, low speed;
Anfangs- —, initial velocity;
Betriebs- —, (commercial) working speed;
End- —, final velocity;
Fortpflanzungs- —, velocity of propagation;
Sende- —, Übertragungs- — speed of transmission;
Umfangs- —, peripheral or circumferential speed;
Winkel- —, angular velocity;
Geschwindigkeits-bereich m, speed range;
— -messer m, tachometer, speedometer;
— -regelung f, speed control;
— -regler m, (speed) governor or regulator.
Gesellschaftsleitung f, (multi-)party line;

Gesellschaftsleitung
— für 4 (10) Anschlüsse, four-(ten-) party line;
—, Rückruf auf die, reverting
gesondert, isolated. [call.
Gesperr *n*, escapement;
Anker= —, anchor escapement;
Malteserkreuz= —, Geneva stop mechanism;
— mit fester und loser Klinke, fast and loose escapement.
gesperrt, blocked.
gespiegelte Welle *f*, reflected wave, *L*.
Gespräch *n*, conversation, call, connection;
ein — anmelden, to book, to file, to place a toll call;
ein — einleiten, to originate or initiate a call;
ein — über 3 Minuten verlängern, to extend a call beyond 3 minutes;
ein — zählen, to record a call on the meter, to meter a call;
abgehendes —, out call;
ankommendes —, in call;
dringendes —, express call;
gebührenpflichtiges —, chargeable call;
Dienst= —, service call, official call;
Fern= —, l. d. or toll call;
— — zu bestimmter Zeit, fixed time call;
Orts= —, local call, city or local conversation;
Staats= —, government call;
Vororts =—, suburban call;
Gesprächs=anmeldung *f*, (toll) ticket; das Anmelden: toll recording, booking of a call;
— =blatt *f*, ticket, toll ticket;
— =dauer *f*, duration of a call;
Feststellung *f* der — = —, timing of calls;

— =dichte *f*, frequency of conversations;
— =einheit *f*, conversation or traffic unit;
— =frequenz *f*, frequency of conversations;
— =gebühr *f*, message rate, message fee;
— =gebühren=tarif *m*, measured rate tariff;
— = — =teilnehmer *m*, measured rate subscriber;
— =minuten *pl*, ticket time *F*;
— =verbindung *f*, connection, call;
eine — = — herstellen, to complete a call;
— =zähler *m*, (conversation) meter, service meter;
— =zählung *f*, (call) metering;
— = — durch Zuschalten einer Zählspannung, booster battery metering;
— =zeit *f*, ticket time;
durchschnittliche — = —, average ticket time;
— =zettel *m*, (toll) ticket.
— = — einordnen, to step in tickets.
gespritztes Metall *n*, die-cast metal.
gesprungener Isolator *m*, cracked insulator.
gestaffelt, staggered; graded *A*;
— er Mehrfachtelegraph *m*, series or echelon multiplex telegraph.
Gestalt *f*, shape, form, figure.
gestaltlos, amorphous.
gestalten, to form, to shape.
Gestänge *n*, standard, pole;
Abschnitts= —, transposition pole;
Abspann= —, end or terminal standard;
Dach= —, roof standard;
— = —, Abspann=, roof end standard;
Doppel= —, H-pole.

gestanzt, punched, blanked out from.

Gestell n, rack, bay, shelf, trestle, stand, frame;
freitragendes —, self-supporting rack;
Anrufsucher= —, finder rack A;
Batterie= —, battery stand or rack or frame;
Boden= —, floor stand;
Einheits= —, unit rack;
Fächer =—, rack shelving;
Leitungsrelais= —, line relay rack F;
Leitungswähler= —, LW= —, final switch rack;
Nachbildungs= —, (balancing) network rack V, F;
Pult= —, desk;
Relais= —, relay rack;
Röhren= —, power tube rack R;
Sicherungs= —, fuse rack;
Verbindungs= —, connecting rack;
Schutz= —, barrier guard B;
Verstärker= —, repeater rack;
Versuchs= —, test rack, test
Vorwähler= —, line- [stand; switchboard;
Wähler= —, (auto) switch-rack, selector rack;
Zähler= —, service meter rack;
Zusatz= —, additional rack;
— =abteilung f, bay.

gestört, faulty, out of order (ab: o. o. o.);
— melden, to report faulty.

Gestört=Summerzeichen n, o. o. o. tone F.

Gesuch n, application.

getastet, keyed R.

geteert, tarred;
— er Hanf m, tarred hemp.

geteilt, divided; split;
— e Batterie f, split battery.

getränkt, impregnated, soaked.

getrennt, disconnected, ab: dis (Leitung, line).

getreue Wiedergabe f, faithful reproduction.

Getriebe n, gear, gearing, machine;
Kegelräder= —, bevel gearing, mitre wheel gearing;
Planeten= —, epicyclic train of gear;
Reduktions= —, reduction gear;
Schnecken= —, worm gear;
— mit Pfeilverzahnung, herring-bone gearing;
— — Schrägverzahnung, helical gearing;
— — Winkelverzahnung, double helical gearing.

geübt, skilled.

gewachst, waxed.

gewährleisten, to guarantee, to warrant.

gewalzt, rolled;
blank=—, bright rolled;
kalt —, cold rolled.

Gewebe n, cloth, gauze;
Draht= —, wire cloth;
Kupfer= —, copper gauze.

gewellt, corrugated.

Gewicht n, weight; gravity;
des Pendels: bob;
spezifisches —, specific gravity;
verschiebbares —, sliding weight;
Antriebs= —, driving weight;
Äquivalent= —, equivalent weight;
Gleit= —, sliding weight;
— des Kupferleiters/ Gewicht der Guttapercha je Seemeile in engl. Pfund, ... lbs./....lbs. (Seekabel, submarine cable);

Gewichtsprozent n, percent by weight.

gewickelt, wound;
auf n Ohm —, wound to a resistance of n Ohms.

Gewinde *n*, thread; — ſchneiden, to tap, to thread;
mit — verſehen, screwed, threaded;
Flach- —, square thread;
Links- —, left-handed thread;
Rechts- —, right-handed thread;
Schrauben- —, screw thread;
— -backen *pl*, screw dies *pl*;
— -bohrer *m*, screw tap;
— -buchſe *f*, screw socket;
— -eiſen *n*, screw plate;
— -gang *m*, thread, 10 — -gänge je cm, 10 threads per centimetre;
— -kaliber *n*, — -lehre *f*, thread gauge;
— -ring *m*, threaded ring;
— -ſchneiden *n* tapping, threading;
— -ſteigung *f*, pitch of a screw;
— -ſtift *m*, headless screw, grub screw.

Gewinn *m*, gain.

Gewitter *n*, thunderstorm;
elektriſches —, electric storm;
magnetiſches —, magnetic storm.

geworfen, warped (Holz, wood).

gewunden, wound.

gezähnt, toothed;
— e Stange *f*, toothed rack; ratch.

gezogen, drawn;
hart—, hard drawn.

gießen, to pour (out); Eiſen: to cast.

Gieß-form *f*, mould;
— -löffel *m*, casting ladle.

Gilbert *n*, gilbert (Einheit der M. M. K., unit of m. m. f.).

giltig, valid.

Gips *m*, gypsum, plaster.

gipſen, to plaster.

Gitter *n*, grid, mesh, grating;
engmaſchiges —, fine grid;
weitmaſchiges —, open grid;
Spannung am iſolierten —, zero grid potential V;
Anodenſchutz- —, anode-screening grid;
Blei- —, lead grid;
Raumlade- —, space-charge grid, filament-screening grid;
Schutz- —, screening grid; barrier guard B;
Steuer- —, control grid;
— -ableitung *f*, grid leak (resistance);
— -batterie *f*, grid battery, C-battery;
— -beſprechung *f*, grid modulation, talking to the grid;
— -blech *n*, perforated sheet;
— -blockkondenſator *m*, grid blocking condenser;
— -kondenſator *m*, grid condenser;
— -kreis *m*, grid (-filament) circuit;
— - — -abſtimmung *f*, grid tuning;
— - — -impedanz *f*, input impedance V;
— - — -kapazität *f*, input capacity V;
— - — -kopplung *f*, grid coupling;
— - — -reaktanz *f*, input reactance V;
— -maſche *f*, grid mesh;
— -maſt *m*, lattice mast or pole, girder pole;
Eiſen- — - —, steel lattice mast;
Holz- — - —, wood lattice mast;
— -modulation *f*, grid modulation;
— -nebenſchluß *m*, grid leak (resistance);
— -platte *f*, grid plate (Sammler, storage cell);
— -ſchauzeichen *n*, grid indicator F;

Gitter
- -**spannung** *f*, grid potential;
- -**ständer** *m*, lattice(d) pole;
- -**steuerung** *f*, grid control;
- -**strom** *m*, grid current;
- - -, negativer, reverse grid current;
- -**tastung** *f*, grid control;
- -**träger** *m*, lattice girder;
- -**vorspannung** *f*, priming or biasing or initial grid voltage, grid bias;
- -**widerstand** *m*, internal input resistance; grid leak resistance.

Glanz *m*, polish;
- -**garn** *n*, glace cotton, glazed cotton, glazed yarn;
- - — -**umklöppelung** *f*, glazed cotton braiding.

Glas *n*, glass;
Element- —, battery jar;
- -**ballon** *m*, carboy, demijohn; Meidingerelement: bell jar;
- -**birne** *f*, glass bulb;
- -**deckel** *m*, glass cover, glass top;
mit — - —**versehen**, glass-topped;
- -**elektrizität** *f*, vitreous electricity;
- -**glocke** *f*, bell jar;
- -**kolben** *m*, (glass) bulb;
- -**papier** *n*, glass paper;
- -**perle** *f*, (glass) bead;
mit — - —n **isoliert**, beaded;
- -**rohr** *n*, glass tube;
- - — -**scheider** *m*, glass rod separator;
- - — -**sicherung** *f*, glass tube fuse;
- -**wandung** *f*, glass walls *pl*.

Glaserkitt *m*, putty.
glasiert, glazed, vitrified.
Glasur *f*, glazing.
glatt, smooth, plain.
Glätte *f*, polish.

glätten, to smooth, tho equalize, to plane.
gleichartig, homogeneous.
Gleichartigkeit *f*, homogeneity.
gleichbelastet, equally loaded;
- e **Phasen** *pl*, balanced phases.

Gleichfeld *n*, (magnetisches), constant or steady (magnetic) field.
gleichförmig, uniform, steady; smooth;
- e **Leitung** *f*, smooth line;
- er **Strom** *m*, steady current.

Gleichförmigkeit *f*, uniformity, equality, regularity.
Gleichgang *m*, unison; synchronism.
gleichgehend, unisonant; synchronous.
Gleichgewicht *n*, balance, equilibrium;
aus dem — **bringen**, to unbalance;
ins — **bringen**, to balance, to equilibrate;
im — **erhalten**, im — **sein**, to equilibrate;
Brücken- —, bridge balance;
Gleichgewichts-Fehler *m*, unbalance;
- -**lage** *f*, position of equilibrium;
- -**zustand** *m*, state of equilibrium, balanced condition.

Gleichheit *f*, equality.
Gleichklang *m*, unison, resonance.
Gleichlauf *m*, synchronism;
- -**impuls** *m*, correcting impulse *T*;
- -**magnet** *m*, correcting magnet *T*;
- -**relais** *n*, correcting or corrector relay *T*;
- -**ring** *m*, correcting ring *T*;
- -**segment** *n*, correcting segment *T*;

Gleichlauf
— =störung f, synchronization troubles pl T;
— =ströme pl, correcting currents pl T;
— =stromstoß m, correcting or governing impulse, unison impulse T;
— =verlust m, loss of synchronism T;
— =zeichen pl, correcting currents, unison signals, phasing or idle signals, pl T.

gleichlaufend, synchronous; paralleling M.
gleichmachen, to equalize; to steady.
Gleichmacher m, equalizer.
Gleichmachung f, equalization.
gleichmäßig, uniform, steady, constant, smooth;
— erhalten, to keep constant, to steady;
— machen, to steady, to equalize.
Gleichmäßigkeit f, uniformity, steadiness, constancy, regularity.
gleichnamig, similar, like.
gleichphasig, co-phasal.
Gleichpol= ..., homopolar.
gleichpolig, homopolar.
gleichrichten, to rectify, to redress.
Gleichrichter m, rectifier;
Elektrolyt= —, electrolytic rectifier;
Hochvakuum= —, kenotron, vacuum tube rectifier;
Kristall= —, crystal rectifier;
Lichtbogen= —, arc rectifier;
Pendel= —, vibrating rectifier;
Quecksilberdampf= —, mercury arc rectifier, mercury vapour rectifier;

Röhren= —, vacuum tube rectifier;
— =Elektronenröhre f, kenotron, rectifier valve;
— =kolben m, Quecksilberdampf=, mercury vapour lamp;
— =kristall m, rectifying crystal;
— =röhre f, rectifying valve; detector valve;
— = —, Empfangs=, audion, detector valve;
— = —, Hochleistungs=, power rectifying valve;
— =wirkung f, rectifying action.
Gleichrichtung f, rectification;
— beider Halbwellen, double-wave rectification;
— einer Halbwelle, half-wave rectification.
gleichschenklig, isosceles.
gleich schnell, isochronous.
gleichseitig, equilateral;
— es Dreieck n, equilateral triangle.
gleichsetzen, to compare (to).
gleichsinnig, aiding;
— parallel, parallel-aiding;
— in Reihe, series-aiding;
zwei Spulen — in Reihe schalten, to connect two coils in series aiding.
Gleich=spannung f, direct voltage, continuous voltage, continuous e. m. f., d. c. potential;
wellige —, ripple voltage;
— =spannungsquelle f, d.c.supply, constant potential supply;
Gleichstrom m, direct current (ab: d. c.), continuous current (ab: c. c.), steady current, zero frequency current;
— =Dosenwecker m, circular trembler;
— =dynamo f, d. c. dynamo, d. c. generator;

Gleichstrom
— **-Kapazitätsmessung** *f*, d. c. charging test method;
— **-komponente** *f*, d. c. component, direct current component;
— **-Meßbrücke** *f*, direct current bridge;
— **-messung** *f*, direct current measurement;
— **-netz** *n*, d. c. supply, d. c. mains *pl*;
— **-quelle** *f*, direct current source;
— **-speisung** *f*, d. c. or c. c. supply (to).

gleichstromüberlagert, superposed on d. c.;
— **-e Welle** *f*, pulsating wave.

Gleichstrom-unterbrecher *m*, trembler, interrupter;
— **-wecker** *m*, trembling bell, trembler;
— **-widerstand** *m*, direct current resistance, steady (current) resistance.

gleichtönend, unisonant.

Gleichung *f*, equation;
die — gilt, the equation holds;
eine —nach *n* **auflösen**, to solve an equation with respect to *n*;
einer — entsprechen, to fit an equation;
allgemeine —, general equation;
quadratische —, quadratic equation;
Wurzel einer —, root of an equation;
Bestimmungs- —, defining equation;
Differential- —, differential equation;
Grund- —, fundamental equation. [pollent.

gleichwertig, equivalent, equi-

Gleichwertigkeit *f*, equivalence.
gleichzeitig, simultaneous.
Gleichzeitigkeit *f*, simultaneity.
Gleiskontakt *m*, rail contact.
Gleit-antrieb *m*, friction drive, slipping or yielding drive;
— **-draht** *m*, (differential) slide wire (der Meßbrücke, of the bridge);
— — **-brücke** *f*, slide wire bridge.
gleiten, to slide, to slip.
Gleiten *n*, sliding.
Gleit-kontakt *m*, sliding contact; (contact) slider;
Spule mit einem —, single slider coil *R*;
— — **zwei —en**, double slider coil *R*;
— **-schienen** *pl*, slide rails, sliding bars, *pl*;
— **-stück** *n*, slide(r).

Glied *n*, section, mesh; limb; link;
Binde- —, link;
Dreiecks- —, delta circuit *L*;
Ketten- —, chain link;
Kettenleiter- —, network mesh, network section;
Längs- —, series element *L*;
Quer- —, shunt element *L*;
Zwischen- —, link.

Glieder-kette *f*, link belt;
— **-werk** *n*, linkwork, linkage.

glimmen, to glow, to blue-glow.
Glimmen *n*, glow(ing), corona.
Glimm-entladung *f*, glow discharge;
— **-lampe** *f*, gaseous conduction lamp;
— **-licht** *n*, blue glow, blue haze;
— — **zeigen**, to blue-glow;
— **-strom** *m*, glow current;
— **-verluste** *pl*, corona losses *pl*.

Glimmer *m*, mica;
— **-kondensator** *m*, mica-dielectric condenser;
— **-platte** *f*, mica sheet.

Glocke f, bell;
  Kelch= —, cup-shaped gong;
  Schalmei= —, sheep gong;
Glocken=halter m, gong support;
— =klöppel m, bell hammer, bell striker;
— =schale f, bell dome, bell gong;
— = —, große flache, gong;
— =zeichen n, bell signal.
Glühelektronen=entladung f, thermionic discharge;
— =strom m, thermionic current.
glühen, to glow; Stahl: to anneal.
Glühen n, glow; annealing; des Fadens: incandescence.
glühend, incandescent;
  rot= —, red hot;
  weiß= —, white hot.
Glüh=faden m, (incandescent) filament;
— = —, 3, Einschmelzstellen pl des, filament seals pl;
— =kathode f, hot or glowing cathode, incandescent filament, hot electrode;
— =kathodenröhre f, thermionic valve;
— =körper m, glower; Nernst needle;
— =lampe f, incandescent lamp;
— =lampenschrank m, lamp switchboard F;
— =verfahren n, annealing process, annealing method.
Glyzerin n, glycerine.
Gold n, gold (Au);
— =blatt n, — =folie f, gold foil;
— =drahtrelais n, gold wire relay (Seekabeltelegraphie, submarine cable telegraphy).
Goldschmidt=Tonrad n, Goldschmidt tone wheel.
Goniometer n, goniometer;
  Radio= —, radio goniometer.
goniometrisch, goniometric(al).
Graben m, ditch, trench;

Kabel= —, cable trench;
— = — herstellen, to trench;
— =bagger m, trenching machine, ditch-digging machine;
— =herstellung f, trenching, ditch-digging;
— =sohle f, bottom of ditch.
Grad m, grade, degree (ab: deg.); gradation;
  n — absolut, n degrees Kelvin, n degrees absolute;
  n — Celsius, n degrees centigrade;
  um 90 — drehen, to turn through 90 degrees;
— =bogen m, arc, bow; Transporteur: protractor.
Gradient m, gradient.
Graduator m, graduator.
graduieren, to graduate.
Gramm n, gram(me);
  Dezi= —, decigram;
  Kilo= —, kilogram;
  Milli= —, milligram;
  Zenti= —, centigram.
Grammophon n, gramophone;
— =nadel f, gramophone needle;
— =platte f, gramophone disc;
— =schallbose f, gramophone sound box.
graphisch, graphic(al);
— darstellen, to represent graphically.
Graphit m, black lead, graphite.
graphitisch, graphitic(al).
Gräting f, grating.
grau, grey(ish).
Gravitation f, gravitation.
greifen, to grip (in), to seize.
Greifen n, gripping(-in), seizure.
Greifer m, grip, grapple;
— =wagen m, pick-up carrier.
Greifklaue f, grip, grapple.
Greifloch n, finger hole A;
Grenz= ...., limiting, cut-off; marginal;

Grenz-bedingungen *pl*, limiting conditions *pl*;
Grenze *f*, limit, limitation, cut-off (point);
  obere —, upper limit;
  untere —, lower limit;
Grenz-frequenz *f*, limiting frequency, cut-off frequency;
  obere (untere) — = —, upper (lower) cut-off frequency (eines Bandfilters, of a band-pass filter);
— -kreisfrequenz *f*, cut-off angular velocity;
— -leistung *f*, limiting output;
— -punkt *m*, cut-off point;
— -strom *m*, marginal current;
— = — -betrieb *m*, marginal operation *F*, *A*;
— = — -relais *n*, marginal operation relay *F A*;
— -wert *m*, limiting value;
— -widerstand *m*, critical resistance.
Griff *m*, handle;
Korbel- —, milled knob handle.
grob, coarse.
Grobeinstellung *f*, coarse adjustment.
grob-körnig, coarse-grained;
— -maschig, coarse-mesh(ed).
Grobsicherung *f*, glass tube fuse.
Größe *f*, size, dimension; magnitude, quantity;
  gerichtete —, directional quantity;
  natürliche —, full size;
— —, doppelte, twice full size;
— —, halbe, one half full size.
Größenordnung *f*, order (of magnitude);
  in der — von, of the order of.
Groß-funkstelle *f*, long-distance radio station, high-power wireless station;

— -stadtgebiet *n*, metropolitan area;
— -station *f*, long-distance station.
Groveelement *n*, Grove cell.
Grube *f*, pit, hole.
Grubenwecker *m*, mining bell.
Grund *m*, bottom, ground;
— -brett *n*, base(board);
— -fläche *f*, base;
— -form *f*, basic form;
— -frequenz *f*, base or fundamental frequency;
  Resonanz- — = —, first resonating frequency;
  Telegraphier- — = —, dot frequency;
— -gleichung *f*, fundamental equation;
— -komponente *f*, fundamental component;
— -kreis *m*, base circle;
— -lage *f*, foundation, basis, base;
grundlegend, fundamental;
Grund-linie *f*, base (line);
— -papier *n*, body paper (der Kondensatoren, of condensers);
— -platte *f*, mounting plate, bed plate, base;
  gemeinsame — = —, common base;
— -regel *f*, basic principle;
— -riß *m*, (ground) plan, floor [plan;
— -satz *m*, principle;
— -schwingung *f*, fundamental oscillation, first harmonic (oscillation), fundamental period;
— -stellung *f*, normal position;
  in die — = — zurückführen, zurückkehren, to return to normal;
— -stück *n*, site;
— -ton *n*, fundamental tone or note;
— -wasser *n*, underground or subsoil water;

Grund-wasser
— — -spiegel m, level of the subsoil water;
— -welle f, fundamental wave;
— -wellenlänge f, fundamental wavelength.
grün, green.
Grünspan m, verdigris.
Gruppe f, group; digit A;
Gruppen-führer m, operator-in-charge, dirigeur T;
— -leitsignal n, pilot signal;
— -meldezeichen n, pilot indicator, pilot signal;
— -schalter m, group switch;
— -wahl f, group selection;
— -wähler m, group selector, intermediate selector;
 I. — - — in sechsstelligen Netzen, code selector, code switch;
 II., III., IV. — - —, tandem selector, numerical switch;
— - — mit Stromstoßübertrager, selector repeater.
gruppieren, to group.
Gruppierung f, grouping, layout.
Gulstadrelais n, vibration relay, Gulstad relay.
Gummi m (n), (india)rubber;
Eisen- —, iron rubber;
Hart- —, ebonite, hard rubber;
Para- —, Para rubber;
Roh- —, raw rubber;
Schwamm- —, sponge rubber;
Weich- —, soft rubber;
— -ader f, rubber-covered wire, rubber-insulated wire;

— -arabikum n, gum arabic;
— -ärmel pl, rubber sleeves pl;
— -buffer m, rubber buffer;
— -decke f, rubber mat;
— -handschuh m, rubber glove;
gummiisoliert, rubber-insulated, rubber-covered;
Gummi-lack m, rubber varnish;
— -Milchsaft m, rubber latex;
— -stopfen m, rubber plug.
Gürtel m, belt;
Sicherheits- —, safety belt.
Guß m, casting; cast...;
schmiedbarer —, malleable cast iron;
Spritz- —, die-cast (metal);
Temper- —, malleable cast iron;
— -bock m, cast-iron frame;
— -eisen n, cast iron;
— - — -sockel m, cast iron base;
gußeisern, cast iron;
Gußmörtel m, concrete;
— -stück n, — -teil n, casting.
Güte f, quality; efficiency; Telegraphierzeichen: definition, legibility; der Röhre: figure of merit.
— des Vakuums, degree of evacuation;
— -verhältnis n, efficiency;
— -ziffer f, figure of merit.
Güter-bahnhof m, goods station;
— -wagen m, truck.
Guttapercha f, guttapercha;
— -schicht f, coat(ing) of guttapercha.

H.

Haar-draht m, Wollaston wire;
— -röhrchen n, capillary tube.
Hacke f, pick, ax.
Hafnium n, hafnium (Hf).

Hagel m, sleet;
— -schlag m, — -sturm m, sleet storm.
Hahn m, cock;

Hahn
Absperr= —, stop cock;
Entwässerungs= —, drain cock.
Haken m, hook, catch, clutch; fastening;
Anker= —, stay hook.
haken=artig, hook-like;
— =förmig, hooked, hook-shaped
Haken=schraube f, hook screw;
— =stütze f, hook-shaped bracket;
— =umschalter m, hook switch, switch hook.
halb=automatisch, semi-automatic(al), semi-mechanical;
— =geschlossen, semi-enclosed;
halbieren, to bisect.
Halbierung f, besection.
Halbkreis m, hemicycle, semicircle;
halb-kreisförmig, hemicyclic(al), semicircular;
— =leitend, semi-conducting;
Halb=leiter m, semi-conductor;
— =messer m, radius;
äußerer —, outer radius;
innerer —, inner radius;
mittlerer —, mean radius;
— =periode f, semi-period, half-period, semi-cycle, half-cycle, semi-oscillation;
halbselbsttätig, semi-automatic(al), semi-mechanical.
Halb=tonklischee n, half-tone block;
— =welle f, half-wave;
— =zylinder m, semi-cylinder.
Hälfte f, half.
Halleffekt m, Hall effect.
Hals m, neck;
— =rille f, neck groove (der Isolatoren, of insulators).
haltbar, durable, lasting.
Haltbarkeit f, durability.
halten, to hold, to support; to keep; anhalten: to stop.
Halter m, fastener, holder, support;

Bürsten= —, brush holder;
Glockenschalen= —, gong support;
Spulen= —, coil holder.
Halte=seil n, guy line;
— =spule f, holding coil;
— =strom m, retaining or holding current;
— = — =kreis m, retaining or holding circuit;
— =wicklung f, holding coil.
Haltzeichen n, stop signal T;
— =geber m, auto-control T.
Hammer m, hammer;
Wagnerscher —, hammer break;
Druck= —, printing hammer;
Holz= —, mallet;
hämmerbar, malleable;
hammerförmig, hammer-shaped;
Hammer=induktor m, hammer break spark coil, trembler coil;
hämmern, to hammer;
Hammer=schlag m, scale;
— =unterbrecher m, hammer break, vibrating break.
Hand= ...., manual, hand ....;
— =amt n, manual exchange;
— = — mit selbsttätiger Schluß=zeichengebung, central battery signalling exchange, c. b. s. exchange;
— =apparat m, telephone handset, microtelephone, combination, hand telephone;
handbedient, hand-operated, manual;
Hand=bedienung f, hand operation, manual operation;
— =betrieb m, manual working, hand work;
handbetriebsmäßig, manual.
Handelszentrum n, commercial centre;
Handfernsprecher m, telephone handset, microtelephone;

Hand
- -fertigkeit f, operating skill;
- -geben n, direct or manual transmission T;
- -geber m, manipulator T;
hand-getastet, key-worked;
Handgriff m, handle; Tätigkeit: manipulation;
handhaben, to handle, to operate, to manipulate;
Hand-habung f, handling, manipulation;
- -lampe f, portable lamp;
- -Morsesystem n, key Morse system;
- -rad n, hand wheel;
- -säge f, hand saw;
- -schuh m, glove;
Gummi- - -, indiarubber
- -tempo n, key speed T; [glove;
- -vermittlungsamt n, manual (telephone) exchange;
- -winde f, hand winch.
Hanf m, hemp;
geteerter -, tarred hemp;
Manila- -, Manil(l)a hemp;
- -seil n, - -tau n, manila rope, hemp rope;
Hänge-isolator m, suspension insulator;
- -wagen m, cableway carriage (für Luftkabel, for aerial cables) B.
hängen, to suspend (an, by, to).
hängt an, Teilnehmer, subscriber clears, subscriber restores the receiver.
Harfenantenne f, harp antenna.
harmonisch, harmonic(al).
Harmonische (meist pl), harmonic(s pl);
in einer - n schwingen, to vibrate to a harmonic;
dreifache oder dritte -, triple harmonics pl;
fünffache oder fünfte -, quintuple harmonics pl;

gerad(zahlige) -, even (higher) harmonics pl;
höhere -, higher harmonics pl;
ungerad(zahlig)e -, odd harmonics pl;
zweite -, second harmonics pl.
hart, hard;
- werden, to harden;
- e Röhre f, hard valve:
härtbar, hardenable;
Hartblei n, hard lead.
Härte f, Stahl, Wasser: hardness, Schärfe: harshness;
- -prüfung f, hardness test;
härten, to harden;
hartgezogen, hard drawn;
Hart-gummi (m), hard rubber, ebonite;
- - -pimpel m, ebonite stud;
- - -platte f, ebonite plate;
- -holz n, hard wood, hardwood;
- -kupfer n, hard (drawn) copper;
- - -draht m, hard-drawn copper wire;
- -lot n, spelter solder, hard solder;
hart löten, to braze (together);
Hartporzellan n, hard porcelain;
Härtung f, hardening;
Einsatz- -, case-hardening.
Harz n, resin;
Fichten- -, pine resin;
- -elektrizität f, resinous electricity;
- -entziehung f, extraction of resin;
- -gehalt m, percentage of resin;
harzig, resinous;
Harz-lot n, resin solder;
- -öl n, resin oil.
Haspel m, reel, drum.

Haube f, cap, hood, cover;
Schutz- —, protecting cap, protecting cover.
häufig, frequent.
Häufigkeit f, frequency.
Haupt..., main, chief;
— -achse f, main shaft;
— -amt n, head office, main office;
— -anschluß m, (Teilnehmer-), subscriber's main station, main set, main telephone station;
— - — -leitung f, exchange line; Gegensatz zur Gesellschaftsleitung: direct line, am.: individual line;
— -auslösetaste f, master release key F;
— -batterie f, main battery;
— -feder f, master spring;
— -fernamt n, main zone centre;
— -gang m, main aisle;
— -kabel n, main (cable);
— -leitung f, main (line);
— -lichtleitung f, lighting mains pl;
— -linie f, primary line, trunk line;
— -masse f, bulk;
— -patent n, parent specification;
— -regler m, master regulator;
hauptsächlich, main;
Haupt-schalter m, main switch, master switch;
— -schlußmotor m, series-wound motor;
— -sicherung f, main fuse;
— -strom, m, main current;
— - — -kreis m, main circuit;
— -taste f, master key;
— -uhr f, master clock;
— -verkehrsstunde f, busy hour, busy period, rush hours pl;
— -verkehrszeit f, busy period;
— -vermittlungsstelle f, main exchange;

— -verteiler m, main distributing) frame, ab: M D F;
— - — -gestell n, main frame.
Haus- ..., house ..., domestic;
— -fernsprecher m, house telephone;
— -fernsprechanlage f, house or domestic telephone plant;
— -rohrpost f, (pneumatic) house tube(s pl);
— -telephon n, domestic telephone;
— -wecker m, domestic electric bell.
Haut f, skin; film, coating; hide;
Oxyd- —, film of oxide;
Roh- —, raw hide;
— -wirkung f, skin effect.
Heavisidenschicht f, Heaviside layer.
Heb-bewegung f, vertical motion A;
— -drehwähler m, vertical and rotary selector, Strowger switch or selector.
Hebe-baum m, crowbar;
— -bock m, lifting jack;
— -daumen m, lift, lifter, lifting cam;
— -klinke f, lifting pawl A.
Hebel m, lever;
drehbar gelagerter —, pivoted lever; zweiarmig: fulcrumed lever;
einarmiger —, one-armed lever;
hin- und hergehender —, oscillating lever, rocking lever;
T-förmiger —, tee lever;
zweiarmiger —, two-armed lever;
Ausrück- —, stopping lever;
Druck- —, printing lever;
Einrück- —, starting lever, tripping lever, engaging lever;

Hebel
Einstell= —, adjusting lever,
Hughesapparat: unison lever;
Null(stell)= —, unison lever,
zero-adjusting lever;
— =arm m, lever arm;
— =schalter m, lever switch;
Doppel= — = —, double lever switch;
— =umschalter m, lever switch;
doppelpoliger — = —, double lever switch;
— =wirkung f, leverage.

Hebemagnet m, vertical magnet, lifting magnet, A;
— =anker m, lifting armature.

heben, to raise, to lift.

Heben n, raising, lift(ing); vertical motion.

Heber m, syringe, syphon;
Batterie= —, battery syringe;
— =Säuremesser m, hydrometer syringe;
— =schreiber m, syphon recorder;
— = — =motor m, mouse mill.

Hebezeug n, lift, lifting jack.

Heck n, stern.

Heimlauf m, back stroke, return (stroke).

heimlaufen, to return.

heiß, hot;
— laufen, to run hot.

heißen to hoist.

Heft n, book; Griff: handle.

Heiz=batterie f, heating battery; filament or A-battery V;
— =draht m, heater, heating wire;

heizen, to heat;

Heiz=faden m, heated filament;
— = — =speisung f, filament supply;
— = — =spannung f, (heated) filament voltage;
— =leistung f, filament power, filament wattage, filament energy consumption;
— =spannung f, filament voltage, filament volts pl;
— =spule f, heating coil, heater;
— =strom m, filament current, heating current;
— = — =änderung f, heating current variation;
— = — =stärke f, heating current intensity;
— =transformator m, heating current transformer;
— =widerstand m, filament rheostat, filament resistance.

Helium n, helium (He);
— =röhre f, helium tube.

hellgrau, bright grey.

hellrot, bright red.
— =glühend, bright red hot.

Hellrotglut f, bright red heat.

hemizyklisch, hemicyclic(al).

hemmen, to impede, to retard, to restrain; Bewegung: to check, to brake, to lock, to trig.

Hemm=rad n, brake wheel; escape wheel, escapement wheel;
— =schuh m, brake, drag.

Hemmung f, check; brake; retardation, impeding, Gesperr: escapement.

Henkel m, bail.

Henry n, henry, secohm, quadrant;
Milli= —, millihenry.

herabführen, to lead down;

Herabführung f, leading-down; downleads pl (einer Antenne, of an aerial);

herab=hängen, to depend (von, from);
— =setzen, to reduce, to bring down, to decrease;

Herabsetzung f, reduction, decrease.

herausnehmen, to remove;

heraus
— =ragen, to protrude, to project;
— =ziehen, to remove (Kabel, cables), to withdraw;

Herausziehen n, withdrawal.

herleiten, to derive (von from), to deduce.

Herleitung f, deduction, derivation.

hermetisch, hermetic(al);
— er Verschluß m, hermetic seal.

herstellen, to make, to manufacture; Anlagen: to erect; to set up, to establish, (eine Verbindung, a connection); einen Stromkreis: to close or make a circuit; Streifen: to prepare the tape T.

Herstellung f, manufacture; erection; setting-up;
wirtschaftliche —, economical manufacture;

Herstellungs=gang m, manufacturing method;
— =kosten pl, cost of construction, first cost pl;
— =verfahren n, manufacturing process, factory process;
— =weise f, manufacturing method.

Hertz n, period, cycle;
n —, n cycles per second;
— scher Oszillator m, Hertzian oscillator.

hervorbringen, to produce, to create, to set up.

Hervorbringung f, production, creation, setting-up.

herzförmig, heart-shaped.

Heterodyn=empfang m, heterodyne reception, beat reception;
— =empfänger m, heterodyne receiver, beat receiver.

heterogen, heterogeneous.

Heterogenität f, heterogeneity, heterogeneousness.

heulen, to howl, V.

Heulen, howl(ing) V, F.

Heuler m, howler F, hooter.

Hf. = Hochfrequenz, high frequency;
— =Sperrkette f, low-pass filter;
— =Sperrkreis m, low-pass selective circuit.

Hilfe f, help, aid;
Ausrüstung für erste —, first aid outfit.

Hilfs=..., auxiliary, subsidiary; Zusatz=...: ancillary;
— =amt n, satellite exchange, sub-office, A;
— =anode f, auxiliary anode;
— =klinke f, ancillary jack F;
— =kreis m, subsidiary circuit;
— =mittel n, auxiliary means pl;
— =pol m, auxiliary pole;
— =wicklung f, auxiliary winding.

hinaufpendeln, to rise in resonance.

Hinaufpendeln n, resonant rise.

hindurchgehen, to pass through, to transit.

hindurchlassen, to let through.

hineinhören, to listen in.

hineinschieben, to slip into.

hinlaufende Welle f, main wave L.

hinten, at the rear.

hintereinander, in series, in tandem;
— =geschaltet, series-connected, serially connected;
— schalten, to connect or join in series;

Hintereinanderschaltung f, series connection.

hinterschnitten, undercut.

hin= und bewegen, (sich), to rock, to shuttle;

Hin- und Herbewegung *f*, to-and-fro motion, reciprocating motion;
hin- und hergehen, to reciprocate, to shuttle;
— = — hergehender Körper *m*, shuttle;
— = — herschwingen, mechanisch: to rock, elektrisch: to surge back and forth.
Hin- und Herschwingen *n*, rocking surging back and forth.
hinweisen, to direct.
Hinweisungsstöpsel *m*, (indicating) peg *F*.
hissen, to hoist, to raise.
Hitzbandstrommesser *m*, hot-band ammeter;
— =draht *m*, hot wire, heated wire;
— = — spannungsmesser *m*, hot-wire voltmeter;
— = — strommesser *m*, hot-wire ammeter;
Hitze *f*, heat;
hitzebeständig, heat-proof, heat-resisting.
Hitzrolle *f*, heat coil, *ab*: h. c.;
— mit Gleitstift, collapsable heat coil.
H-Leitung *f*, H-circuit, I-circuit.
Hobel *m*, plane;
— =eisen *n*, bit;
— =maschine *f*, planing machine.
hobeln, to plane.
hoch, high.
Hochantenne *f*, high antenna;
hoch-beheizte Röhre *f*, bright valve;
— =empfindlich, highly sensitive, supersensitive;
— =frequent, high frequent, high frequency...;
Hochfrequenz *f*, high frequency, *ab*: h. f.; radio frequency;
— =drossel *f*, high frequency choke coil;
— =erzeuger *m*, high frequency generator;
— =generator *m*, h. f. generator, h. f. alternator; radio alternator;
— =kabel *n*, radio cable *R*;
— =kreis *m*, h. f. circuit;
— =leiter *m*, radio cable;
— =leitung *f*, h. f. circuit;
— =litze *f*, spiralweave cable, litzendraht;
— =maschine *f*, h. f. alternator, h. f. generator;
— =Mehrfachfernsprechen *n*, h. f. multiple telephony;
— =sicherung *f*, h. f. fuse, h. f. cut-out;
— =Siebgebilde *n*, high-pass selective circuit;
— =Sperrkette *f*, low-pass filter, infra filter, higher limiting filter;
— =telephonie *f*, h. f. telephony;
— = —, leitungsgerichtete, h. f. telephony along lines;
— =transformator *m*, h. f. transformer;
— =unterbrecher *m*, h. f. interrupter, h. f. commutator;
— =verstärker *m*, high frequency amplifier, radio frequency amplifier;
— =verstärkung *f*, h. f. amplification;
hoch-gesättigt, highly saturated;
— =kant, edgewise, on edge;
— = — gewickelte Bandspule *f*, edgewise wound ribbon coil;
Hochleistungs-..., high-power(ed); high-capacity...;
— =Gleichrichterröhre *f*, power rectifying valve;
— =röhre *f*, power tube;
hoch-leitfähig, highly conductive, high-conductivity...;
— =ohmig, highly resistive, high-resistance...;

hoch
— =richten, to hoist, to erect, B;
Hochspannung f, high tension, high pressure, ab: h. t.
Hochspannungs=gleichstrom m, high-tension direct current, ab: h. t. d. c.;
— =linie f, high voltage line;
— =pfeil m, danger arrow;
— =seite f, high tension side, h. t. side;
Höchst=...., maximum, peak;
— =geschwindigkeit f, top speed;
— =grad m, maximum;
— =leistung f, maximum output; peak power;
— =spannung f, super tension, extra high tension;
— =strom m, maximum current, peak current;
— = =ausschalter m, maximum cut-out;
— =wert m, maximum value;
Hoch= und Niederfrequenzverstärkung f, high and low frequency amplification;
— = — —, gleichzeitige, dual or reflex amplification;
— =vakuum=Elektronenröhre f, high vacuum electron valve;
— = =Gitterröhre f, pliotron;
— = =gleichrichterröhre f, high vacuum rectifier valve, kenotron;
— = =glühkathodengleichrichterröhre f, kenotron;
— = =röhre f, high vacuum valve or tube.
hochwertig, high-grade.
hochwinden, to jack up, to hoist.
Höhe f, height, altitude (auch M); level;
wirksame —, effective height R;
Sonnen= —, sun's altitude;
Höhenlinie f, level;
Energie= —, power level;

Höhenlinien=darstellung f, — =diagramm n, level diagram, level chart;
Höhen=reihe f, level A;
— =schritt m, vertical step A; level A;
— = — =vielfach(feld) n, level multiple A.
hoher Ton m, high-pitched note.
hohl, hollow;
— er Kern m, hollow core, tubular core;
Hohl=kern m, tubular core, hollow core;
— =leiter m, tubular conductor;
— =raum m, void, cavity;
— = — =kabel n, air-space cable;
— = — —, Papier=, dry core cable, air-space paper-core cable;
— =trieb n, lantern pinion;
Höhlung f, cavity, hollow;
hohlwandig, hollow-walled;
Hohlwelle f, tubular shaft.
Höllenstein m, nitrate of silver ($AgNO_3$).
Holundermarkkugel f, pith ball.
Holz n, wood, Bauholz: timber;
— = ... wood(en)
weiches, saftreiches —, sappy wood;
Hartholz, hardwood;
— =(bohr)käfer m, wood (boring) beetle;
hölzern, wood..., wooden;
Holz=gittermast m, wood lattice mast;
— =hammer m, mallet;
— =kasten m, wood box, wood tank, wooden cabinet;
— =kohle f, charcoal;
— =kohleneisen n, charcoal iron, Norway iron;
— =kohlenpfanne f, charcoal brazier;
— =latte f, wooden lath;
— =papier n, wood pulp paper;

8*

Holz
— -rinne f, wood trough;
— -scheider m, wood separator;
— -schliff m, wood pulp;
— -schraube f, wood screw;
— -stange f, wood(en) pole;
— - —, rohe oder unzubereitete, untreated wooden pole;
— -teer m, wood tar;
— -werk n, woodwork.

Homodyn-empfang m, homodyne reception, zero-beat reception;
— -empfänger m, homodyne or zero beat receiver.

homogen, homogeneous; smooth;
— e Leitung f, smooth or homogeneous line L.

Homogenität f, homogeneity, homogeneousness.

Honigwabenspule f, honeycomb coil.

hörbar, audible;
— e Anzeigung f, audible detection;

Hörbarkeit f, audibility;
Hörbarkeits-faktor m, audibility factor;
— -grenze f, limit of audibility;
— - —, obere (untere) upper(lower) limit of audibility;

Hörempfang m, audible reception;
Hörempfindung f, auditory sensation;
Schwellenwert m schmerzhafter —, threshold of feeling.

hören, to hear, lauschen: to listen (in);

Hören n, listening; audition.
Hörer m, listener(-in); Fernhörer: telephone, receiver, phone;
Fern- —, (telephone) receiver;
O. B.- —, l. b. receiver;

Z. B.- —, c. b. receiver;
hörfrequent, audible, audio- ...;
über- —, ultra-audible, superaudible;
unter- —, sub-audible, infraaudible;

Hör-frequenz f, audio frequency, audible frequency, acoustic frequency; ab: a. f.;
über- — - , super-audible frequency, ultra-audio frequency;
Unter- — - —, sub-audio frequency;
— - — -kreis m, audio circuit;
— - — -verstärker m, note amplifier, note magnifier.

horizontal, horizontal;
Horizontale f, horizontal (line);
Horizontal-intensität f, horizontal intensity, horizontal force;
— -komponente f, horizontal component;
— - — des Erdmagnetfeldes, earth's horizontal field.

Hörmuschel f, earpiece, cap.
Horn n, horn.
hornartig, horny.
Hörner-ausschalter m, horn (type) switch;
— -blitzableiter m, horn-shaped lightning arrester;
— -pol m, horn-shaped pole;
— - — -relais n, horn-type pole relay.

hornig, horny.
Hör-schwelle f, threshold of audibility;
— -stellung f, listening position;
— -taste f, listening key;
Sprech- — - —, talk-listen button;
— -weite f, hearing distance.

Hub m, stroke, Pendel: throw;
Anker- —, armature stroke, play of tongue T;   [A.
— -magnet m, vertical magnet

Hufeisenmagnet *m*, horse-shoe magnet.
Hügel *m*, hill.
hügelig, hilly.
Hülle *f*, envelope sheath(ing), jacket, Schiff: hull.
Hülse *f*, tube, collar, collet, bush(ing), sleeve, barrel; mit einer – versehen, to sleeve, to jacket;
verschiebbare –, sliding sleeve;
Fiber= –, fibre sleeve *F*;
Stecker= –, plug sleeve, cover or handle of plugs, *F*;
Stöpsel= –, plug cover;
– = – n=gewinde *n*, plug cover thread;
Verbindungs= –, jointing sleeve *B*;
Hülsen=bund *m*, sleeve joint *B*;
– =verbinder *m*, jointing sleeve *B*.
Hunderterstufe *f*, hundreds digit *A*.
Hupe *f*, hooter, horn;
elektrische –, electric horn.
Hütchen *n*, cup;
Achat= –, agate cup.

Hydrant *m*, hydrant.
hydraulisch, hydraulic(al);
– e Presse *f*, hydraulic press.
hydroelektrisch, hydroelectric(al).
Hydroxyd *n*, hydroxide.
Hygrometer *n*, hygrometer.
hygroskopisch, hygroscopic(al).
Hyperbel *f*, hyperbola;
– =funktion *f*, hyperbolic function.
hyperbolisch, hyperbolic(al);
– e Funktion *f*, hyperbolic function;
– er Sinus *m*, hyperbolic sine, sinh.
Hypotenuse *f*, hypotenuse.
Hysterese *f*, **Hysteresis** *f*, hysteresis;
magnetische –, magnetic hysteresis;
Hysteresis=schleife *f*, hysteresis loop, hysteretic loop, magnetic cycle.
– =verlust *m*, hysteretic loss, hysteresis loss;
– = – =zahl *f*, coefficient of hysteresis (watt/cm^3, per).
hysteretisch, hysteretic(al);
– e Nacheilung *f*, hysteretic lag.

J.

I=Anker *m*, shuttle armature, H-armature.
identisch, identical.
Identität *f*, identity.
I=Eisen *n*, double T-iron, I-iron;
– = – (=stange *f*), I-beam.
Igeltransformator *m*, hedgehog transformer.
imaginär, imaginary;
– e Einheit *f*, imaginary unit;
– e Zahl *f*, imaginary number.
Impedanz *f*, impedance;
– einer mit halber Spule beginnenden Leitung, mid-load impedance *K*;
Blindkomponente der –, reactive impedance;
Wirkkomponente der –, dissipative impedance;
Abschluß= –, terminating or end impedance, load impedance;
Eingangskreis= –, input impedance;
End= –, terminating or end impedance, load impedance;

Impedanz
Gitterkreis- —, input impedance V;
Kurzschluß- —, short circuit impedance, closed end impedance;
Längs- —, series impedance;
Leerlauf- —, open-end impedance, no-load impedance, open-circuit impedance;
Parallel- —, leak impedance;
Quer- —, leak impedance, shunt impedance, $(g + j\omega c)$;
— - —**glied** n, shunt impedance element;
Reihen- —, (line) series impedance, $(r + j\omega l)$;
— - —**glied** n, series impedance element;
Streu- —, leakage impedance.

impfen, to inject B.
Impfung f, injection B.
imprägnieren, to impregnate, to soak.
Imprägnierung f, impregnation.
Impuls m, impulse, pulse;
— **e erteilen** to impulse;
Impuls-..., impulse..., impulsive;
Öffnungs- —, break impulse;
Schließungs- —, make impulse;
Strom- —, current (im)pulse;
Wähl- —, dialling impulse A;
— -**dauer** f, duration of impulse A;
— -**frequenz**, impulse frequency A;
— -**gabe** f, impulsing A;
— -**geber** m, impulse sender A;
Tasten- — - —, impulse sending key A;
— -**periode** f, impulse period A;
— -**reihe** f, succession of impulses, train of impulses;
— -**relais** n, impulsing relay, impulse relay A;

— -**speicher** m, impulse storing device, (digit-storing) register, A;
— -**teilung** f, break-to-impulse duration ratio A;
— -**übertrager** m, impulse repeater A;
— -**verhältnis** n, impulse ratio, break-to-make ratio, A;
— -**verzerrung** f, impulse distortion A;
— -**wahl** f, impulse action A.

Inbetriebsetzung f, starting.
Inkrement n, increment;
logarithmisches —, (äquivalentes), (equivalent) logarithmic increment.
Index m, index, pl: indices; suffix.
indifferent, neutral.
Indifferenzzone f, neutral zone.
Induktanz f, inductance, inductive reactance, positive reactance;
— -**spule** f, inductance coil, inductor, retard(ation) coil, reactor, graduator;
— - — **mit Eisenkern**, ferric inductance coil;
— - — — **Luftkern**, air core inductance coil.
Induktion f, induction;
elektromagnetische —, electromagnetic induction;
elektrostatische —, electrostatic induction;
gegenseitige —, mutual induction;
magnetische —, magnetic induction;
Elektro-—., electro-induction;
Magnet- —, magnetic induction;
Seiten- —, inductive interference;
— - —**s** -**schutz** m, anti-induction device;

Induktion
Zahn- —, tooth induction;
Induktions-feld n, induction field;
— -fluß m, induction flux;
induktionsfrei, non-inductive;
nicht —, inductive;
— es Kabel n, screened conductor cable, anti-induction cable;
— er Widerstand m, plain resistance, non-inductive resistance;
Induktions-geräusch n, induced noise;
— -motor m, induction motor;
— -schutz m, anti-induction device;
— -spule f, induction coil, inductor (coil);
— -störung f, inductive interference, inductive trouble;
— -strom m, induced current.
induktiv, inductive;
elektro- —, electro-inductive;
nicht —, non-inductive;
— gekoppelt, inductively coupled;
— geladen, inductively loaded;
— e Kopplung f, inductive coupling, magnetic coupling.
Induktivität f, inductance;
durch gemeinsame — gekoppelt, auto-inductively coupled;
Einheit f der —, unit (of) inductance;
Leitung f mit erhöhter —, loaded circuit;
natürliche —, natural inductance (eines Kreises, of a circuit);
verteilte —, distributed inductance;
gleichmäßig oder stetig — —, evenly or uniformly or continuously distributed inductance;

punktförmig — —, lumped inductance;
Gegen- —, mutual inductance;
— - -s -koeffizient m, coefficient of mutual inductance;
Reihen- —, series inductance;
Streu- —; stray inductance;
Induktivitäts-symmetrie f, inductance balance (von Spulen, of coils).
Induktor m, inductor;
Funken- —, spark coil, induction coil;
Hammer- —, hammer break spark coil, trembler coil;
Kurbel- —, magneto generator, magneto, hand generator;
Umschalter m am — - —, automatic cut-out of magneto;
Magnet- —, magneto generator;
— -anruf m, generator call;
— -apparat m, magneto telephone station;
— -generator m, inductor alternator R;
— -maschine f, inductor type generator R;
— -rad n, inductor wheel R;
— -schlußzeichen n, ring-off signal.
induktorisch, inductive;
— e Beeinflussung f, inductive interference, inductive trouble.
Induktorium n, induction coil, spark coil.
induzieren, to induce.
induzierender Magnetpol m, inducing magnetic pole;
— Strom m, inducing current.
induzierter Magnet m, induced magnet;
— Strom m, induced current.
ineinandergreifen, to intermesh.

Ineinandergreifen n, intermeshing.

inert, inert.

Infinitesimal- ...., infinitesimal;

— **-rechnung** f, infinitesimal calculus.

Influenz f (electro)static induction, influence;

— **-maschine** f, influence machine.

influenzieren, to influence.

Infusorienerde f, infusorial earth.

Ingenieur m, engineer.

Inhalt m, content; volume.

Inklination f, inclination.

inklinieren, to incline.

inkonstant, inconstant.

Inkonstanz f, inconstancy.

Inkrement n, increment.

Inkrement- ...., incremental.

Inland- ..., inland...;

— **-telegramm** n, inland message.

Innen- ...., internal;

— **-anlage** f, internal plant;

— **-dienst** m, internal service; Störungssucher im — - —, internal faultsman;

— **-durchmesser** m, inner or internal diameter;

— **-leitung** f, internal wiring;

— **-polbdynamo** f, internal pole dynamo, external armature generator;

— **-seite** f, inner side;

— **- — des Hauptverteilers**, exchange side of M D F;

— **-verzahnung** f, internal teeth pl;

— **-widerstand** m, internal resistance.

innere(r), internal, inner.

instabil, instable.

Instabilität f, instability.

instandhalten, to maintain, to keep up.

Instandhaltung f, maintenance, upkeep.

Instandhaltungsarbeiten pl, maintenance work, routine work.

instandsetzen, to repair.

Instandsetzung f, repair, restoration.

Instandsetzungs-arbeiten pl, repair work;

laufende — - —, routine repair work;

— **-trupp** m, repair gang B.

Instrument n, instrument; **Blas-** —, wind instrument; brass instrument; **Musik-** —, musical instrument; **Saiten-** —, stringed musical instrument.

Integral n, integral; **Linien-** —, line integral.

Integration f, integration.

integrieren, to integrate (über eine Periode, over a cycle).

Intensität f, intensity; **Feld-** —, field intensity; **Horizontal-** —, horizontal force, horizontal intensity; **Zeichen-** —, signal strength, signal intensity;

Intensitätsschwankung f, variation in intensity.

Interferenz f, interference;

— **-** ...., heterodyne, beat, R;

— **-empfang** m, beat reception, heterodyne (beat) reception;

— **-empfänger** m, beat receiver, heterodyne receiver;

— **-erscheinung** f, interference phenomenon;

— **-frequenz** f, beat frequency, frequency of beats;

— **-vorgang** m, beating effect;

— **-wirkung** f, interference effect.

interkristallin, intercrystalline.

intermittieren, to intermit.

intermittierend, intermittent, interrupted, tapping;
— **e Erdverbindung** *f*, interrupted earth, tapping earth.

Interpolation *f*, interpolation.

Interpolator *m*, interpolator.

interpolieren, to interpolate.

Interpunktionszeichen *n*, punctuation mark.

Interpretation *f*, interpretation.

interpretieren, to interpret.

Intervall *n*, interval.

Invariant *n*, Invariant (47% Ni, 53% Fe).

Ion *n*, ion;
basisches —, basic ion;
Säure- —, acid ion;

Ionen-bewegung *f*, migration of ions;
— **-strom** *m*, stream of ions;
— **-wanderung** *f*, travelling of ions, migration of ions.

Ionisation *f*, ionisation;
Stoß- —, ionisation by impact, ionisation by collision;

Ionisations-kammer *f*, ionisation chamber;
— **-spannung** *f*, ionising or ionisation potential;
— **-strom** *m*, ionisation current.

ionisieren, to ionise.

Ionisierung *f*, ionisation.

irden, earthenware;
— **es Formstück** *n*, earthenware block *B*.

Iridium *n*, iridium (Ir).

irrational, irrational.

Irrationalität *f*, irrationality.

irreversibel, irreversible.

Irrtum *m*, error.

Irrung *f*, erasure, rub-out, *T*; erase signal *T*;

Irrungs-taste *f*, erase key *T*;
— **-zeichen** *n*, erase signal, rub-out signal, *T*.

isochron, isochronous.

Isochronismus *m*, isochronism.

Isolation *f*, insulation;
niedrige —, low insulation;
— **gegen Erde**, insulation against ground;

Isolations-fehler *m*, insulation failure, insulation fault;
— **-messer** *m*, insulation tester, megger;
— **-prüfer** *m*, insulation tester;
— **-prüfung** *f*, insulation test;
— **-widerstand** *m*, insulation (resistance), dielectric resistance;
— **-zustand** *m*, state of insulation.

Isolator *m*, insulator;
gesprungener —, cracked insulator;
Abspann- —, terminal insulator;
Bradfield- —, Bradfield insulator;
Doppelglocken- —, double-shed or double-cup or double-petticoat insulator;
Durchführungs- —, wall tube insulator, leading-in insulator;
Einführungs- —, leading-in insulator;
— — **mit Vergußkammer**, pothead insulator *B*;
Hänge- —, suspension insulator;
Kreuzungs- —, transposition insulator *B*;
Pilz- —, mushroom or umbrella insulator;
Porzellan- —, porcelain insulator;
Schirm- —, umbrella insulator;
Stütz- —, pin type insulator;
Überführungs- — **mit Vergußkammer**, pothead insulator;
— **mit eingebauter Sicherung**, fuse insulator;

Isolator
— mit gerader Stütze, pin type insulator;
— -stütze f, bracket, insulator spindle;
gerade — - —, insulator pin.
Isolier-band n, (adhesive) insulating tape;
mit — - — umwickeln, to serve with insulating tape;
— -buchse f, insulating bush;
— -fähigkeit f, insulating property;
— -formstück n, moulded insulation;
— -klemme f, insulating clamp;
— -lack m, insulating varnish, isolac;
mit - —— überzogen, isolac...;
— -masse f, insulating compound;
gepreßte — - —, moulded insulation;
mit — - — getränkt, compounded;
— -matte f, insulating mat;
— -perle f, bead;
— -rohr n, insulating tube;
— -stöpsel m, insulating plug;
— -zwischenlage f, insulating separator.
isolieren, to insulate, to seal; rein darstellen: to isolate.
isolierend, insulating;
— -e Zwischenlage f, insulating layer.
isoliert, insulated, sealed; isolated;
— es Ende n einer Leitung, sealed end of a line.
Isolierung f, insulation; covering.
J-Stütze f, I-bolt B..

Jahres-gebühr f, annual subscription (rate);
— -zeit f, season.
jährlich, annual;
— -e Schwankungen pl, annual variations pl.
Japan-lack m, japan;
mit — - — überzogen, japanned;
— -papier n, Japanese paper.
Jigger m, jigger.
Joch n, yoke.
justieren, to adjust, to set.
Justierung f, adjustment, setting.
Jute f, jute;
mit — umwickelt, jute served;
mit Tannin getränkte —, tanned jute;
— -garn n, jute yarn;
— -packung f, jute packing;
— -umwicklung f, jute serving;
— -wicklung f, wrapping of jute.

K.

Kabel n, cable;
ein — aufnehmen, to take up, to pick up a cable;
ein — aufspleißen, to fan out a cable;
ein — ausformen, to form out a cable;
ein — auslegen, to lay a cable;
ein — einziehen, to draw in a cable;
ein — herausziehen, to remove a cable;
ein — (an)landen, to land a cable;
ein — schneiden, to cut a cable;
Schlag eines —s, turn of a cable;

Kabel
achterverseiltes —, quadruple pair cable;
belastetes —, loaded cable;
— —, stetig, continuously loaded cable;
— —, punktförmig, lump-loaded cable;
doppeladriges —, twin cable, bifilar cable;
dreiadriges —, triple core cable;
dickdrähtiges —, heavy gauge wire cable;
dünndrähtiges —, small gauge wire cable;
einadriges —, single core(d) cable;
getränktes —, imprägniertes —, impregnated cable;
induktionsfreies —, anti-induction cable, sceened conductor cable;
konzentrisches —, concentric cable;
künstliches —, artificial cable;
— —, aus reinen Widerständen gebildetes, non-reactive artificial cable;
verseiltes —, twisted cable;
— —, fest, tight cable;
— —, lose, loose cable;
— —, stern=, spiral(led) four cable, spiral quad cable;
versenktes —, underground cable, buried cable;
vieladriges —, multiple (conductor) cable, multicore cable;
vielpaariges —, multi-pair cable; multiple twin cable, m. t. cable;
vieradriges —, four-wire cable;
viererpupinisiertes —, composite loaded cable;
viererverseiltes —, duplex (telephone) cable, phantom cable, quadded cable;
zweiadriges —, two wire core cable, twin core cable, bifilar cable;
zweipaariges —, two pair core cable;
Abschluß= —, terminal cable;
Amts= —, office cable;
Band= —, ribbon(-shaped) cable;
Baumwoll= —, cotton-covered cable;
Baumwoll=Seiden= —, silk and cotton-insulated cable;
Blei= —, lead (-covered) cable;
— = — mit Hohlraumisolierung, airspace lead-covered cable;
Bleimantel= —, lead (-covered) cable;
Bleirohr= —, zweiadriges, two-wire lead (-covered) cable;
— = — =ende n an Kabelaufführungspunkten, tail end, pothead tail;
Dieselhorst=Martin= —, D.=M.= —, (paper core) multiple twin cable, (p. c.) m. t. cable;
Dreileiter= —, triple core cable;
Einführungs= —, leading-in cable, terminal cable;
— = — für Spulenkästen, stub cable;
Einleiter= —, single core(d) cable;
Erd= —, buried cable, underground cable;
Faserstoff= —, fibre-covered cable;
Fern= —, long-distance cable; trunk (telephone) cable,
— = — =linie f, toll line (route);
— = — =netz n, long-distance cable system;
Fernleitungs=bezirks= —, trunk zone cable;
— =end= —, toll entrance cable;

Kabel

Fernleitungs-zwischen- —, toll intermediate cable;
Fernsprech- —, telephone cable;
Flach- —, flat cable, ribbon-shaped cable;
Fluß- —, river cable, subfluvial cable;
Haupt- —, main cable, main;
Hochfrequenz- —, radio cable, h. f. cable;
Hohlraum- —, air-space cable;
— = —, vielpaariges, air-space multiple-twin cable, a. s. m. t. cable;
— = —, Papier-, dry-core cable;
Kontaktsatz- —, bank (-to-bank) cable A;
Küsten- —, shore-end cable, shallow-water cable;
Leit- —, Lotsen- —, pilot cable;
Luft- —, aerial cable;
Luft-Hohlraum- —, air-space cable;
Mehrfach-Zwillings- —, multiple twin cable, m. t. cable;
Normalfern- —, standard l. d. cable;
Not- —, interruption cable;
Ozean- —, ocean cable;
Papier- —, paper (core) cable, p. c. cable, dry-core cable;
— = — mit abgeschirmten Leitern, screened (conductor) paper core cable;
Papierhohlraum- —, dry-core cable, air-space paper-core cable;
Pupin- —, coil (-loaded) cable;
Querverbindungs- —, tie cable;
Röhren- —, conduit cable;
Schrank- —, switchboard cable;
See- —, submarine cable, ocean cable;

Seiden- —, silk-covered cable;
Spulen- —, coil-loaded cable
Standard- —, standard cable;
— = —, ... Meilen, ... miles of standard cable;
Sternvierer- —, spiral(led) four cable, spiral quad cable;
Stich- —, branch cable; tie cable;
System- —, switchboard cable, multiple cable;
— = —, 63adriges, 63 wire multiple cable, 63 wire switchboard cable;
Telegraphen- —, telegraph cable;
Thomson- —, non-loaded (submarine) telegraph cable;
Tiefsee- —, deep sea cable;
Überlandfern- —, overland l. d. cable;
Verbindungsleitungs- —, junction cable;
Vielfach(feld)- —, multiple cable;
— = —, Wähler-, bank (multiple) cable;
Vierfach-Zwillings- —, quadruple twin cable;
Zweig- —, branch cable;
Zweileiter- —, twin core cable;
Zwillings- —, twin cable;
— = —, Vielfach-, multiple twin cable, m. t. cable;
— = —, Vierfach-, quadruple pair cable;
Zwischen- —, intermediate cable;
— mit abgeschirmten Leitern, screened conductor cable;
— — verschiedenen Aderstärken, composite cable;
— — Belastung der Stamm- und Viererleitungen, composite-loaded cable;
— — erhöhter Induktivität, loaded cable;

Kabel
- mit Viererverseilung, phantom cable, duplex cable;
- - 4 (7) Vierern, four- (seven-) quad cable;
- -abschluß *m*, cable terminal, cable termination;
- -abzweig *m*, cable joint, cable tap;
- - - -kasten *m*, cable joint box;
- - - -muffe *f*, cable distribution plug;
- - - -punkt *m*, u. g. (= underground) distributing point;
- -ader *f*, core; cable line;
- - - -ausgleich *m*, cable balancing K;
- -aderpaar *n*, cable pair;
- -alphabet *n*, cable Morse code;
- -anlage *f*, cable plant, cable system;
- -aufhänger *m*, cable bearer, hanger;
- -aufhängung *f*, cable suspension;
- -aufnahme *f*, removal of a cable;
- -ausgleich *m*, cable balancing K;
- -ausgleichsverfahren *n*, cable balancing method K;
- -auslegung *f*, laying of cable;
- -brett *n*, cable shelf;
- -bruch *m*, cable break;
- -brunnen *m*, (cable) manhole, cable pit; joint box, jointing chamber;
- - -, kleiner, cable jointing chamber, flush box;
- -buchstabe *m*, bei dem positive und negative Stromstöße abwechseln, cross letter T;
- -dampfer *m*, cable steamer;
- -einziehstrumpf *m*, cable grip B;
- -endmuffe *f*, pothead terminal;
- -endverschluß *m*, terminal box, box head, cable terminal;
- -endverzweiger *m*, cable distribution head;
- -fehler *m*, cable fault;
- -führung *f*, running of cable(s);
- -gerüst *n*, cable rack;
- -gesellschaft *f*, cable company;
- -gestell *n*, cable rack, cable shelf;
- -graben *m*, trench, ditch; einen - - herstellen, to trench;
- -haus *n*, cable house;
- -hütte *f*, cable hut;
- -kanal *m*, cable conduit, cable way, cable duct; electrical subway;
- - - -führung *f*, duct route;
- -kasten *m*, cable trough; kleiner Kabelbrunnen: flush box;
- -keller *m*, cable cellar;
- -länge *f*, cable length:
- -legung *f*, laying of a cable;
- -linie *f*, cable line;
Fern- - -, toll line (route);
- -lötbrunnen *m*, cable joint box, jointing chamber;
- -löter *m*, (cable) jointer, splicer, cable solderer;
Blei- - -, plumber jointer;
- -lötstelle *f*, cable joint;
- -lötung *f*, cable jointing;
- -mantel *m*, cable sheathing;
- - - -korrosion *f*, corrosion of the lead sheathing;
- -maschine *f*, cable making machine;
mehrspulige - - -, multiheaded cable machine;
- -meßkarren *m*, cable testing car;
- -muffe *f*, cable sleeve, joint box, splice box;

Kabel=muffe
ausgegossene – = –, filled joint;
— =muffenverzweigung f, multiple joint;
— =netz n, cable system, cable plant;
— =öse f, cable eye;
— =plan m, cable lay-out;
— =Reparaturschiff n, restorer;
— =rinne f, cable trough(ing), cable channel;
U=förmige – = –, U-troughing;
— =rohr n, cable pipe, cable duct;
irdenes – = –, earthenware cable duct;
Fiber= – = –, fibre cable pipe;
— = –strang m, cable conduit, cable duct;
— =rost m, cable shelf, cable rack;
— =schacht m, cable chute;
— =schiff m, cable ship;
— =schuh m, cable eye, (connector) lug, cable socket, thimble;
offener – = –, spade terminal;
— =seele f, cable core;
— =seite f, cable side;
— =spleißung f, cable joint, cable splice;
— =stein m, cable tile;
— =stumpf m, stub cable;
— =stütze f, cable bearer, cable bracket;
— =tank m, cable tank;
— =taste f, cable key;
— =träger m, cable bearer, cable bracket, cable shelf;
— =traggerüst n, cable rack, cable support;
— =trommel f, cable reel, cable drum;
— =überführungskasten m, cable distributing box;
— =unterbrechung f, cable break, cable interruption;
— =verteilungssystem n, cabling system;
offenes – = –, tapering cabling system;
— =verzweiger m, cable connection box, distribution box or case;
— =verzweigung f, cable branching;
— =verzweigungsmuffe f, distribution plug;
— =werk n, cable work(s pl);
— =winde f, cable winch;
— =zopf m, cable form;
Herstellung f eines – = –es, forming-out of a cable.
Kadmium n, cadmium (Cd).
Käfer m, beetle, chafer;
Bohr= –, boring beetle.
Käfig m, cage;
— =anker m, squirrel cage rotor;
— =antenne f, cage antenna;
käfigförmig, cage-like;
Käfigspule f, cage coil.
Kalander m, calender.
kalandrieren, to calender.
Kaliber n, calibre, gauge; gage (am.);
Gewinde= –, thread gauge.
kalibrieren, to calibrate, to gauge, gage (am.).
Kalibrierung f, calibration.
Kali n, kohlensaures, pottassium carbonate (K_2CO_3).
Kaliko m, cloth.
Kalium n, potassium (K);
— =perchlorat n, chlorate of potassium ($KClO_4$).
Kalk m, lime;
gelöschter –, slaked quicklime;
Ätz= –, quicklime.
Kalkulagraph m, calculagraph.
Kalorie f, calory.
Kalorimeter n, calorimeter.

Kalorimetrie *f*, calorimetry.
kalorimetrisch, kalorimetrisch(al).
kalt, cold;
— e **Elektrode** *f*, cold electrode;
— **=brüchig,** cold short brittle;
— **gewalzt,** cold rolled.
Kälte *f*, frost.
Kalzium *n*, calcium (Ca);
Chlor= —, calcium chloride ($CaCl_2$);
— **=chlorid** *n*, calcium chloride ($CaCl_2$);
— **=oxyd** *n*, calcium oxide, lime (CaO).
Kamm *m*, comb; am **Lötösen**streifen: fanning strip;
— **=lager** *n*, thrust bearing;
— **=rad** *n*, cog wheel.
Kammer *f*, chamber;
Dämpfer= —, damping chamber.
Kanal *m*, channel, duct, conduit, trough;
einzügiger —, single-way duct *B*;
mehrzügiger —, multiple-way duct *B*;
Formstück= —, block conduit;
— **=** —, **irdener,** earthenware block conduit;
— **=** —, **Zement=,** concrete block conduit;
Kabel= —, cable conduit, cable way; electrical subway;
Rohr= —, pipe line;
— **=öffnung** *f*, pipe.
Kanne *f*, can;
Öl= —, oil can.
Kanonenmetall *n*, gunmetal.
Kante *f*, edge;
Kanten-länge *f*, edge length;
von .... cm — **=** —, ... cm in edge;
— **=wirkung** *f*, edge effect.
Kapazitanz *f*, capacitance, negative reactance;
Kapazität *f*, capacity; dielectric capacity; als **Leitungskon**stante: (line shunt) capacity, capacitance;
elektrostatische —, electrostatic capacity;
gegenseitige —, mutual capacity;
gemeinsame —, joint capacity;
punktförmige —, concentrated capacity;
verteilte —, distributed capacity;
— —, **punktförmig,** lumped capacity;
— —, **stetig,** continuously distributed capacity;
Ausgleichs= —, balancing capacity; counterpoise *R*;
Betriebs= —, mutual capacity wire-to-wire capacity, *K*;
— **=** —, **Vierer=,** pair-to-pair capacity;
Eigen= —, self-capacity;
Elektroden= —en *pl*, inter-electrode capacities *pl V*;
Erd= —, (wire-to-) earth capacity (von **Leitungen,** of lines);
Gitterkreis= —, input capacity *V*;
Kopplungs= —, coupling capacity;
Leistungs= —, watt-hour capacity (von **Sammlern,** of storage cells);
Leitungs= —, wire-to-earth capacity, (line) shunt capacity;
Röhren= —en *pl*, inter-electrode capacities *pl V*;
Spulen= —, coil capacity;
Schleifen= —, wire-to-wire capacity;
— **=** —, **Vierer=,** side-to-side capacity, phantom capacity;
Streu= —en *pl*, spurious capacities *pl*;

Kapazität
Vierer- —, pair-to-pair capacity, side-to-side capacity;
Windungs- —, internal capacity (einer Spule, of a coil);
— gegen Erde, earth capacity, capacity to earth;
— zwischen den Transformatorwicklungen, inter-winding capacity of a transformer;
Kapazitäts-ausgleich m, capacity balance;
— = — durch Abernkreuzen, test-balancing method, test-splicing method, K;
— = — durch Zusatzkondensatoren, condenser balancing method K;
— =ausgleichsverfahren n, capacity balancing method;
— =brücke f, capacity bridge;
kapazitätsfrei, non-capacitive;
Kapazitäts-kopplung f, capacity coupling;
— =meßbrücke f, capacity bridge;
— =messung f, capacity test;
Gleichstrom- — = —, d. c. charging test;
— =probe f, capacity test (der Sammler, of storage cells);
— =reaktanz f, capacity reactance;
— =symmetrie f, capacity balance;
— =ungleichheit f, capacity unbalance;
— =unsymmetrie f, capacity unbalance;
— =widerstand m, capacitance, capacitive resistance.
kapazitiv, capacitive, capacitative;
—er Blindwiderstand m, capacity reactance;
— e Kopplung f, capacity coupling;
ungewollte —e —, spurious capacities pl;
— e Reaktanz f, capacity reactance, negative reactance;
— er Schirm m, electrostatic shield;
— e Schirmung f, electrostatic shielding.
Kapillar- …., capillary;
Kapillarität f, capillarity.
Kapillar-rohr n, capillary tube;
— =wirkung f, capillary action.
Kapital n, capital, fund; stock;
— =anlage f, investment.
kapitalisieren, to capitalize.
Kapitalisierung f, capitalization.
Kappe f, cover, hood, cap;
mit einer — versehen, to cap;
geschlitzte —, split cover;
Aufsteck- —, slip-on cap;
Schutz- —, safety cap;
Staub- —, dust cover.
Kapsel f, capsule, shell, kleine —: button;
Mikrophon- —, transmitter inset, microphone button, capsule;
— =mikrophon n, inset transmitter, button transmitter;
Doppel- — = —, double button transmitter.
kapseln, to enclose.
Karbolineum n, carbolineum.
Karborund m, carborundum (SiC).
Karbonat n, carbonate.
Kardan-gelenk n, Hooke's joint;
— =getriebe n, cardan gear.
kardanische Aufhängung f, cardanic suspension.
Karren m, car, cart;
Apparate- —, instrument cart;
Dynamo- —, supply cart;
Kabelmeß- —, cable testing cart;

Karren
  Maschinen= —, engine cart;
  Mast= —, mast cart;
  — =station *f*, cart type radio station, waggon radio set.
Karte *f*, card; chart, map;
  Generalstabs= —, ordnance
  Land= —, map;   [map;
  See= —, chart;
  — in großem Maßstabe, large scale map.
Kartei *f*, card index;
  — =system *n*, record card system.
Karton *m*, cardboard; Kasten: carton.
Kaskade *f*, cascade;
  in —, in cascade;
  — — schalten, to (join in) cascade;
Kaskaden=umformer *m*, cascade converter;
  — =verstärker *m*, cascade amplifier.
Kassiervorrichtung *f*, coin collector F.
Kastanie *f*, chestnut.
Kasten *m*, case, box, tank;
  Eisen= —, iron case;
  Holz= —, wood box, wood tank;
  Kabel= —, cable troughing;
  — = —, eiserner, iron cable troughing;
  Schutz= —, protecting case, (protecting) cover;
  Untersuchungs= —, test case;
  — mit Schirmwänden, screening box;
  — =deckel *m*, boxlid;
  — =untersuchungsstelle *f*, test case.
Katheten *pl*, perpendicular sides *pl*, smaller sides *pl* of right-angled triangle.
Kathode *f*, cathode, negative electrode;
  Faden= —, filamentary cathode;
  Glüh= —, glowing cathode, hot cathode;
  Oxyd= —, oxide cathode, Wehnelt cathode; oxide-coated filament;
  Wehnelt= —, Wehnelt cathode;
Kathoden=dunkelraum *m*, dark space round the cathode;
  — =fall *m*, cathode fall;
  — =röhre *f*, cathode tube;
  — =strahlen *pl*, cathode rays *pl*;
  — = — =oszillograph *m*, Braun tube oscillograph, cathode ray oscillograph;
  — =röhre *f*, thermionic valve;
kathodisch, cathodic;
  — e Entladung *f*, cathodic discharge.
Kation *n*, cation.
Kausche *f*, thimble.
Kautschuk *m*, rubber, caoutchouc, gum elastic;
  vulkanisierter —, vulcanized caoutchouc.
Kegel *m*, cone; Kegel= ...., conical;
  — =antenne *f*, cone antenna;
  Doppel= — = —, double cone antenna;
  — =form *f*, conicalness;
kegelförmig, conical;
Kegel=rad *n*, conical wheel, bevel wheel, mitre wheel;
  — =rädergetriebe *n*, mitre (wheel) gearing, bevel gearing;
  — =radübertragung *f* 1:1, equal ratio bevel gear.
Kehrpunkt *m*, cusp (einer Kurve, of a curve).
Keil *m*, wedge, key;
  Feder= —, feather key;
  — =form *f*, wedge shape;
keilförmig, wedge-shaped;
Keilnute *f*, key way.
Kelchglocke *f*, cup-shaped gong.

Keller *m*, cellar;
  Kabel= —, cable cellar.
Kenngröße *f*, characteristic;
  Übertragungs= — *n pl*, transmission characteristics *pl* (eines Stromkreises, of a circuit).
Kennleitwert *m*, indicial admittance *L*.
Kennlinie *f*, characteristic (curve);
  Frequenz(abhängigkeits)= —, frequency characteristic;
  Maschinen= —, speed-load characteristic;
  Röhren= —, valve characteristic;
  — = —, **geradliniger Teil der**, straight portion of the valve characteristic;
  — = —, **oberer (unterer) Knick der**, upper (lower) bend of the valve characteristic;
  — = —, **Steilheit** *f* **der**, slope of the valve characteristic;
  Strom=Spannungs= —, current-voltage characteristic;
  Tourenzahl=Belastungs= —, speed-load characteristic.
kennzeichnen, to characterize (durch by).
kennzeichnend, characteristic.
Kennzeichnung *f*, characterization.
Kenotron *n*, kenotron.
Kerbe *f*, kerf, notch, groove,
Kern *m*, core;   [slot, nick.
  geblätterter —, laminated core;
  geschlitzter —, split core;
  geschlossener —, closed core;
  offener —, open core;
  ohne —, coreless;
  unterteilter —, **(fein)**, (finely) subdivided core;
  Blätter= —, laminated core;
  Blech= —, laminated core, stamped sheet core;
  Draht= —, wire core;
  — = — **=spule** *f*, wire core coil;
  Eisen= —, iron core;
  Flach= —, flat core;
  Hohl= —, hollow core, tubular core;
  Luft= —, air core;
  Masse= —, compressed iron powder core *K*;
  Pulver= —, **(gepreßter)**, (compressed) powder core;
  Rund= —, round core;
  Staub= —, **(gepreßter)**, (compressed) dust core;
  Tauch= —, plunger;
  Weicheisen= —, soft iron core;
kernlos, coreless;
Kernplatte *f*, core plate (der Sammler, of storage cells);
  — **=scheibe** *f*, core disc;
  — **=transformator** *m*, core transformer;
  — **=verluste** *pl*, core losses *pl*;
  — **=vierer** *m*, central quad *K*.
Kerze *f*, candle;
Kerzenstärke *f*, candle power.
Kessel *m*, tank, vessel;
  Tränk= —, impregnating vessel;
  Vulkanisier= —, vulcanizing pan;
  — **=pauke** *f*, tymbal, kettledrum.
Kette *f*, chain elektr.: network, circuit, filter;
  durchlässige —, transmission filter, acceptor circuit;
  endlose —, endless chain;
  Dreiecksglied= —, π-network, π-circuit;
  Entzerrer= —, corrective network, correcting network;
  Gelenk= —, link chain;
  Glieder= —, link belt;
  Kondensator= —, high-pass filter, infra filter, lower-limiting filter;
  Relais= —, relay chain *A*;

Kette
Sperr= —, rejector network;
— = —, Hf.=, low-pass filter, higher-limiting filter;
Spulen= —, low-pass filter, ultra-filter;
Sternglied= —, T-network, T-circuit;
Verzögerer= —, delay network, delay circuit;
Ketten=glied n, chain mesh, chain link; Filter: network mesh, filter section;
— =leiter m, network, filter, periodic recurrent structure, chain, chain system, wave filter;
— = — **I. Art**, π-network;
— = — **II. Art**, T-network;
— = —, aus gekoppelten Kreisen bestehender, coupled circuit chain;
— = —, aus Impedanzen gebildeter, impedance network;
— = —, mehrgliedriger, multi-mesh network;
— = —, mit einem halben Längsglied (Querglied) beginnender, wave filter terminated at mid-series (mid-shunt);
— = — =abschluß m durch ein halbes Längsglied (Querglied); mid-series (mid-shunt) termination of a network;
— = — =glied n, network mesh, filter section;
— =linie f, catenary.
Kf.=Leitung f, transfer circuit.
Kiefer f, fir.
Kies m, shingle.
Kiesel m, pebble; Kiesel= ....
— =erde f, silica; [pebbly];
— =grund m, pebbly bottom, shingle bottom;
— =gur f, infusorial earth.
kieselig, pebbly;
kieselsauer, silicated;

Kieselstein m, pebble.
Kilogrammeter n, kilogrammetre;
Kilo=hertz n, kilo-cycle, kilocycle, k. c.;
— =meter n, kilometre (statute mile = 1.609 km);
— =ohm n, kilohm;
— =volt n (ab: kV), kilovolt;
— =watt n (ab: kW), kilowatt, k. W.
— = — =stunde f, (ab: kWh) kilowatt hour.
kinetisch, kinetic;
— e Energie f, kinetic energy, vis viva.
Kiosk m, kiosk;
Straßen= —, street kiosk.
kippen, to tilt, to upset.
Kippschalter m, switching key.
kirschrot, cherry red.
Kissen n, cushion, pad;
Fernhörer= —, receiver cushion;
Stoß= —, pad.
Kitt m, cement; putty;
Glaser= —, putty.
kitten, to cement; to putty.
Klammer f, clamp, clip, bracket (auch M); parenthesis M;
eckige —n pl, square brackets pl M;
Feder= —, clamping spring, spring clip;
U= —, clevis, U-link.
Klampe f, cleat.
Klang m, musical sound;
zusammengesetzter —, complex musical sound;
— =farbe f, timbre, stamp, tone (colour);
— =feder f, coiled wire gong (der Münzfernsprecher, of coin collector stations);
— =platte f, sounding plate;
— =plattenklopfer m, plate sounder T.

9*

Klapp- ...., hinged;
— **-rahmen** m, hinged frame;
— **-schalttafel** f, hinged switchboard;
— **-transformator** m, rotatable coaxial coil transformer;
— **-variometer** n, hinged coil variometer.

Klappe f, lid; trap door; flap; shutter, indicator F;
 die — aufrichten, to restore the shutter;
 einschenklige —, single-coil indicator;
 selbsthebende —, self-restoring indicator;
 zweischenklige —, two-coil indicator;
 Anruf- —, calling indicator;
 Fall- —, drop indicator, drop shutter;
 — - — mit elektrischer (mechanischer) Rückstellung, electrical (mechanical) replacement indicator;
 Mantel- —, tubular indicator;
 Nebenstellen- —, extension indicator;
 Rohrpost- —, flap;
 Rückstell- —, plug-restored indicator, selfrestoring indicator;
 Schluß- —, ring-off indicator;
— mit Topfmagnet, tubular indicator;

Klappenschrank m, switchboard, board;
 aufklappbarer —, hinged switchboard;
 schnurloser —, cordless switchboard;
 Einschnur- —, single cord switchboard;
 Fernsprech- —, telephone switchboard;
 O. B.- —, magneto board, l.b. switchboard;
 Rückstell- —, switchboard with plug-restored drop indicators;
 Vielfach- —, multiple switchboard;
 Z. B.- —, c. b. board;
 Zweischnur- —, double-cord switchboard;
 — für Induktoranruf, magneto switchboard;
 — zu 6 Leitungen, six-line switchboard;

Klappenstreifen m, strip of indicators.
klappern, to chatter, to rattle.
Klappern n, chatter(ing), rattle, rattling.
klar, clear, distinct; Sprache: articulate.
Klarheit f, clearness, distinctness; articulation.
klassifizieren, to classify.
Klassifizierung f, classification.
Klaue f, claw, catch, jaw, pawl;
Klauenkupplung f, claw clutch.
Klebebeamter m, gummer T.
kleben, to gum (auf ein Formular, to a form); Anker: to stick, to freeze.
Kleben n, gumming;
— des Ankers, sticking of the armature.
Kleber m, gummer T.
Kleb-kraft f, elektrische, electrostatic adhesion, electrostatic retentive force;
— -stift m, stop, distance piece.
— -stoff m, gum.
Kleister m, paste, gum.
Klemme f, clamp; binding post, terminal, connector;
 Anschluß- —, binding post, terminal;
 Ausgangs- —n pl, output terminals;

Klemme
Doppel= —, double terminal;
Eingangs= —n *pl*, input terminals *pl*;
Entnahme= —n *pl*, output terminals *pl*;
Erd= —, earth(y) terminal, ground terminal;
Feder= —, spring clip;
Frosch= —, draw-vice, eccentric grip;
Kniehebel= —, come-along, drawing tongs *pl*;
Messing= —, brass terminal;
Pol= —, pole terminal;
Prüf= —, test clip;
Schnur= —, cord fastener;
Schraub= —, screw terminal;
Verbindungs= —, connector.

klemmen, to clamp; **sich** —, to jam (tight).

Klemmen-kasten *m*, terminal block, connection box;
— =leiste *f*, connection strip, terminal strip;
— =platte *f*, connection plate;
— =spannung *f*, terminal voltage;
— =streifen *m*, connection strip.

Klemm=platte *f*, clamping plate;
— =ring *m*, clamping ring;
— =schraube *f*, clamping screw; binding post.

Klettereisen *n*, climbing iron.

Klima *n*, climate.

klimatisch, climatic.

Klinge *f*, blade.

Klingel *f*, bell;
— =leitung *f*, bell circuit;
— =transformator *m*, bell transformer.

klingen, to sound.

klingend, sounding; Sprache: sonorous.

Klinke *f*, jack, switch spring, spring jack; latch, pawl;
an —n endigend, terminated on jacks;

besetzte —, engaged jack;
dreiteilige —, three-way jack, three-point jack;
freie —, disengaged jack, idle jack;
fünfteilige —, five-point jack;
unbesetzte —, disengaged jack;
vierteilige.—, four-point jack;
zweiteilige —, two-point jack, two-way jack;
Abfrage= —, answering jack, home jack;
Abzweig= —, branching jack;
Amts= —, exchange jack;
Anschalte= —, operator's jack, service jack;
— = —, zweite, ancillary service jack;
Apparat= —, instrument jack *T*;
Batterie= —, battery jack, power jack;
Doppelsperr=—, double detent, double pawls *pl*, double dog *A*;
Doppelunterbrechungs= —, double break jack;
Dreh= —, turning pawl;
Feder= —, spring catch;
Fern(leitungs)= —, toll line jack;
Hebe= —, lifting pawl;
Hilfs= —, ancillary jack;
Leitungs= —, line jack;
Mithör= —, monitor(ing) jack;
Nebenstellen= —, extension (line) jack;
Parallel= —, parallel jack, branching jack; duplicate jack;
— = —, Vielfach=, parallel multiple jack;
Prüf= —, test jack;
— = —, Schnur=, cord testing jack;
Sperr= —, pawl, holding pawl, lock(ing) pawl, stop pawl;

Klinke
Stoß= —, driving pawl, propelling pawl; thrust pawl;
Teilnehmer= —, subscriber's jack;
Trenn= —, break jack, interrupt jack;
Unterbrechungs= —, break jack;
— = —, Vielfach=, series multiple jack;
Verbindungsleitungs= —, junction jack;
— = —, abgehende, out junction jack;
Vielfach= —, multiple jack;
Vorschub= —, feed pawl T;
Wiederholungs= —, ancillary jack;
Zwillings= —, pair of jacks;
Klinken-auflager n, spring rest;
— =brett n, jack panel;
— =buchse f, jack bush;
— =feder f, jack spring (v. Feder)
— =feld n, jack panel, jacks pl;
— = —, Teilnehmer=, subscriber's multiple;
— = —, Vielfach=, (jack) multiple;
— =hülse f, jack barrel, jack bush;
— =körper m, jack body, jack socket;
— =streifen m, strip of jacks, jack strip;
— =sucher m, jack finder A;
— =umschalter m, jack switchboard, line switchboard;
— =vielfachfeld n, jack multiple.
Klippe f, rock.
Klippklapptelegraph m (bei dem abwechselnd ein Buchstabe gesandt und empfangen wird), flipflap telegraph, to-and-fro telegraph.
Klischee n, block.
klopfen, to tap; to sound T;
Klopfen n, tapping; sounding T.

Klopfer m, sounder; am Fritter: tapper;
Klangscheiben= —, plate sounder;
Übertragungs= —, relaying sounder, uprighting sounder;
die — = — abstellen, to silence T
— mit trägem Rade, Hughes silencer;
— mit zwei Klangscheiben, double plate sounder;
— =taste f, sounder key.
Klöppel m, hammer, striker;
des Wheatstonelochers: mallet, punching handle;
Glocken= —, bell hammer, bell striker.
— =kugel f, bob;
— =maschine f, braider.
klöppeln, to braid.
Klotz m, block;
Anker= —, stay block B.
Klümpchen n, nodule.
Klumpen m, lump;
Stoff= —, lump of matter.
Kluppe f, pliers pl B;
Draht= —, pliers pl B;
Schneid= —, screw stock.
knacken, to crackle, to click, to crack;
Knacken n, click, crack(s pl), crackling;
— bei der Besetztprüfung, engaged click F.
Knagge f, tappet, cam, stop
Knall m, bang; [cam.
— =gas n, oxyhydrogen;
— = — =gebläse n, oxyhydrogen blow pipe.
knarren, to jar.
Knarren n, jar.
knarren, to crack.
Knattern n, cracks pl.
Knauf m, knob;
mit einem — versehen, knobbed.
Knebel(griff) m, capstan, tommy;

Knebel
— -schraube *f*, tommy screw, capstan head screw.

kneten, to knead.

Knick *m*, bend, knee; im Draht: kink;
oberer (unterer) — der Röhren-kennlinie, upper (lower) bend of valve characteristic.

Knie *n*, knee, bend, elbow;
oberes —, top bend, upper bend, *V*;
unteres —, bottom bend, lower bend, *V*.
Rohr- —, pipe elbow, pipe bend;
— -gelenk *n*, knuckle joint;
— -hebelklemme *f*, come-along, drawing tongs *pl*, B.

knirschen, to grind.

Knirschen *n*, grinding; grinders *pl R*.

knistern, to sizzle, to crack.

Knistern *n*, sizzle, cracks *pl*.

Knochen *m*, bone;
— -öl *n*, bone oil.

Knopf *m*, button, knob;
mit einem — versehen, knobbed;
Druck- —, press button, push, push button.

Knötchen *n*, nodule.

Knoten *m*, node; Längenmaß: knot (= 1.853 km);
Schwingungs- —, interference point, nodal point of vibration, vibration node, null point;
Spannungs- —, potential node;
Strom- —, current node;

Knoten- ..., centre ...; nodal;
— -amt *n*, chief centre office, main centre office; repeating centre, main office, *T*;
— - -system *n*, zone system, repeating centre system, *T*;
— -punkt *m*, centre; nodal point.

koagulieren, to coagulate.

Koagulierung *f*, coagulation.

koaxial, coaxial.

Kobalt *m*, cobalt (Co).

kochen, to boil; Sammler: to gas.

Kochen *n*, boiling.

kochendes Geräusch *n*, boiling noise.

Kochsalz *n*, sodium chloride (NaCl).

Kode *m*, code;
Telegraphen- —, telegraph code.

Koeffizient *m*, coefficient.

Koerzitivkraft *f*, coercive force, retentivity.

Kohäsion *f*, cohesion.

Kohle *f*, coal, carbon;
Bogenlampen- —, carbon;
Galvano- —, coppered carbon;
Mikrophon- —, microphonic carbon;
Retorten- —, retort carbon;
— -faden *m*, carbon filament;
— - -lampe *f*, carbon filament lamp;
— -blitzableiter *m*, carbon protector;
— -gehalt *m*, carbon content (des Stahls, of steel);
— - —, Stahl *m* mit hohem (niedrigem), high (low) carbon steel;
— -korn *n*, carbon granule;
— -membran *f*, carbon diaphragm;
— -papier *n*, carbon paper;
— -pol *m*, carbon pole, carbon terminal;

Kohlen-beutel-elektrode *f*, carbon bag electrode;
— - — -mikrophon *n*, carbon bag transmitter;
— -dioxyd *n*, carbon dioxyde (CO_2);

Kohlen
— -körnerkammer f, carbon granule chamber;
— -körnermikrophon n, carbon granule transmitter;
— -mikrophon n, carbon transmitter;
— -pfanne f, charcoal brazier B;
— -pulvermikrophon n, carbon powder transmitter;
— -säure f, carbonic acid gas (CO_2);
— -staubmikrophon n, carbon dust transmitter;
— -stoff m, carbon (C);
— -teer m, coal tar;
— -verbrauch m, carbon consumption;
— -vorschub m, feed(ing) of the carbons.
Koinzidenz f, coincidence.
Kokon m, cocoon;
— -faden m, silk fibre.
Kolben m, piston; Glasbirne: bulb;
Glas- —, bulb, glass bulb;
Gleichrichter- —, rectifier lamp;
Pumpen- —, bucket;
— -ventil n, piston valve.
Ko.-Leitung f, trunk junction circuit, toll switching trunk (am.) F.
Kollektor m, collector;
— -geräusch n, commutator noise;
— -lamelle f, commutator bar;
— -ring m, collector ring;
— -stab m, commutator bar, commutator segment.
kollidieren, to collide.
Kollision f, collision.
Kolonne f, gang;
Bau- —, construction gang.
Kolophonium n, colophony.
Kombination f, combination;

Kombinations-frequenz f, combination frequency;
— -ton m, combination tone.
Kombinator m, combiner T;
— -scheibe f, combiner wheel T.
kombinieren, to combine.
kombinierter Widerstand m, joint resistance.
Komma n im Dezimalbruch, decimal point.
kommerziell, commercial.
Kommutator m, commutator;
— -brand m, burning of commutator;
— -motor m, commutator motor;
— -unterbrecher m, commutator interrupter, commutator break.
kommutieren, to commutate.
kommutierter Strom m, commutated current.
Kommutierung f, commutation;
geräuschlose —, noiseless commutation;
Kommutierungs-frequenz f, ripple frequency;
— -wellen pl, commutation ripple, commutator ripples pl.
Kompaß m, compass, magnetic compass;
Funk- —, radio compass, wireless compass;
Magnet- —, magnetic compass;
— -strich m, rhumb.
Kompensation f, compensation;
Kompensations-schaltung f, compensating circuit, compensating network;
— -strom m, compensating current;
— -widerstand m, compensating resistance.
Kompensator m, compensator.
kompensieren, to compensate (for).

Kompenfierung *f*, compensation (for).
Komplementwinkel *m*, complementary angle.
komplex, complex;
 konjugiert —e Größen *pl*, conjugate complex quantities *pl*.
Komponente *f*, component;
 flüchtige —, transient component;
 imaginäre —, imaginary component;
 periodische —, periodic component;
 reelle —, real component;
 Blind= —, wattless component, reactance component;
 Ein= und Ausschwingungs= —, transient component;
 Energie= —, energy component;
 Grund= —, fundamental component;
 Schein= —, apparent component;
 Spannungs= —, voltage component;
 Strom= —, current component;
 Widerstands= —, resistance component;
 Wirk= —, energy component.
kompoundieren, to compound.
kompoundiert, compound(ed).
Kompoundmotor *m*, compound (-wound) motor.
Kompression *f*, compression.
Kompressor *m*, compressor.
komprimieren, to compress.
Kondensanz *f*, condensance, capacitance.
Kondensator *m*, condenser;
 mit Ableitung behafteter —, leaky condenser;
 einstellbarer —, adjustable condenser;
 großer —, large-capacity condenser;
 variabler —, variable condenser, adjustable condenser;
 Abflachungs= —, smoothing condenser;
 Abstimm= —, tuning condenser;
 — = —, Luftdraht=, aerial tuning condenser, *ab*: a. t. c.;
 Antennen=Verkürzungs= —, aerial series condenser, short wave condenser;
 Ausgleichs= —, balancing condenser, compensating condenser;
 Block= —, blocking condenser, stopping condenser;
 — = —en *pl*, **Doppel=,** double block (condensers *pl*) *T*;
 — = —, Gitter=, grid blocking condenser;
 — = —, Luftdraht=, aerial blocking condenser;
 Differential= —, differential (twin) condenser, twin condenser;
 Dreh= —, disc condenser;
 Drehplatten= —, rotating plate condenser;
 Druckluft= —, compressed air condenser;
 Eich= —, calibration condenser;
 Funkenlösch= —, spark quench condenser;
 Gitter= —, grid (blocking) condenser;
 Glimmer= —, mica (dielectric) condenser;
 Kopplungs= —, coupling condenser;
 Luft= —, air condenser;
 Meß= —, measuring condenser;
 Öl= —, oil (dielectric) condenser;

Kondensator
  Platten= —, plate condenser, disc condenser;
  Quer= —, bridging condenser, by-pass condenser;
  Sende= —en pl, signalling condensers, sending condensers pl (in den Brückenarmen, in the bridge arms) T;
  Speicher= —, reservoir condenser;
  Spritzguß= —, die-cast condenser;
  Überbrückungs= —, bridging condenser, by-pass condenser;
  Verkürzungs= —, Luftdraht=, aerial series condenser, shortwave condenser;
  Vorrats= —, reservoir condenser;
  Wellplatten= —, corrugated plate condenser;
  Wickel= —, roll type condenser;
  — für große Leistung, power condenser;
  — =mit Parallelwiderstand, shunted condenser T;
  — =mit geringen dielektrischen Verlusten, low loss condenser;
  —,en=batterie f, condenser bank;
  — =belegung f, condenser armature, condenser coating;
  — en=kasten m, condenser box, condenser pot, K;
  — =kette f, high-pass filter, lower limiting filter, infra filter;
  — =kreis m, condenser circuit;
  — =mikrophon n, condenser transmitter;
  — en=muffe f, condenser sleeve K;
  — =telephon n, condenser telephone;
  — -wickel m, condenser reel.
kondensieren, to condense;
Kondensierung f, condensation.

Konduktanz f, conductance.
konduktiv, conductive.
Konfetti pl perforations pl T.
kongruent, congruent;
  — e Dreiecke pl, congruent triangles.
Kongruenz f, congruence, congruency.
Königswasser n, aqua regia.
konisch, conical, coned, taper(ed), bevelled;
  — er Stift m, taper pin.
Konizität f, conicalness.
konjugiert, conjugate;
  — komplex, conjugate complex.
konkav, concave.
Konservierung f, preservation (von Holz, of wood).
konsistent, consistent.
Konsistenz f, consistence, consistency.
Konsole f, bracket, cantilever.
Konsonant m, consonant;
  explosiver —, explosive consonant;
  Nasal= —, nasal consonant.
konstant, constant;
  — halten, to keep constant, to maintain constant.
Konstantan n, constantan.
Konstante f, constant.
Konstanz f, constancy;
  Ton= —, constancy of pitch;
  Touren= —, constancy of speed;
  Wellen(längen)= —, steadiness of the wave.
konstruieren, to design, to construct.
Konstruierung f, designation.
Konstrukteur m, designer.
Konstruktion f, design, construction;
  mechanische —, mechanical
Kontakt m, contact; [design.
  die —e enger (weiter) stellen, to close up (open) the contacts;

Kontakt
die —e reinigen, to clean contacts;
— machen, to make contact, to make connection (auf, an, with, on);
ausgefressener —, pitted contact, worn contact;
federnder —, spring contact, flexible contact;
inniger —, intimate contact;
intermittierender —, tapping contact, intermittent contact; ticker R;
kleiner —, shortened segment T;
schlechter —, poor contact;
weicher —, spring contact, flexible contact;
Amboß= —, anvil contact, buffer contact;
Arbeits= —, make contact, operating contact; marking contact T;
Auslöse= —, release contact A;
Bank= —, bank contact;
Buffer= —, buffer contact;
Doppel= —, double contact, collateral contact;
Doppelarbeits= —, double make contact;
Doppelschließ= —, double make contact;
Doppeltrenn= —, double break contact;
Feder= —, spring contact, flexible contact;
Folge= —, make-before-break contact, continuity-preserving contact;
Gleis= —, rail contact;
Gleit= —, sliding contact; (contact) slider;
Kopf= —, mechanical contact, off-normal (contact), A;

Messer= —, knife blade contact, switch jack;
Nocken= —, cam contact;
Relais= — e pl, relay contacts, relay points pl;
Ruhe= —, rest(ing) contact, non-operative contact; spacing contact T;
Schiebe= —, sliding contact, (contact) slider;
Schienen= —, rail contact;
Schlepp= —, make-before-break contact, continuity-preserving contact;
Schließ(ungs)= —, make contact;
Selbstunterbrecher= —, trembler (bell) contact, vibrating contact;
Trenn= —, spacing contact T; break contact;
Tür= —, door push;
Unterbrecher= —, vibrating contact;
Unterbrechungs= —, beak contact;
Wackel= —, variable contact, defective contact, loose contact;
Wechsel= —, make and break contact;
Wellen= —, cam contact, cam springs pl, shaft contact, A;
Zeichen= —, marking contact T;
— =abstand m, contact clearance;
— =arm m, wiper A;
— = — =satz m, wiper set, wiper assembly, A;
— =backen pl, contact jaws pl;
— =bank f, contact bank, (selector) bank, A;
— = — =drähte pl, bank wires pl A;

Kontakt
— ꞊ — ꞊**verkabelung** f, bank-to-bank cabling A;
— ꞊**bürste** f, contact brush; wiper A;
— ꞊**druck** m, contact pressure;
— ꞊**feder** f, contact spring;
— ꞊ — **des Stromstoßgebers**, impulse spring A;
— ꞊**feile** f, contact file;
— ꞊**feld** n, (**Vielfach**꞊), (multiple) contact bank A;
— ꞊**finger** m, contact finger, wiper;
— ꞊**fläche** f, contact (sur)face;
— ꞊**hebel** m, contact lever;
— ꞊**kamm** m, contact comb;
— ꞊**mikrophon** n, contact transmitter, contact microphone;
— ꞊**prellen** n, contact chatter;
— ꞊**rahmen** m, contact carriage;
— ꞊**satz** m, contact bank;
 a꞊b꞊ — ꞊ —, line contact bank A;
 c꞊ — ꞊ —, local contact bank, private (contact) bank, testing and guarding bank, A;
 Leitungs꞊ — ꞊ —, line contact bank A;
— ꞊ — ꞊**vielfachkabel** n, bank cable A;
— ꞊ — ꞊**vielfachverdrahtung** f, bank wires pl A;
— ꞊**schiene** f, contact bar;
 vordere (hintere) — ꞊ — **der Taste**, front (back) contact of the key;
— ꞊**schlitten** m, contact carriage;
— ꞊**spannung** f, contact potential;
— ꞊**spitze** f, contact point;
— ꞊**stift** m, contact pin; **kurzer:** contact stud;
— ꞊**stöpsel** m, contact plug;
— ꞊**stück** n, contact block, (contact) stud;
— ꞊**weite** f, contact clearance.

Konto n, account;
 ein — **belasten**, to debit;
— ꞊**einlage** f, deposit.
Kontra꞊alt m, contralto;
— ꞊**baß** m, contrabass.
Kontroll꞊druck m, home record T;
— ꞊**drucker** m, — ꞊**empfänger** m, control printer T;
— ꞊**platz** m, supervisor's position, monitor's position;
— ꞊**relais** n, supervisory relay;
— ꞊**lampe** f, supervisory lamp;
 Zähler꞊ — ꞊ —, meter lamp;
— ꞊**schrift** f, home record;
— ꞊**streifen** m, home record, control slip;
— ꞊**stromkreis** m, checking circuit;
— ꞊**taste** f, check key;
— ꞊**ton** m, check tone.
Kontrolle f, checking.
Kontermutter f, lock nut.
kontinuierlich, continuous.
Kontraktion f, contraction.
Konus m, cone;
— ꞊**antenne** f, cone aerial.
Konvektionsstrom m, convection current.
konvektiv, convective.
konvergent, convergent.
Konvergenz f, convergence, convergency, (**zu, in, nach**, into).
konvergieren, to converge.
Konzentration f, concentration.
konzentrieren, to concentrate; to lump.
konzentrisch, concentric(al) (**mit, zu**, to, with);
— **es Kabel** n, concentric cable.
Konzession f, concession.
Koordinaten pl, coördinates pl;
 rechtwinklige —, rectangular coördinates;
 Polar꞊ —, polar coördinates;
— ꞊**papier** n, coördinate paper, ruled paper;

Koordinaten-papier, logarithmisches, logarithmic coordinate
paper;
— -system *n*, coordinate system.
Kopal-firnis *m*, copal varnish;
— -lack *m*, copal lacquer, copal
varnish.
Kopf *m*, head, top (end); eines
Telegramms: preface, pre-
amble;
Flach- —, flat head;
Rund- —, round head;
Schrauben- —, screw head;
Sechskant- —, hexagonal head;
Torsions- —, torsion head;
Vierkant- —, square head;
Zeichen- —, signal head *TL*;
— -bügel *m*, head band;
— -fernhörer *m*, head receiver,
headgear receiver, h. g. receiver, headphone;
Doppel- — - —, headphones
pl;
— -kontakt *m*, off--normal (contact), mechanical contact,
A;
— -rille *f*, top groove *B*.
Kopier-presse *f*, copying press;
— -telegraph *m*, copying telegraph.
koppeln, to couple.
Koppler *m*, coupler.
Kopplung *f*, coupling;
autoinduktive —, autoinductive coupling;
direkte —, direct coupling,
conductive coupling;
feste —, tight coupling;
galvanische —, galvanic coupling;
induktive —, inductive coupling, inductance coupling,
magnetic coupling;
kapazitive —, capacity coupling, condenser coupling,
capacitive coupling;

— -en *pl*, ungewollte, spurious
couplings *pl*;
kritische —, critical coupling;
lose —, loose coupling;
magnetische —, magnetic coupling, inductive coupling;
reaktionslose —, non-reactive
coupling;
statische —, (electro)static
coupling;
Gitterkreis- —, grid coupling;
Kapazitäts- —, capacity coupling;
Mitsprech- —, phantom-to-
side unbalance *K*;
Nebensprech- -en *pl*, crosstalk couplings *pl*, crosstalk
paths *pl*, *K*;
Röhren- —, intervalve coupling, intervalve linkage *V*;
Übersprech- —, crosstalk couplings *pl*, zwischen den Stäm-
men eines Vierers: side-to-side
unbalance;
Widerstands- —, resistance
coupling, resistive coupling;
— zwischen zwei Röhren, intervalve coupling, intervalve
linkage;
— durch gemeinsame Induktivität, autoinductive coupling;
— — — Kapazität, auto-capacity
coupling;
Kopplungs-faktor *m*, coupling
factor, coupling coefficient;
Rück- — - —, regenerative
coefficient of coupling;
— -kapazität *f*, coupling capacity;
— -koeffizient *m*, coupling factor;
— -kondensator *m*, coupling condenser;
— -regler *m*, coupling control;
— -schleife *f*, search coil, expor-
ing coil;
— -spule *f*, coupling coil, coupler;

**Kopplungs-spule
veränderliche** — = —, vario-coupler;
— =**transformator** m, repeating transformer, coupling transformer, jigger;
— =**wechsler** m, poling switch, coupling changer, KV;
— =**welle** f, partial wave;
— = — **n** pl, coupling waves pl;
— =**widerstand** m, coupling resistance;
— =**ziffer** f, coupling coefficient.
Koralle f, coral.
Korallen= ...., coralline.
Korb m, basket;
— =**bodenspule** f, basket (type) coil, spider web coil;
— =**deckelspule** f, basket-wound coil;
— =**flasche** f, demijohn, carboy.
Kordel=griff m, milled knob handle;
— =**knopf** m, milled knob;
— =**(kopf)schraube** f, thumb screw;
— =**mutter** f, milled nut;
— =**rolle** f, milled roller;
— =**schraube** f, knurled screw.
kordeln, to mill.
Kork m, cork;
— =**klotz** m, cork pad (am Hughes=regler, of the Hughes governor).
Korn n, granule, grain;
— =**größe** f, grain size;
Körner pl, granules pl;
Kohlen=—, carbon granules;
— = — =**mikrophon** n, carbon granule transmitter;
— =**fritter** n, granular coherer;
— =**mikrophon** n, granular transmitter.
körnig, grained, granular.
Körnung f, granulation.
Korona f, corona;

— =**verluste** pl corona(l) losses pl.
Körper m, body;
fester —, solid body;
Apparat= —, body (portion);
Klinken= —, jack socket;
Stecker= —, plug body;
Tasten= —, centre of the key;
— =**schluß** m, body contact.
Körperchen n, corpuscle.
Korrektion f, correction;
Korrektions=daumen m, correcting cam;
— =**faktor** m, correction factor;
— =**rad** n, correcting wheel, correction wheel, corrector wheel.
korrigieren, to correct.
korrodieren, to corrode.
Korrosion f, corrosion;
Kabelmantel= —, corrosion of the lead sheathing.
Kosten pl, cost pl, expense;
Anlage= —, first cost, cost of construction, purchasing cost;
Anschaffungs= —, prime cost;
Bedienungs =—, cost of attendance;
Betriebs= —, operating expense, working cost;
Durchschnitts= —, average cost;
Errichtungs= —, cost, cost of construction, purchasing cost; [struction;
Herstellungs= —, cost of con-
Un= —, cost, expense;
— = —, **laufende**, running cost;
Unterhaltungs= —, cost of upkeep, maintenance cost;
Wartungs= —, cost of attendance;
— =**anschlag** m, estimate, tender;
— =**berechnung** f, calculation of cost;
— =**rechnung** f, bill. [ent, cot.
Kotangente f (ab: cot), cotang-

KR = Kapazität × Widerstand, KR = capacity × resistance,
- =Gesetz *n*, KR-law *T*.
Kraft *f*, force;
  elektromotorische —, (*ab*: EMK) electromotive force, *ab*: e. m. f.
  lebendige —, momentum, vis viva;
  magnetische —, magnetic force;
  magnetisierende —, magnetizing force;
  Einheit der —, unit force;
  Ablenkungs= —, deflecting force;
  Abstoßungs= —, repulsive force, force of repulsion;
  Antriebs= —, motive force;
  Anziehungs= —, force of attraction, attractive force;
  Gegen= —, opposing force;
  Kleb= —, retentive force;
  — =—, elektrische, electrostatic adhesion;
- =anlage *f*, power plant, power system;
- =ausbeute *f*, power output;
- =leitung *f*, power line, power circuit;
- =linie *f*, line of force; magnetische — = —, magnetic line of force, line of magnetic force;
- =linien schneiden, to cut or intersect lines of force;
- = — =richtung *f*, direction of lines of force;
- = — =weg *m*, magnetischer, magnetic flux path;
- = — — =, mittlerer, mean path of lines of force;
- = — =zahl *f*, number of lines of force;
- = — = —, gesamte, total flux;
- =maschine *f*, prime mover;
- =messer *m*, dynamometer;
- =quelle *f*, source of power;
- =röhre *f*, tube of force; große Senderöhre: power tube;
- =schalttafel *f*, power (switch-) board, power panel;
- =übertragung *f*, transmission of power;
- = —s =anlage *f*, power transmission system;
- =wagen *m*, motor car, automobile;
  Last= — = —, motor truck;
- =werk *n*, power station.
kräftig, powerful.
Kragen *m*, collet, collar.
Kramme *f*, Krampe *f*, staple;
Krampf *m*, cramp *T*;
Kranbaum *m*, derrick;
- zum Aufrichten der Stangen, pole setting derrick *B*.
Kranz *m*, rim.
Krarup=ader *f*, continuously loaded conductor, iron whipped core, Krarup conductor;
- =bandumspinnung *f*, iron tape winding;
- =drahtumspinnung *f*, iron wire whipping;
- =kabel *n*, continuously loaded cable;
- =leiter *m*, continuously loaded conductor;
- =umspinnung *f*, iron whipping, Krarup winding.
krarupisieren, to load continously, to krarupize.
Krarupisierung *f*, Krarup loading, continuous loading, krarupization.
Krater *m*, crater;
- =fläche *f*, crater area (des Lichtbogens, of the arc).
kratzen, to scratch, to scrape.
Kratzen *n*, scratching, scraping; Geräusch: scratchy noise, jar, scratch.
kratzendes Geräusch *n*, scratchy noise.

Kratzer *m*, scraper; Geräusch: jar, scratch.
Kräuselung *f*, ripple;
— des Gleichstromes, d. c. ripple.
Krauskopf *m*, rose bit, countersink.
Kreide *f*, chalk.
Kreis *m*, circle; Stromkreis: circuit;
abgestimmter —, tuned circuit;
eingeschriebener —, inscribed circle;
gekoppelte —e *pl*, coupled circuits *pl*;
magnetischer —, magnetic circuit;
selektiver —, selective circuit;
Anoden- —, plate (-filament) circuit;
Ausgangs- —, output circuit; plate circuit, V;
Eingangs- —, input circuit;
Eisen- —, ferric circuit;
Empfänger- —, receiver circuit;
Entladungs- —, discharge circuit;
Entnahme- —, output circuit, load circuit;
Erreger- —, exciting circuit;
Meß- —, measuring circuit;
Orts- —, local circuit;
Resonanz- —, (series-) resonant circuit;
Selektiv- —, selecting circuit, selective circuit;
Sieb- —, selective circuit;
Stoß- —, impulsing circuit;
Teil- —, divided circle, graduated circle;
Verbraucher- —, load circuit;
Zwischen- —, intermediate circuit;
— - —, abgestimmter, tuned intermediate circuit;
— -bewegung *f*, circular motion; schnelle — - —, gyration;
— -bogen *m*, arc;
— -durchmesser *m*, diameter;
— -fläche *f*, circular surface; area of a circle M.
kreisen, to circulate, to rotate; schnell: to gyrate, to eddy;
kreisend, circulating, rotatory; gyratory, eddy.
Kreisfluß *m*, circular flux;
kreisförmig, circular;
Kreis-frequenz *f*, angular velocity, frequency in radians;
Grenz- — - —, cut-off angular velocity;
— -funktion *f*, circular function;
— -halbmesser *m*, radius;
— -lauf *m*, cycle;
— - — -Selbstanschlußsystem *n*, by-pass automatic telephone system;
— -nut *f*, recess;
— -prozeß *m*, cycle; [cycle; magnetischer — - —, magnetic
kreisrund, circular;
— er Querschnitt, circular cross-section;
Kreis-segment *n*, segment;
— -sektor *m*, sector;
— -skala *f*, circular scale;
— -strom *m*, circular current;
— -teilung *f*, circular scale, divided or gratuated circle;
— -umfang *m*, circumference.
Kreosot *n*, creosote;
mit — tränken, to creosote;
— — getränkte Stange *f*, creosote(d) pole;
— -tränkung *f*, creosoting.
Krug *m*, jar.
Krümmung *f*, curvature, bend.
Kreuz *n*, cross;
Malteser —, maltese cross;
— -kopf *m*, cross-head;
— -lochschraube *f*, capstan head screw;
— -rahmenantenne *f*, cross-coil aerial.

Kreuz
—-spulmeßgerät n, crossed-coil measuring instrument;
— -support m, cross slide;
— -verbindung f, cross-connection.
kreuzen, to cross, to traverse; Leitungen: to cross-connect, am Gestänge: to transpose.
Kreuzen n, crossing, cross-connecting; transposing.
Kreuzung f, cross(ing); von Leitungen: cross-connection, am Gestänge: transposition;
Draht- —, wire crossing;
Fluß- —, river crossing;
Leitungs- —, transposition; line crossing;
Starkstrom- —, power line crossing;
Über- —, crossing (-over);
Kreuzungs-abschnitt m, transposition section B;
— -festpunkt m, transposition section point B;
— -isolator m, transposition insulator;
— -punkt m, (point of) intersection; crossing B;
— -stange f, transposition pole B;
— -system n, cross (-over) system, transposition system, B.
kreuzweise, crossways, traverse.
kriechen, to creep; to leak;
Kriechen n, creeping; des Stromes: surface leakage.
Kriech-strom m, surface leakage current;
— -weg m, surface leakage path.
Kristall m, crystal;
eingelöteter —, solder-mounted crystal;
Gleichrichter- —, rectifying crystal;

— -bildung f, formation of crystals, crystallisation;
— -detektor m, crystal detector;
— -gefüge n, crystal structure;
— -gleichrichter m, crystal rectifier.
kristallin(isch), crystalline.
Kristallisation f, crystallisation.
kristallisieren, to crystallise.
kritisch, critical;
— e Kopplung f, critical coupling
— er Wert m, critical value.
Kronrad n, crown wheel.
Kröpfung f, bend, shoulder.
Krücke f, crutch.
Krügerelement n, Krueger cell.
krumm, buckled, bent;
sich — ziehen, to buckle (Stangen, poles).
krümmen, (sich), to bend, to curve, to buckle.
Krümmen n, bending, buckling.
Krümmer m, bend;
Rohr- —, bend.
krummlinig, curved, curvilinear.
Krummwerden n, buckling.
Krümmung f, bend, curve, curvature, crank;
Erd- —, curvature of the earth;
Krümmungs-halbmesser m,
— -radius m, radius of curvature.
Kruste f, crust;
mit einer — überziehen, to incrust;
Erd- —, earth's crust.
Kübel m, tray.
Kubik-inhalt m, cubic contents pl;
— -meter n, $(ab: m^3)$ cubic metre $(=35.317$ cub. ft.);
— -wurzel f, cube root;
— -zahl f, cube;
— -zentimeter n, $(ab:$ cm^3, ccm), cubic centimetre $(=0.061\,026$ cub. in.).

Kubus *m*, cube.
Kugel *f*, ball, globe; sphere *M*;
 Klöppel= —, bob;
— =fläche *f*, spherical surface;
— =form *f*, ball shape, spherical shape;
kugelförmig, ball-shaped, spherical;
Kugel-funkenstrecke *f*, sphere gap, balls *pl*;
— =gelenk *n*, ball joint, Hooke's joint;
kugelig, spherical;
Kugel-lager *n*, ball bearing;
— =panzergalvanometer *n*, ball shield galvanometer;
— =variometer *n*, ball variometer;
— =welle *f*, spherical wave.
kühlen, to cool.
Kühl-flansch *m*, cooling flange;
— =flügel *m*, cooling vane;
— =flüssigkeit *f*, cooling fluid;
— =gefäß *n*, cooling tank;
— =mantel *m*, cooled jacket;
— =rippe *f*, cooling vane, cooling flange, fin;
— =schlange *f*, cooling coil.
Kühlung *f*, cooling.
Kühl-wasser *n*, cooling water;
— = — =teich *m*, cooling pond.
Kumulation *f*, cumulation.
kumulativ, cumulative.
Kunst *f*, art;
— =leitung *f*, artificial line, artificial cable, artificial circuit; balancing network *V*;
— =schaltung *f*, network;
 End= — =—, terminal network;
— =seide *f*, imitation silk.
künstlich, artificial.
Kupfer *n*, copper;
 Elektrolyt= —, electrolytic copper;
— =band *n*, copper band, copper tape;

— =dämpfung *f*, copper damping;
— =draht *m*, copper wire;
 Hart= — = —, hard-drawn copper wire;
— = — =geflecht *n*, copper mesh;
— =element *n*, copper-zinc cell;
— =gaze *f*, copper gauze;
— =gewebe *n*, copper mesh copper gauze;
— = — =bürste *f*, copper gauze brush;
— =gewicht *n*, copper weight;
— =kies *m*, (Eisen=), chalcopyrite, iron copper sulphide $(Cu_2S + Fe_2S_3)$;
— =klotz *m*, copper block;
— =lahn *m*, copper tinsel;
— =Leitfähigkeitsnormal *n*, copper conductivity standard;
— =litze *f*, copper stranded wire;
— =mantel *m*, copper collar, copper jacket;
 mit einem — = — **versehen**, copper jacketed;
— = — =relais *n*, coppered relay, copper jacketed relay;
— =nickel *n*, copper nickel alloy;
kupferplattieren, to copperplate;
kupferplattiert, copper-plate(d);
Kupfer-pol *m*, copper pole, *ab*: C;
— =pyrit *m*, iron copper sulphide, chalcopyrite, $(Cu_2S + Fe_2S_3)$;
— =ring *m*, copper collar;
— = — =relais *n*, copper collar relay;
— =röhre *f*, copper tube; copper sleeve *B*;
— = —n =verbindung *f*, copper sleeve joint, twisted sleeve joint, *B*;
— =streifen *m*, copper strip;
— =verbindungshülse *f*, copper jointing sleeve *B*;

Kupfer
— =verluste *pl*, copper losses *pl*;
— =vitriol *n*, blue vitriol, copper sulphate, (CuSO$_4$);
mit — = — tränken, to boucherize;
Tränkung *f*, mit — = —, boucherization;
— =wirkungsgrad *m*, copper efficiency.

kuppeln, to clutch, to couple, (mit, to).

Kupplung *f*, coupling, clutch; elastische —, flexible coupling; elektromagnetische —, electromagnetic coupling; lösbare —, disengaging coupling; magnetische —, magnetic clutch; Band= —, belt coupling; Ein=Umlauf= —, single-revolution clutch; Klauen= —, claw clutch; Reib= —, friction clutch.

Kupronelement *n*, copper oxide cell, cupron cell.

Kurbel *f*, crank; mit einer — versehen, cranked;
— =griff *m*, crank handle;
— =induktor *m*, hand generator, magneto (generator); dreilamelliger — = —, three-bar magneto;
Umschalter am — = —, automatic cut-out of magneto;
— =schalter *m*, radial arm switch, lever switch;
Doppel= — = —, double-lever switch;
— =scheibe *f*, crank disc.

Kursweiser *m*, radio beacon *R*.

Kurve *f*, curve, bow;
— n aufnehmen, to trace curves, to plot curves;
Anstieg *m* einer —, rise, slope of a curve;

Aufnahme *f* einer —, plotting of a curve;
nach einer logarithmischen —, on a logarithmic curve;
spitze —, peaky curve, peaked curve;
wellige —, corrugated curve;
zweispitzige —, double-peaked curve;
Exponential= —, exponential curve;

Kurven=blatt *n*, curve sheet, graph;
— =darstellung *f*, graph, graphical representation;
— =form *f*, curvature;
in — = — darstellen, to represent graphically;
Darstellung *f* in — = —, graphic representation;
— =lineal *n*, French curve;
— =schar *f*, set of curves;
— =schlitz *m*, cam slot; curved slot;
— =tafel *f*, curve sheet, graph.

kurzschließen, to short-circuit, to short; to close.

Kurzschließer *m*, short-circuiting device.

Kurzschluß *m*, short-circuit, short, (zu, round);
im — arbeiten, to work on short-circuit *T*;
teilweiser —, partial short-circuit;
vollständiger —, dead short-circuit;
— =impedanz *f*, short-circuit impedance, closed-end impedance;
— =schalter *m*, short-circuiting switch;
— =strom *m*, short-circuit current.

Kurzschrift *f*, shorthand.
kurzwellig, short-wave(length)...
kurzzeitig, momentary.
Küste *f*, coast.
Küsten-funkstelle *f*, coastal radio station;

— **-kabel** *n*, shore-end cable, shallow-water cable.
kyanisieren, to cyanise.
Kyanisieren *n*, cyanising.
Kyanisierung *f*, cyanisation.

L.

labil, unstable.
Laboratorium *n*, laboratory;
— **-versuch** *m*, laboratory test.
Lack *m*, varnish, lacquer, lac;
  Email(le)- — *m*, enamel lac;
  Gummi- —, rubber varnish;
  Isolier- —, insulating varnish;
  Japan- —, japan;
  Kopal- —, copal varnish;
— **-firnis** *m*, lac varnish.
lackieren, to varnish, to lacquer; to japan.
Lackierofen *m*, lacquering stove.
lackiert, varnished, japanned.
Lackierung *f*, varnishing; finish;
  schwarze —, black finish.
Lade-baum *m*, derrick;
— **-dynamo** *f*, charging generator;
— **-einrichtung** *f*, charging equipment;
— **-maschine** *f*, (battery) charging generator;
— **-satz** *m*, battery charger, charging set;
— **-schalter** *m*, (battery) charging switch;
— **-schalttafel** *f*, charging switchboard;
— **-spannung** *f*, charging voltage;
— **-stellung** *f*, charging position;
— **-strom** *m*, charging current; einer Leitung: surge current;
— **-stromkreis** *m*, charging circuit.
laden, (sich), to charge (auf *n* Volt, to *n* volts); Leitungen: to load (*cf.* belasten);
  über- —, to overcharge;
  wieder —, to recharge;
  zu wenig —, to undercharge.
Laden *n*, charging; loading.
Ladung *f*, charging, loading; charge, load.
  unter —, under charge;
  elektrische —, electric charge;
  entgegengesetzte —en *pl*, opposite charges *pl*;
  erste — der Sammler, first or initial charge of storage cells;
  freie —, free charge;
  gebundene —, bound charge;
  gleichförmige —, uniformly or continuously distributed loading K;
  induzierte —, induced charge;
  punktförmige —, lumped loading K;
  ruhende —, static charge;
  ungebundene —, free charge;
  Einheits- —, unit charge;
  Nach- —, additional charge (der Sammler, of storage cells);
  Raum- —, space charge V;
Ladungs-abschnitt *m*, loading coil section, pupinization section, K;
— **-dichte** *f*, density of charge;
— **-frequenz** *f*, charge frequency (eines Kondensators, of a condenser);

Labungs
— =vermögen, capacity;
elektrisches — = —, electrostatic capacity.

Lage f, coat(ing), layer; lamina (pl laminae); Standort: position, Peilung: bearing;
zwei —n Papier stark, two thicknesses of paper;
— n-drall m, twist of the wire layers (eines Kabels, of a cable);
— =plan m, general plan.

Lager n, d. Maschine: bearing; Fundament: bed(ding); f. Vorräte: store, stock;
auf —, in stock;
ein — anlegen, to lay out a store;
mit einem — versehene Drehspule f, unipivot moving coil;
Achs= —, bearing;
Auf= —, back (der Klinkenfeder: of the jack spring);
Fuß=—, footstep bearing;
Kamm= —, thrust bearing;
Kugel= —, ball bearing;
Rollen= —, roller bearing;
Spur= —, thrust bearing;
Steh= —, vertical bearing;
Stein= —, jewelled bearing;
Walzen= —, roller bearing;
— mit Kugelbewegung, spherical bearing;
— =büchse f, hub, brass;
— =deckel m, cap;
— =metall n, anti-friction metal, bearing metal;
— =schale f, brass;
— =schiene f, fulcrum bar;
— =schild m, bearing bracket;
— =spitze f, pivot point;
— =stein m, jewel cup;
— =stift m, hinge pin;
— =stütze f, bearing bracket;
— =typ m, stock type.

lagern, to (be) store(d);
drehbar —, to pivot (in, on).
Lagerung f, bedding;
Spitzen= —, pivot suspension;
Stift —, hinge pin bearing;
— in Steinen, jewelled bearing.
Lahn m, tinsel;
Kupfer= —, copper tinsel;
— =litze f, tinsel (cord);
— =litzenschnur f, tinsel cord.
Lamelle f, lamination, lamina pl: laminae, bar;
Kollektor= —, commutator bar, commutator segment.
Lamellenmagnet m, compound magnet, laminated magnet.
lamellieren, to laminate.
lamelliert, laminated.
Lampe f, lamp, pilot F;
weiße —, clear pilot;
Anruf= —, line lamp; calling (-in) lamp;
Ballast= —, ballast lamp;
Besetzt= —, engaged lamp;
Blink= —, flash lamp;
Glimm= —, gaseous conduction lamp;
Hand= —, portable lamp;
Kohlefaden= —, carbon filament lamp;
Kontroll= —, supervisory lamp, pilot lamp;
Löt= —, blow lamp, soldering lamp;
Melde= —, alarm lamp, pilot lamp;
Metallfaden= —, metal filament lamp, tungsten lamp;
Morse= —, flash lamp;
Neon= —, neon lamp;
Platz= —, pilot lamp, pilot indicator;
Quecksilberdampf= —, mercury vapour lamp;
Ruf= —, calling lamp, calling pilot;

Lampe
Schluß= —, clearing lamp, supervisory lamp; am B=Platz: junction clearing lamp;
— = — für den rufenden Teilnehmer, answering supervisory lamp;
— = — — — angerufenen Teilnehmer, calling supervisory lamp;
Signal= —, flash lamp; pilot lamp;
Steck= —, jack(s) lamp;
Taschen= —, pocket lamp, flash lamp;
Überwachungs= —, supervisory lamp, pilot lamp;
Verstärker= —, amplifier valve, lamp, three-electrode lamp;
runde — = —, amplifier bulb;
Widerstands= —, resistance lamp;
Wolfram= —, tungsten lamp;
Zählerkontroll= —, meter lamp F;

Lampen=arm m, lamp bracket;
— =fassung f, lamp holder;
— =feld n, lamps p', lamp panel;
— =kappe f, lamp cap F;
— =schirm m, lamp shade;
— =sockel m, lamp socket;
— =stempel m, (lamp) squash, press;
— =widerstand m, lamp resistance;

Land n, land; country;
Aus= —, foreign counrty;
Bestimmungs= —, country of destination;
In= —, inland;
Ursprungs= —, country of origin;
— =briefträger m, rural postman.

landen, to land (Kabel, cables).

Land=karte f, map;
— =linie f, land line;
— =straße f, road.

Landung f, landing.

Landungsstelle f, landing place.

lang, long;
elektrisch —, electrically long L;
unendlich —, indefinitely long L.

Länge f, length; geographisch: longitude, ...° westlicher (östlicher) —, long. ....° W (E);
axiale —, axial length;
elektrische —, electric length;
mittlere —, mean length;
Fabrikations= —, manufacturing length, factory length, drum length, (der Kabel, of cables);
Probe= —, sample length;
Trommel= —, drum length.

Längen=einheit f, unit length;
für die — = —, per unit length;
— =maß n, linear measure; eines Telegraphenkabels total distortion of a telegraph cable
$$\left(\frac{R}{2}\sqrt{\frac{C}{L}} - \frac{G}{2}\sqrt{\frac{L}{C}}\right).$$

länglich, elongated, Öffnung: slotted.

langsam, slow;
— abfallendes Relais, slow-to-release relay;
— ansprechend, slow operating;
— laufend, slow speed ...., low speed ....;
— lösend, slow release ....;
— wirkend, slow acting;
— er werden, to slow down.

Langsamkeit f, slowness; Ansprechen: sluggishness.

Längs=feld n, longitudinal field;
— =glied n einer Kette, series element of a network;
in $1/2$ — = — endendes Filter n, wave filter terminated at mid-series (position);

Längsglied
— = —, Abschluß *m* eines Filters durch ¹/₂, termination of a filter at mid-series position, mid-series termination;
— =impedanz *f*, series impedance.
— =lochstreifen *m*, lengthways perforated tape *T*;
— =magnetisierung *f*, longitudinal magnetization:
— =schnitt *m*, longitudinal section;
— =schwingungen *pl*, longitudinal vibrations *pl*;
— =widerstand *m*, series resistance.
langwellig, long wave . . . ., high wavelength . . . .
L=Antenne *f*, (inverted) L-aerial.
Lappen *m*, lug;
vorspringender —, projecting lug.
Lasche *f*, bond, fishplate.
Last *f*, load, charge:
mit voller —, on full load;
zur — schreiben, to debit;
Blind= —, reactive load;
Bruch= —, breaking load;
Schnee= —, snow load;
Wirk= —, active load;
— =kraftwagen *m*, motor truck;
— =wagen *m*, truck, waggon.
Laterne *f*, lantern.
Latte *f*, lath;
Holz= —, wooden lath.
Lauf —, running:
stoßfreier —, smooth running.
laufen, to run.
Laufen *n*, running.
laufend, running;
— e Instandsetzungsarbeiten *pl*, routine repair work;
— e Nummer *f*, consecutive number;
— e Überwachung *f*, routining, constant supervision.

Läufer *m*, rotor;
— der Funkenstrecke, spark gap rotor.
Lauf=geschwindigkeit *f*, speed; einer Welle: velocity of progression;
— =nummer *f*, consecutive number;
— = — =stempel *m*, numbering machine:
— =ring *m*, ball race (am Kugellager, of the ball bearing);
— =zeit *f*, time of transit, duration of transmission;
— = — des Stromes über eine Leitung, line lag.
Lauschmikrophon *n*, pick-up transmitter.
laut, loud; noisy.
Laut *m*, sound;
Summer= —, buzzing sound;
Zisch= —, sibilant (sound), hissing sound;
— =minimum *n*, (position of) silence;
— =sprecher *m*, loudspeaker, megaphone, talker;
Trichter= — = —, horn type loudspeaker;
— =stärke *f*, loudness, intensity of sound, volume of sound;
— = — der Sprache, volume of speech;
Verhältnis *n* der — = — zu den Störern, signal-to-noise ratio;
— = — =n =anzeiger *m*, volume indicator;
— = — =messer *m*, sound measuring device, audibility meter;
— = — =messung *f*, sound measurement, audibility test; telephonometry;
— = — =regelung *f*, volume control;
— = — =regler *m*, volume regulator; [fier.
— =verstärker *m*, sound ampli-

läuten, to ring.
Läuten *n*, ringing, ring.
Leatheroid *n*, leatheroid.
Lebensdauer *f*, (length of) life.
leck, leaky;
— sein, to leak.
Leck *n*, leak;
— strom *m*, leakance current, leak(age) current.
lecken, to leak.
Leclanchéelement *n*, Leclanché
Leder *n*, leather, hide. [cell.
— riemen *m*, leather belt.
leer, empty; Raum: void; Papierblatt: blank;
luft= —, vacuous;
— es Blatt *n*, (paper) blank.
Leerlauf= ...., no load...;
— arbeit *f*, no-load work;
— impedanz *f*, open-circuit impedance, no-load impedance;
— spannung *f*, no-load voltage, open-circuit voltage;
— strom *m*, no-load current;
Leerscheibe *f*, idler wheel.
legen, to put; Kabel: to lay (cables).
legieren, to alloy, to compound.
legiert, alloyed;
— er Stahl *m*, alloy steel.
Legierung *f*, alloy, composition;
Nickel-Eisen= —, nickel-iron composition.
Legung *f*, laying (eines Kabels, of a cable);
Kabel= —, cable laying.
Lehm *m*, clay.
Lehre *f*, gauge (*am*: gage) calibre; zum Zusammenbau: assembling jig, fixture;
Bohr= —, drilled jig;
Draht= —, wire gauge;
Gewinde= —, thread gauge;
Rachen= —, gap gauge;
Schub= —, slide gauge.
Leim *m*, glue;
Marine= —, marine glue.

leimen, to glue, to cement.
Leinen *n*, linen, cloth;
Öl= —, oiled linen;
Paus= —, tracing cloth;
— garn *n*, flax yarn.
Leinöl *n*, linseed oil;
— firnis *m*, boiled oil;
— wand *f*, linen, cloth;
geölte —, oiled cloth.
Leiste *f*, strip, ledge;
Abstands= —, spacing strip;
Fuß= —, skirting;
Klemmen= —, terminal strip, terminal block;
Scheuer= —, skirting;
Sicherungs= —, fuse strip.
Leistung *f*, power; in Watt: wattage;
abgegebene —, power output;
aufgenommene —, power input;
besondere — en *pl*, additional service *T*;
entnommene —, outoput;
hohe —, high power;
—r —, von, high-power(ed);
kleine —, low power;
—r —, von, low-power(ed);
zugeführte —, input;
Empfangs= —, received power, incoming power;
Grenz= —, limiting output;
Heiz= —, filament power, filament wattage;
Höchst= —, maximum output;
Schein= —, apparent power;
Sende= —, sending power;
Spitzen= —, peak power;
Strahlungs= —, radiated power;
Verkehrs= —, traffic capacity;
— = — in einer Richtung, one-way traffic capacity;
— = — in beiden Richtungen, two-way or duplex traffic capacity;

Leistung
 Wirk= —, real power;
Leistungs=abgabe f, power delivery;
— =aufwand m, power input;
— =einheit f, unit (of) power;
— =empfindlichkeit f, power sensitivity;
— =faktor m, power factor;
— =entnahme f, taking of power;
— =kapazität f eines Sammlers, watt-hour capacity of a storage cell;
— =messung f, power measurement;
— =verbrauch m, power dissipation;
— =verhältnis n, power ratio;
— =verstärkung f, power amplification, power magnification;
— = —s =verhältnis n, power amplification ratio;
— =zerstreuung f, power dissipation.
Leitauge n, guide.
Leitbuchstaben pl, office code A;
 Gruppenwähler mit —, code selector A;
— =speicher m, office code register A;
leiten, to conduct, to lead, to guide, to pilot; (Telegramme, Gespräche: to route; über eine Linie, over a trunk);
Leiten n, routing.
leitend, conductive;
— machen, to render conducting.
Leiter m, conductor, leader, lead; pilot;
— f, ladder;
guter — m, good conductor;
schlechter — m, poor conductor;
mit negativem Widerstand behafteter — m, third class conductor;

Außen= — m, outer main;
Erd= — m, earth wire, ground wire;
— = —, Starkstrom=, power ground wire;
Halb= — m, semi-conductor;
Hohl= — m, tubular conductor;
Mittel= — m, neutral wire;
Null= — m, neutral wire;
Roll= — f, rolling ladder;
Schiebe= — f, travelling ladder;
Starkstrom= — m, power wire;
Wärme= — m, heat conductor;
— =bündel n, bunched conductors pl, bunch of conductors;
— =durchmesser m, conductor diameter;
— =querschnitt m, cross-sectional area of conductor;
— =stärke f, conductor diameter;
— =widerstand m, conductor resistence.
leitfähig, conductive.
Leitfähigkeit f, conductance, conductivity;
asymmetrische —, asymmetric conductance;
magnetische —, permeance, permeability;
negative —, negative conductance;
spezifische —, conductivity;
unipolare —, unilateral conductivity, unidirectional conductivity;
Wärme= —, heat conductance, heat conductivity, thermal conductivity;
— =s =normal n, conductivity standard;-
— =s — des Kupfers, copper conductivity standard.
Leit=kabel n, pilot cable;
— =rolle f, guide roller; guide pulley;
— =signal n, pilot indicator;

Leit-signal
Gruppen- — - —, pilot signal;
— **-stelle** f, routing desk;
— **-vermögen** n, conductance;
richtungsabhängiges — - —, asymmetric conductance;
spezifisches — - —, conductivity;
— **-wert** m, conductance;
negativer — - —, negative conductance;
spezifischer — - —, conductivity;
— **-widerstand** m, conduction resistance.

Leitung f, line, circuit, wire, lead; conductance; conduction (der Gase, through gases);
in der — bleiben, to hold the line F;
eine — oberirdisch führen, to run a line overhead;
an der Stangenspitze geführte —, saddle wire B;
eine — isolieren, to insulate, to disconnect a line;
zwei —en kreuzen, to cross, to transpose two wires;
mehrere Ämter in einer —, several stations upon a circuit;
— **mit erhöhter Induktivität**, (inductively) loaded circuit;
— **mit punktförmiger Ladung**, lump-loaded circuit;
— **von endlicher Länge**, finite line;
— **en am Verteiler schalten**, to cross-connect lines;
— **en vertauschen**, to cross lines;
äquivalente —, equivalent circuit;
belastete —, loaded circuit;
— —, **gleichförmig oder stetig**, continuously loaded circuit;
— —, **punktförmig**, lump-loaded circuit;   [line;
besetzte —, busy line, engaged

betriebsfähige —, perfect circuit;
dünndrähtige —, small-gauge line;
durchgehende —, through circuit;
endigende —, terminating circuit;
endliche —, finite line L;
gegabelte —, forked circuit;
gekreuzte —, transposed line;
gestörte —, faulty circuit;
gleichförmige —, smooth line;
gummiisolierte —, rubber-insulated leader;
homogene —, smooth or uniform or homogeneous line;
künstliche —, artificial line, balancing network KV;
lange —, elektrisch, electrically long line;
— **r — en, Theorie** f, long line theory;
— —, **unendlich**, infinite line L;
— —, **quasi-unendlich**, semi-infinite line, quasi-infinite line;
natürliche —, actual line;
oberirdische—, aerial line, overhead line;
starkdrähtige —, heavy (-gauge line;
unipolare —, unidirectional conductance;
unterirdische —, underground circuit;
verdrallte —, twisted line B;
verlustlose —, line of no loss L;
verzerrungsfreie —, distortionless circuit L;
wirkliche —, real line;
zusammengesetzte —, composite circuit;
a- —, a-line;
Abgreif- —, tapping lead;

Leitung
Amts= —, exchange line; junction from p. b. x. to exchange;
Anmelde= —, recording trunk F;
Ausgleichs= —, (artificial) balancing line, balancing network, compensation circuit;
Auslands= —, international (trunk) line;
Außen= —, external leads pl;
Automaten= —, coin box circuit F;
b= —, b-line;
c= —, c-wire, S-wire, third conductor, test wire, F;
Dienst= —, order wire, ab: o. w., F; service circuit, speaker wire; call wire; Kf.=L.: transfer circuit;
— = —, unmittelbare, straight order wire;
— = —, Sammel=, split order wire;
— = — zwischen zwei Plätzen desselben Amtes, transfer circuit F;
Doppel= —, metallic (return) circuit, two-wire circuit, twin conductor(s pl), looped circuit;
— = —, für ein km, per loop kilometre;
— = —, gekreuzte, transposed pair, crossed pair;
— = —, verdrallte, twisted pair;
Doppelsprech= —, duplex telephone circuit;
Drahtfunk= —, carrier line, high frequency carrier circuit;
Duplex= —, duplex circuit;
Durchgangs= —, transit circuit, through (am: thru) circuit;
Einzel= —, grounded line;

Ersatz= —, spare circuit, reserve circuit; äquivalente Leitung: equivalent circuit.
Einfach= —, simplex circuit;
Fern= —, toll circuit (engl.) long-distance circuit (am.);
— = —, mit Verstärkern ausgerüstete, repeatered toll circuit;
— = — ohne Verstärker, non-repeatered toll circuit;
— = —, Durchgangs=, through trunk, through toll line;
Fernkabel= —, toll cable circuit;
— = — mit Sprechstromverstärkern, repeatered toll cable circuit;
Fernmelde= —, communication line;
Fernsprech= —, telephone circuit;
Fernvermittlungs= —, trunk junction circuit, toll switching trunk (am.);
Feuermelder= —, fire alarm circuit;
Fort= —, conduction;
Frei= —, open line;
Gegensprech= —, duplex circuit;
Gemeinschafts= —, YQ-circuit (engl.), omnibus circuit, way circuit (am.);
Gesellschafts= —, (multi-) party line;
— = — mit 4 (10) Anschlüssen, four-(ten-)party line;
H= —, H-circuit, I-circuit;
Haupt= —, main (line);
Haupt=Licht= —, lighting mains pl;
Hauptanschluß= —, direct line (engl.), individual line (Gegensatz zu Gesellschaftsleitung); subscriber's line;

Leitung, Innen- 156 Leitung, Verbindungs-

Leitung
Innen- —, interior wiring, internal wiring; zwischen Wählern: trunk, link, A;
Kf- —, transfer circuit;
Klingel- —, bell circuit;
Ko- —, trunk junction circuit;
Kraft- —, power line;
Kraftübertragungs- —, power transmission circuit;
Kunst- —, artificial line, artificial circuit, balancing network VK;
Licht- —, lighting circuit;
— = —, Haupt-, lighting mains pl;
Luft- —, aerial line, aerial circuit, overhead line, open (-wire) circuit;
Melde- —, (Fernamts-), record(ing) trunk, record junction circuit;
Miet- —, leased wire, rented wire;
Münzsprecher- —, coin box circuit;
Nachbildungs- —, balancing network, artificial line;
Nebenanschluß- —, extension line;
Nebenstellen- —, extension circuit;
Omnibus- —, omnibus circuit, YQ-circuit (engl.), way circuit (am.);
Ortsverbindungs- —, junction, trunk A;
Privatfernsprech- —, private telephone line;
Privatnebenstellen- —, private wire;
Privattelegraphen- —, private telegraph line;
Prüf- —, test(ing) circuit, holding wire F; pilot wire K;

Pupin- —, lump-loaded circuit, coil-loaded line;
Rohr- —, piping, tubing, conduit;
Rück- —, return wire;
Schwachstrom- —, signalling circuit, communication circuit;
Simplex- —, simplex circuit;
Simultan- —, plus circuit, composite circuit;
Simultantelegraphen- —, telegraph superposed circuit;
Sp- —, (rural) phonogram circuit, rural telephone circuit;
Speise- —, feeder (circuit), supply circuit; power circuit;
— = —, Einphasenbahn-, single phase electric railway power circuit;
Sprech- —, speaker wire;
Staffel- —, series circuit, echelon circuit, T;
Stamm- —, side circuit, physical circuit, transformer circuit, combining circuit, component line;
Starkstrom- —, power (-transmission) line, power wire;
Stich- —, tie line;
Tast- —, keying circuit T;
Teilnehmer- —, subscriber's line, subscriber's loop;
— = —, kurze, zero loop;
Telegraphen- —, telegraph line;
— = — mit Sprechbetrieb, phonogram circuit;
— = —, Simultan- —, telegraph superposed circuit;
Übertragungs- —, transmission line;
Verbindungs- —, junction circuit (engl.), trunk (am.), junction, trunking circuit;
— = — zwischen Wählern desselben Amts, trunk, link;

Leitung, Verbindungs=
— = — — zweier Ämter, (interoffice) trunk (circuit);
— = —, ankommende (abgehende), incoming (outgoing) junction or trunk; in (out) junction;
— = — für Verkehr in beiden Richtungen, both-way junction;
— = —, doppeltgerichtete, bothway junction, two-way trunk circuit;
— = —, Vororts= —, surburban junction;
Vergleichs= —, (standard) reference circuit;
Verlängerungs= —, extension circuit, pad, K;
Vierdraht= —, four-wire circuit K;
Vierer= —, phantom circuit, plus circuit, duplex circuit, phantom pair, compound circuit combined circuit;
Vorort= —, suburban junction F;
Vorschalte= —, trunk junction circuit;
Wasser= —, water pipe(s pl);
— = —s =hahn m, water tap;
Zeitungs= —, news circuit;
Zimmer= —, office wiring, office cabling;
Zweidraht= —, two-wire circuit;
Leitungs=abschluß m, circuit termination (durch Übertrager, in repeating coils);
— =anlage f, wire plant;
— =äquivalent n, line equivalent;
— = — in Meilen Standardkabel, standard cable equivalent (ab: s. c. e.), line equivalent in miles of standard cable (ab: in m. s. c.);

— =aufseher m, lineman, linesman;
— =ausgleich m, line balance;
— =berührung f, contact; zeitweise — = —, tapping contact, intermittent contact;
— =bündel n, bunch of circuits, bundle of trunks; ankommendes (abgehendes) — = —, bunch of incoming (outgoing) trunks;
— =dämpfung f, line loss, line attenuation, transmission equivalent;
— =draht m, line wire;
— =eigenschaft f, line constant, line characteristic;
— =ergänzung f, extension circuit, (artificial) extension line, excess network, K;
— =feder (der Klinke), line spring (of jacks);
— =fehler m, line failure, line fault;
— =führung f, wiring, running of wires; offene — = —, open wiring; verdeckte — = —, concealed wiring;
— =geräusch n, line noise;
leitungsgerichtet, along lines;
— e Trägerwellentelegraphie f, wired wave telegraphy, highfrequency telegraphy along lines;
Leitungs=kapazität f, line (shunt) capacity;
— = — gegen Erde, wire-to-earth capacity;
— =klinke f, line jack, line switch spring;
— =konstante f, line constant, conductor constant;
— = —n pl, circuit constants pl;
— =kontaktsatz m, line contact bank A;

Leitungs
— **-kreuzung** f, line crossing; Platzwechsel: crossing or transposition of wires B;
— **-länge** f, circuit length;
— **-nachbildung** f, line balance; artificial balancing line, balancing network;
— **-netz** n, line system, network, outdoor plant;
— **-nummer** f, circuit number;
— **-plan** m, circuit plan, map of network;
— **-relais** n, line relay;
— — **-gestell** n, line relay rack;
— **-schleife** f, loop, metallic circuit;
— **-seite** f, des Hauptverteilers, line side of the main distributing frame;
— **-spule** f (des Relais), line coil (of the relay) T;
— **-theorie** f, line theory;
— **-umschaltung** f, line change;
— **-unterbrechung** f, disconnection;
— **-verlängerung** f, excess network, extension circuit, artificial extension line, K;
— **-verluste** pl, line losses pl;
— **-verzögerung** f, line lag;
— **-wähler** m, connector, final selector, final switch, A;
— — —, Folgeschalter und Relaissatz für, final sequence switch and relay set A;
— — — mit Frequenzwahl (für Gesellschaftsleitungen) frequency selecting connector A;
— — —gestell n, final switch rack;
— — —vielfach(feld) n, final selector (bank) multiple A;
— **-wasser** n, tap water;
— **-wicklung** f, line winding (des Differentialrelais, of differential relay);

— **-zeit** f, line time, circuit time;
— **-zustand** m, line conditions pl.

Leitvermögen n, conductance; magnetisches —, magnetic conductance; spezifisches —, conductivity, specific conductance.

Leitweg m, route; einen — geben, to route (über, over).

Leitwert m, admittance; conductance; susceptance; kapazitiver —, capacity susceptance; Blind- —, susceptance; Kenn- —, indicial admittance; Schein- —, admittance; — — einer Doppelleitung, line shunt admittance; Wirk- —, conductance;
— **-messer** m, conductometer.

lesbar, legible.
Lesbarkeit f, legibility; definition T.
Letten m, clay.
leuchten, to light.
Leucht-gas n, illuminating gas, coal gas;
— **-wirkungen** pl, luminous effects pl.
Leydener Flasche f, Leyden jar;
— **Flaschenbatterie** f, battery of Leyden jars.
Libelle f, level, spirit level, bubble level.
Licht n, light; **Büschel- —**, brush light;
— **-bild** n, photo, photograph;
— **-bogen** m, arc; **Bildung** f eines — —s, formation of an arc, arcing; **Entstehung** f eines — —s, arcing; **Poulsen̈scher** — —, Poulsen arc; **singender** — —, singing arc;

Lichtbogen
sprechender – = –, speaking arc;
Voltascher – = –, voltaic arc;
Gleichstrom= – = –, direct current arc;
Wechselstrom= – = –, alternating arc;
– = –**entladung** f, arc discharge;
– = –**generator** m, oscillating arc, arc generator, arc converter, R;
– = –**gleichrichter** m, arc rectifier;
– = –**löschung** f, arc extinction, arc extinguishing;
– = –**schwingungen** pl, arc oscillations pl;
– = – – I, II, III **Art**, type 1, II, III arc oscillations;
– = –**schwingungserzeuger** m, arc generator;
– = –**sender** m, arc transmitter;
– = –**sicherheit** f, non-arcing property (von Isoliermaterial, of insulating material).
– = – **zündung** f, arc ignition.

lichtelektrisch, photoelectric(al);
– **e Zelle** f, photoelectric cell.

lichtempfindlich, (photo-) sensitive, light-reactive;
Papier – **machen**, to sensitise paper;
– **e Zelle** f, light-reactive cell.

Licht-empfindlichkeit f, photo-sensitivity;
– =**fleck** m, spot of light;
– =**geschwindigkeit** f, velocity of light, electromagnetic wave velocity;
– =**leitung** f, lighting circuit, lighting mains pl;
– =**pause** f, blue print;
– =**schreiber** m, photographical recorder;
– =**sicherung** f, lamp fuse;
– =**signal** n, luminous signal;
– =**steckdose** f, light socket;
– =**strahl** m, beam (of light), light ray, luminous ray;
– =**tableau** n, luminous indicator board;
– =**zeichen** n, luminous signal;
– =**zeiger** m, spot of light.

Liebhaber m, amateur;
Funk= –, radio amateur;
– =**lizenz** f, amateur license.

liefern, to deliver, to render.
Lieferung f, delivery.
Lineal n, ruler;
Kurven= –, French curve.
linear, linear;
– **es Ansprechen** n, linearity of response.

Linie f, line;
eine – **abpfählen**, to stake out, to peg out a line;
in gerader – **mit**, in alignment with;
in eine – **bringen**, to line, to align with;
– **mit gekreuzten Leitungen**, transposition line B;
ausgezogene –, **voll**, solid line, full line, whole line;
dünne –, light line;
gestrichelte –, dashed line;
oberirdische –, overhead line;
punktierte –, dotted line;
strichpunktierte –, chain-dotted line, dash-dotted line;
versenkte –, underground line, covered line;
Außen= –, contour;
Fern= –, trunk (line), toll line;
Fernkabel= –, long-distance cable line;
Haupt= –, primary line;
Hochspannungs= –, high-voltage line;
Kenn= –, characteristic line;
Kraft= –, line of force;

Linie
 Land= —, land line;
 Neben= —, secondary line, side line;
 Null= —, zero line; zero axis;
 Orts= —, local line;
 Stangen= —, pole line;
Linien-anker m, longitudinal stay B
 — =integral n, line integral;
 — =material n, line material;
 — =relais n, line relay, main (line) relay;
 — =strom m, line signal T;
 — =umschalter m, line switchboard T;
 — =wähler m, selector; connector, final selector, A;
 Stöpsel= — = —, plug selector;
 — =zeit f, line time T;
 — =zug m, trunk route.
linieren, to line, to rule.
Linksdrall m, left-handed lay or twist.
linksdrehend, counter-clockwise, anti-clockwise.
Linksdrehung f, counter-clockwise rotation, rotation to the left.
linksgängig, left-handed, counter-clockwise.
Linksgewinde n, left-handed thread.
Linse f, lens.
Lipowitzmetall n, Lipowitz alloy (27.7 Pb, 13.3 Sn, 50 Bi, 10Cd 74.5⁰).
Lippe f, lip.
Liter n, litre (1 quart = 1.1359 litres).
Litze f, strand, stranded wire, flexible;
 Draht= —, strand;
 Emaille= —, enamelled strand;
 Hochfrequenz= —, radio cable, litzendraht; [wire;
 Kupfer= —, copper stranded
 Lahn= —, tinsel;
 — = —n =schnur f, tinsel cord;
 Litzen=draht m, composite wire litzendraht; flexible, stranded wire;
 ....=litzig, ...strand;
 sieben= —, seven strand....
Lizenz f, license;
 Amateur= —, amateur license;
 Experimentier= —, experimenter's license;
 Liebhaber= —, amateur license;
 Versuchs= —, experimenter's license;
 — =inhaber m, — =nehmer m, licensee.
Loch n, hole, aperture; pit;
 ein — herstellen, to hole;
 längliches —, slotted hole;
 Schau= —, sight opening, inspection hole;
 Stangen= —, pole hole;
 — =breite f eines Frequenzsiebes, transmitted band of frequencies, transmission range, band width of a filter;
 Bandfilter n von großer — = —, broad band filter;
 — =eisen n, hollow punch;
 — =lage f eines Filters, position of the transmission range of a filter;
 — =räumer m, drift B;
 — =stempel m, punch, die;
 — = — =satz m, gang of punches;
 — =streifen m, perforated tape or slip;
 Längs= — = —, lengthwise perforated tape;
 Quer= — = —, cross-perforated tape;
 — = — =empfänger m, receiving perforator, reperforator;
 — = — =sender m, tape transmitter; [mission.
 — = — =sendung f, tape trans-

lochen, to perforate, to punch.
Lochen *n*, perforation;
Neu= —, reperforation.
Locher *m*, puncher, perforator;
Empfangs= —, receiving perforator, reperforator;
Hand= —, puncher;
Maschinen= —, reperforator;
Tasten= —, keyboard perforator;
Wheatstone= —, Wheatstone perforator.
löcherig, pitted.
Lochung *f*, perforation.
Empfangs= —, **Neu**= —, reperforation.
locker, loose;
— **werden,** to get loose, to loosen.
lockern, (sich), to loosen, to slacken, to work loose.
Lockklingel *f*, call bell.
Löffel *m*, ladle;
Gieß= —, casting ladle.
Logarithmenpapier *n*, logarithmic (cross-section) paper.
Logarithmierung *f*, logarithmation.
logarithmisch, logarithmic(al);
— **e Spirale** *f*, equiangular spiral, logarithmic spiral.
Logarithmus *m*, logarithm;
Basis des —, base of logarithms;
gemeiner —, common logarithm;
natürlicher —, natural logarithm.
Lohn *m*, wages *pl*;
Boten= —, porterage;
Fest= —, fixed wages;
Stück= —, piece wages.
lokalisieren, to locate.
Longitudinalschwingungen *pl*, longitudinal vibrations *pl*.
lösbar, soluble; detachable.
Lösbarkeit *f*, solubility.
Lösch=decke *f*, fire-extinguishing cover;

— **=drossel** *f*, quenching choke.
löschen, to extinguish, **Lampen:** to darken; **Funken:** to quench;
Lösch=funken *m*, quenched spark;
— = **=sender** *m*, quenched spark transmitter, quenched gap transmitter;
— = — **=strecke** *f*, quenched spark gap, quench(ing) gap;
— **=gerät** *n*, fire-extinguishing appliances *pl*;
— **=kreis** *m*, quenching circuit;
— **=wirkung** *f*, quenching action.
Löschung *f*, extinction, darkening; quenching.
Funken= —, quenching of sparks;
Lichtbogen= —, arc extinction.
lose, loose, slack;
— **gekoppelt,** loosely coupled;
— **Kopplung** *f*, loose coupling;
— **werden,** to slacken, to get loose;
— **Zuführung** *f*, wandering lead.
Lose *f*, slack.
lösen, losmachen: to loosen; **ab=nehmen:** to detach, to disengage; **aufdrehen:** to untwist; *M*: to solve; in **Wasser:** to dissolve.
loslassen, to release (dem **Anker,** the armature).
loslöten, to unsolder.
losschrauben, to unscrew.
Lösung *f*, loosening, detachment; untwisting; solution, dissolution, (*cf.* **lösen**);
feste —, solid solution;
wässerige —, aqueous solution;
Lösungs=druck *m*, solution pressure;
— **=mittel** *n*, solvent.
Lot *n*, solder; vertical *M*; **Richt=lot:** plumb;
Blei= —, lead solder;
Hart= —, spelter solder, hard solder;

Lot
 Harz= —, resin solder;
 Richt= —, plumb;
 Schlag= —, spelter solder;
 Weich= —, Zinn= —, tin solder, soft solder.
Lötbrunnen m, cable joint(ing) box, jointing chamber.
loten, to plumb; Wassertiefen: to sound.
löten, to solder.
Loten n, plumbing; sounding.
Löten n, soldering.
Löter m, solderer;
 Blei= —, plumber;
 Bleikabel= —, plumber jointer;
 Kabel= —, jointer, splicer;
 — =zelt n, wireman's tent.
lotfrei, solderless.
Löt=klemme f, tag;
 — =kolben m, soldering iron;
 — =lampe f, blow lamp, soldering lamp; kleine blow torch;
 Spiritus= — —, alcohol blow torch, spirit lamp;
Lotmaschine f, sounding machine.
Löt=naht f, soldered seam, soldered joint;
 — =ofen m, charcoal brazier;
 — =öse f, tag, (flat type) soldering tab;
 mit — =' —n versehen, tagged;
 Streifen m mit 80 — =—n, 80 tag strip;
 — = —n =streifen m, tag strip, terminal strip, connection strip.
lotrecht, plumb, vertical.
Lötrohr n, blowpipe;
 — =säure f, killed spirit $(ZnCl_2 + H_2O)$;
Lotsenkabel n, pilot cable.
Löt=stelle f, soldered joint, soldered junction, splice;
 schlechte — = —, dry joint;
 Auskreuz= — = —, test-splice K;

 Kabel= — = —, cable joint, cable splice;
 Wickel= — = —, Britannia joint;
 — =stift m, tag, soldering pin, (wire type) soldering tab;
 — =topf m, charcoal brazier.
Lotung f, sounding;
 Echo= —, echo-sounding;
Lötung f, soldering.
Lotungs=gerät n, sounding apparatus.
Löt=verbindung f, soldered joint, soldered junction;
 — =wasser n, killed spirit, soldering fluid, $(ZnCl_2 + H_2O)$;
 — =wulst m (f), wipe, plumber's wiped joint.
Lücke f, interstice, void.
Luft f, air; **Luft=** .... air, atmospheric;
 komprimierte —, compressed air;
 verdünnte —, rarefied air;
 Betriebs= —, air draught;
 Druck= —, compressed air, pressure;
 — = — (=strom m) forced draught.
 Saug= —, vacuum;
 — =abzug m, vent;
 — =blase f, air bubble;
 eingeschlossene — = —, air cavity;
 — =dämpfung f, air damping.
luftdicht, air-tight;
 — verschlossen, hermetically sealed;
 —er Verschluß m, hermetical seal.
Luftdraht m, aerial, antenna;
 — =abstimm-kondensator m, aerial tuning condenser, ab: a. t. c.;
 — = —=spule f, aerial tuning inductance, ab: a. t. i.;
 — =abstimmung f, antenna tuning;

Luftdraht
— **=amperemeter** *n*, aerial ammeter;
— **=Blockkondensator** *m*, aerial blocking condenser;
— **=gebilde** *n*, aerial structure, radiating system;
— **=haspel** *m*, aerial winch;
— **=kreis** *m*, aerial circuit;
— **=spule** *f*, antenna helix, aerial inductance;
— **=träger** *m*, antenna support;
— **=umschalter** *m*, aerial changeover switch;
— **=verkürzungskondensator** *m*, shortening condenser, shortwave condenser;
— **=verlängerungsspule** *f*, aerial loading inductance, antenna load coil;
— **=winde** *f*, aerial winch;
— **=zuleitung** *f*, aerial feeder.

Luft=drossel *f*, air core choke;
— **=druck** *m*, atmospheric pressure, barometric pressure;
— **= — =....**, pneumatical;
— **=einschluß** *m*, air cavity, air bubble;

luftelektrisch, static(al);
— e Störungen *pl*, statics *pl*, atmospherics *pl*, *ab*: X.'s *pl*.

Luftelektrizität *f*, atmospheric electriciy.

lüften, to ventilate.

luftgekühlt, air-cooled.

Luft=Hohlraumkabel *n*, air-space cable;
— **=kabel** *n*, aerial cable;
— **=kanal** *m*, air duct;
— **=kern** *m*, air core;
— **=kompressor** *m*, air compressor;
— **=kondensator** *m*, air condenser.

luftleer, vacuous, evacuated;
— er Raum *m*, vacuous space.

Luftleerblitzableiter *m*, vacuum lightning arrester;

— **=leere** *f*, vacuum, evacuated state;
— **=leerfunkenstrecke** *f*, vacuum spark gap;
— **=leiter** *m*, aerial, antenna; (einseitig) gerichteter — = —, (uni-) directional aerial;
— **= — =gebilde** *n*, aerial structure, aerial network;
— **= — =system** *n*, aerial system;
— **= — =widerstand**, aerial resistance;
— **=leitung** *f*, open-wire circuit, open line, overhead line;
— **=linie** *f*, aerial line, overhead line;
— **=puffer** *m*, dash-pot;
— **=pumpe** *f*, air pump;
Quecksilber= — = —, mercurial air pump;
— **=reibungsverluste** *pl*, windage losses *pl*;
— **=rückstand** *m*, residual air;
— **=schicht** *f*, atmosphere; stratum of air;
— **=spalt** *m*, air gap, entrefer;
— **= —, Generator** *m* mit gleichförmigem, non-salient pole generator;
— **= —, Generator** *m*, mit nach den Polkanten erweitertem, salient pole generator;
— **=spule** *f*, air (core) coil, air core solenoid;
— **=störung** *f*, atmospheric disturbance, atmospherics *pl*, X.'s *pl*;
Beseitigung *f* von — = —en, X. stopping, elimination of statics;
Einrichtung *f* zur Ausscheidung von — = —en, anti-atmospheric device, X. stopper;
Verhältnis *n* der Zeichen zu den — = — en, signal-to-static ratio;
— **= —ß =....**, parasitic;
— **=strom** *m*, air blast,

Luft
— **-transformator** *m*, air core transformer.
Lüftung *f*, ventilation.
Lüftungsrohr *n*, ventilation pipe.

Luft-weg *m*, air path;
— **-widerstand** *m*, air resistance;
— **-zug** *m*, air draught.
L.W. (= **Leitungswähler**), final selector *A*;
— **-gestell** *n*, final switch rack.

M.

Made(nschraube) *f*, headless screw, grub screw.
Magnesia *f*, oxide of magnesium, magnesia (MgO).
Magnesium *n*, magnesium (Mg);
— **-oxyd** *n*, oxide of magnesium (MgO);
— **-sulfat** *n*, sulphate of magnesium (MgSO$_4$).
Magnet *m*, magnet;
induzierender —, inducing magnet;
induzierter —, induced magnet.
natürlicher —, natural magnet, loadstone, lodestone;
permanenter —, permanent magnet;
starker —, powerful magnet;
zusammengesetzter —, battery of magnets, compound magnet;
Anlaß- —, start(ing) magnet;
Anrufer- —, silencer magnet *T*;
Antriebs- —, drive magnet, driving magnet;
Auslöse- —, z. Einrücken: trigger magnet, trip magnet; z. Rückstellen: release magnet;
Brems- —, brake magnet;
Dauer- —, permanent magnet;
Dreh- —, rotary magnet *A*;
Druck- —, printing magnet, printer magnet;
Druck-Auslöse- —, printing trip magnet;

Einrück- —, trigger magnet, trip magnet, starting magnet;
Einstell- —, setting magnet;
Feld- —, field magnet;
Gleichlauf- —, correcting magnet;
Heb- —, vertical magnet *A*; lifting magnet;
Hochleistungs- —, power magnet;
Hub- —, vertical magnet (des Strowgerwählers, of the Strowger switch); stepping magnet;
Hufeisen- —, horse-shoe magnet;
Kupplungs- —, clutch magnet;
Lamellar- —, compound magnet;
Molekular- —, molecular magnet;
Polarisations- —, polarizing magnet;
Richt- —, controlling magnet; setting magnet;
Ring- —, annular magnet;
Rückstell- —, resetting magnet; release magnet;
Schalt- —, driving magnet, stepping magnet; switching magnet;
Sperr- —, locking magnet;
Stanz- —, punch(ing) magnet;

Magnet
Topf= —, pot-shaped magnet, iron-clad magnet;
Vorschub= —, feeding magnet, spacing magnet;
Wähl= —, selecting magnet;
— =achse f, magnetic axis;
— =anker m armature; des Dauermagneten: keeper;
Auslöse= — = —, release armature A;
Hebe= — = —, lifting armature A;
— =ausschalter m, field break switch;
— =detektor m, magnetic detector;
— =eisenerz n, magnetite (Fe_3O_4)
— =eisenstein m, magnetite, lodestone, black oxide of iron, (Fe_3O_4);
magnetelektrisch, magneto-electrical;
— e Maschine f, magneto (-electric machine);
Magnet=feld n, magnetic field;
ein — = — erzeugen, to produce a magnetic field;
transversales = — —, transverse magnetic field;
Erd= — = —, earth's magnetic field;
— =fluß m, magnetic flux;
— =induktion f, magnetic induction;
— =induktor m, magneto, hand generator;
— =joch n, yoke of a magnet;
— =kompaß m, magnetic compass;
— =kreis m, magnetic circuit;
— =nadel f, magnetic needle;
— =rad n, magnet wheel;
— =magazin n, battery of magnets;
— =schenkel m, magnet limb, leg;

— =spule f, coil, winding, solenoid;
— =stahl m, magnet steel;
— =zahn m, field projection.
magnetisch, magnetic(al);
erd= —, earthmagnetic;
ferro= —, ferromagnetic;
nord= —, north-magnetic;
süd= —, south-magnetic;
un= —, non-magnetic, unmagnetized;
— es Gewitter n, magnetic storm;
— es Feld n, magnetic field;
— e Feldstärke f, magnetic field intensity;
— er Fluß m, magnetic flux;
— es Gleichfeld n, constant magnetic field;
— e Kopplung f, magnetic coupling;
— e Kraft f, magnetic force, magnetic intensity;
— e Kraftlinie f, line of magnetic force, magnetic line of force;
— er Kreis m, magnetic circuit;
— er Kreisprozeß m, magnetic cycle;
— e Leitfähigkeit f, permeance, magnetic conductance;
— es Magazin n, compound magnet;
— e Masse f, magnetic substance;
— es Moment n, spezifisches, specific magnetic moment;
— er Nebenschluß m, magnetic leak;
— e Permeabilität f, magnetic permeability;
— e Polstärke f, magnetic pole strength;
— es Potential n, magnetic potential;
— e Reibung f, magnetic friction;
— e Schicht f, magnetic layer;

magnetisch
— er Schirm m, magnetic shield;
— e Schirmung f, magnetic shielding;
— e Streuung f, magnetic dispersion; [bias;
— e Vorspannung f, magnetic
— e Wage f, magnetic balance;
— es Wechselfeld n, alternating magnetic field;
— er Widerstand m, magnetic resistance, reluctance;
— e Wirkung f, magnetic effect;
— er Zyklus m, magnetic cycle.

magnetisierbar, magnetizable.
Magnetisierbarkeit f, magnetizability;
magnetisieren, to magnetize.
magnetisierende Kraft f, magnetizing force.
Magnetisierung f, magnetization;
überlagerte —, superposed magnetization;
Dauer= —, permanent magnetization;
Längs= —, longitudinal magnetization;
Quer= —, transverse magnetization;
Vor= —, superposed magnetization, magnetic bias.
Magnetisierungskurve f, magnetization curve, magnetization characteristics pl;
— =stärke f, intensity of magnetization;
— =strom m, magnetizing current, polarizing current;
— =zyklus m, magnetization cycle.

Magnetismus m, magnetism;
remanenter —, residual magnetism;
Erd= —, terrestrial magnetism;
Ferro= —, ferromagnetism.

Magnetometer n, magnetometer.
magnetometrisch, magnetometrical.
magnetomotorisch, magnetomotive;
— e Kraft f (ab: MMK), magnetomotive force, ab: m. m. f.
Magnetoskop n, magnetoscope;
magnetoskopisch, magnetoscopical.
Magnifier m, magnifier.
Mahagoniholz n, mahogany.
Makadam m, macadam.
makadamisieren, to macadamize.
Malteserkreuz n, maltese cross;
— =gesperre n, Geneva stop mechanism.
Mangan n, manganese (Mn);
— =chlorid n, manganise chloride ($MnCl_2$);
— =dioxyd n, manganese dioxide (MnO_2);
— =oxyd n, — =sesquioxyd n, manganese sesquioxide (Mn_2O_3);
— =stahl m, manganese steel;
— =superoxyd n, pebble manganese, manganese dioxide, (MnO_2).
Manganin n, manganin (84 Cu, 12 Mn, 4 Ni).
Mangel m, Fehlen: deficiency (in, bei, in; an, of), lack, shortage; Fehler: defect.
Manila-hanf m, manila hemp;
— = =seil n, manila rope;
— =papier n, Manil(l)a paper.
Mannloch n, manhole;
— =deckel m, manhole cover.
Manometer n, pressure gauge.
Mantel m, shell, jacket, Kabel: sheath(ing);
nahtloser —, seamless sheathing;
Blei= —, lead sheath;
Kühl= —, cooled jacket;
Kupfer= —, copper jacket;

**Mantel, Kupfer-
mit einem − = −versehen,** copper-jacketed;
− =**baumen** m, edge cam;
− =**klappe** f, tubular indicator;
− =**transformator** m, shell transformer, ironclad transformer.
manuell, manual.
Marineleim m, marine glue.
Marke f, label, mark.
markieren, to mark.
Markierpfahl m, marking post,
Marmor m, marble; [peg.
− =**platte** f, marble slab;
− =**schalttafel** f, marble switchboard.
− =**tafel** f, marble slab.
Masche f, mesh, interstice;
Gitter= −, grid mesh;
Kettenleiter= −, mesh of network;
Maschenwerk n, meshed network.
Maschine f, machine, engine;
mit −n herstellen oder bearbeiten, to machine;
− **schreiben,** to type, to typewrite;
Bohr= −, drilling machine;
Fräs= −, milling machine;
Hobel= −, planing machine;
Kraft= −, prime mover;
Schleif= −, grinding machine;
Maschinen-antrieb m, machine drive;
mit − = −, machine driven;
− =**charakteristik** f, speed-load characteristic;
− =**einheit** f, machine unit;
− =**elemente** pl, machine elements pl;
− =**geber** m, automatic transmitter T;
− =**geräusch** n, generator hum
− =**haus** n power house; [F;
− =**karren** m, engine cart;
− =**raum** m, power room;

− **schreiben** n, typing, typewriting;
blindes − = −, touch-typing;
− =**schreiber** m, typist;
− =**sender** m, machine transmitter; automatic transmitter T;
− =**sendung** f, automatic transmission T;
− =**telegraph** m, automatic telegraph, machine telegraph;
− =**wählersystem** n, machine switching system A.
Maschinerie f, machinery.
maskieren, to mask.
Maskierung f, masking;
Gehör= −, auditory masking.
Maß n, measure; dimension; rate;
− =**einheit** f, unit (of measure), standard of measurement;
abgeleitete − = −, derived unit;
absolute − = −, absolute unit;
Außen= −, overall dimension, outer dimension;
Flächen= −, superficial measure;
Innen= −, inner dimension;
Längen= −, linear measure (cf. **Längen**= ...)
− =**nahme** f, means pl, measure;
− =**stab** m, rule; scale, rate;
in großem − = −**e,** on a large scale;
in kleinem − = −**e,** on a small scale;
ungefährer − = − 1:6, scale about $1/6$;
in vergrößertem − = −**e zeichnen,** to draw to an enlarged scale;
in verkleinertem − = −**e,** on a reduced scale.
Masse f, mass, substance;
Hauptmasse: bulk; **Mischung:** compound, composition;
aktive −, active paste;

Masse
magnetische —, magnetic substance;
Füll= —, filling paste;
Verguß= —, sealing compound;

Massen-anziehung f, gravitation;
— =einheit f, unit of mass.

Masse-kern m, compressed iron powder core;
— =platte f, pasted plate.

Massentelegraph m, high-capacity telegraph.

massiv, solid.

Mast m, mast, pole, post;
freitragender —, tower, self-supporting mast;
Beton= —, concrete pole;
Eisen= —, iron pole;
Funk= —, radio tower, radio mast;
Gitter= —, lattice mast, lattice(d) pole;
— = —, Eisen=, iron lattice pole;
— = —, Holz=, wood lattice pole;
Kuppel= —, coupled poles pl;
Rohr= —, tubular pole, tube pole;
Stahlrohr= —, tubular steel pole, wrought steel pole;
Stumpf= —, stub mast;
Teleskop= —, telescopic mast;
— =fundament n, pole foundation, mast foundation;
— =fuß m, pedestal, pole footing;
— =karren m, mast cart;
— =schalter m, pole switch.

Material n, material;
— =verbrauch m, consumption of materials;

Materie f, matter.

Mathematik f, mathematics pl.
Mathematiker m, mathematician.
mathematisch, mathematical.
Matrize f, matrix, die;

Stanz= —, cutting die plate;
Stempel= —, punching die T.

matt, glanzlos: mat; Licht: dull.
mattieren, to mat.
Matte f, mat.
Mauer f, wall;
— =bohrer m, stone drill;
— =bügel m, (wall) bracket;
— =kanal m, wall channel;
— =stütze f, wall bracket;
— =werk n, masonry, brick work;
Ziegel= — = —, brick work.

mauern, to brick (up).

Maxima und Minima pl, maxima and minima.

Maximal-amplitude f, maximum amplitude;
— =ausbeute f, maximum output;
— =ausschalter m, overload circuit breaker, maximum cutout;
— =wert m, maximum value.

Maximum n, maximum.

Maxwell n, maxwell;
— =anordnung f, shunted condenser, reading condenser, T;
— =erde f, Maxwell earth T;
— =schaltung f, shunted condenser, reading condenser, Maxwell earth, T.

Mechanik f, mechanics pl.
Mechaniker m, mechanic.
mechanisch, mechanic(al);
— e Kraft f, mechanical force;
Einheit der —en —, unit mechanical force;
— es Moment n, momentum.

Mechanismus m, mechanism.
Medium n, medium (pl. media);
Übertragungs= —, transmitting medium.

Megaphon n, megaphone.
Megger m, megger;
Brücken= —, bridge megger.
Megohm n, megohm.

mehrarmig, multi-arm(ed).
mehrdrähtig, multiple-wire....
mehrfach, multiple.
Mehrfach=anschlußbündel *n*, p. b. x. junction group;
— =fernsprechen *n*, multiple telephony;
— =funkenstrecke *f*, multiple spark gap;
— =leitungswähler *m*, private branch exchange final selector, p. b. x. final selector;
— =luftleiter *m* mit abgestimmten Zweigen, multiple tuned aerial;
— =schaltung *f*, multiple connection;
in — = — betreiben, to multiplex;
— =senden *n*, multiple transmission *R*;
— =telegraph *m*, multiplex telegraph, multi-channel telegraph, multiple-way telegraph;
— = — in Gabelschaltung, split multiplex telegraph, forked multiplex telegraph;
— = — in Staffelschaltung, series multiplex telegraph;
— = — mit abgestimmten Wechsel=strömen, harmonic multiple telegraph;
— =telephonie *f* mit hochfrequenten Trägerströmen, h. f. multiple telephony;
— =verteiler *m*, multiplex distributor, multi-channel distributor, *T*;
— =verstärker *m*, multi-stage amplifier;
— =Zwillingskabel *n*, multiple twin cable, m. t. cable.
Mehrgitterröhre *f*, multiple grid valve.
mehrgliedrig, multi-mesh;

— er Kettenleiter *m*, multi-mesh network.
mehrlagig, multi-layer.
Mehrnabeltelegraph *m*, multiple needle telegraph.
mehrphasig, polyphase.
Mehrröhrenverstärker *m*, multi-valve amplifier.
mehrstufig, multistage.
mehrteilig, multisectional.
mehrwegig, multi-way, multi-channel;
— er Schalter *m*, multi-point switch;
— er Telegraph *m*, multi-channel telegraph.
mehrzügiges Formstück *n*, multiple tile *B*;
mehrzügiger Kanal *m*, multiple-way duct *B*.
Meile *f*, mile;
englische —, British mile, statute mile (= 1,60933 km);
See= —, nautical mile (= 1,854965 km).
Meilenlänge *f*, mileage.
Meisel *m*, chisel.
meiseln, to chisel, to chip.
Melde=amt *n*, record section *F*;
— =beamtin *f*, record (table) operator, recorder, recording operator, *F*;
— =lampe *f*, alarm lamp, pilot lamp;
— =leitung *f*, toll record circuit, recording trunk, *F*;
— =platz *m*, record position *F*;
— =relais *n*, pilot relay, supervisory relay;
— =sicherung *f*, alarm type fuse;
— =spitzenplatz *m*, record transfer position *F*;
— = — beamtin *f*, record transfer operator *F*;
— =tisch *m*, record table;
— =zeichen *n*, indicator, alarm signal;

Melde-zeichen
Gruppen- – – –, pilot indicator, pilot signal.
melden, to signal.
Melder *m,* signal, alarm;
  Einbruchs- –, burglar alarm;
  Feuer- –, fire alarm.
Membran *f,* diaphragm, membrane;
  Kohle- –, carbon diaphragm;
  Weicheisen- –, ferrotype diaphragm.
Menge *f,* quantity, portion, amount.
Mennige *f,* minium, red lead, (Pb_3O_4).
Meridian *m,* meridian.
Meridionalebene *f,* meridional plane.
merkbar, appreciable.
Merk-mal *n,* characteristic;
— **-punkt** *m,* point-de-repère, **(Baudot)** *T*;
— **-zeiger** *m,* adjustable index, indicator needle.
merzerisiert, mercerised.
Merzerisierung *f,* mercerisation.
Meß-amt *n,* testing office;
— **-band** *n,* surveyor's chain;
meßbar, measurable;
Meß-batterie *f,* testing battery;
— **-bereich** *m,* (measuring) range;
— **-brücke** *f,* measuring bridge;
  Gleichstrom- – – –, direct current bridge, d. c. bridge;
  Kapazitäts- – – –, capacity bridge;
  Schleifdraht- – – –, (differential) slide wire measuring bridge;
  Wechselstrom- – – –, alternating current bridge, a. c. bridge;
— **-einrichtung** *f,* measuring device;
— **-ergebnis** *n,* test reading;

— **-genauigkeit** *f,* precision of the test;
— **-karren** *m,* testing cart;
— **-kette** *f,* surveyor's chain;
— **-kondensator** *m,* measuring condenser;
— **-kreis** *m,* testing circuit;
— **-methode** *f,* testing method;
— **-röhre** *f,* graduated tube;
— **-spannung** *f,* measuring voltage;
— **-strom** *m,* testing current, measuring current;
— **– – -kreis** *m,* testing circuit;
— **– – -stärke** *f,* testing current intensity;
— **-technik** *f,* testing technique;
  Fernsprech- – – –, telephonometry;
— **-tisch** *m,* test desk;
— **-verfahren** *n,* testing method;
— **-wert** *m,* measured value, test value;
— **-zeit** *f,* testing time.
messen, to test, to measure;
  eine Leitung auf Isolation und Leitfähigkeit –, to test a line for insulation and conductivity.
Messer *n,* knife; — *m* meter, gauge;
  Druck- – *m,* pressure gauge;
  Frequenz- –, frequency meter;
  Geräusch- –, noise measuring set;
  Lautstärke- –, *m,* audibility meter;
  Nebensprech- –, crosstalk meter;
  Wellen- – *m,* wave meter;
— **-kontakt** *m,* knife blade contact, switch jack;
  Wähler *m* **mit – – -en,** jack mounted selector;
— **– – -backen** *pl,* contact jaws *pl*;
— **-schalter** *m,* knife blade switch;

Meſſer
— **ſchneide** *f*, knife edge.

Meſſing, brass;
— **-band** *n*, brass tape;
   mit — - — umwickelt, brass-taped;
— - — **-umlappung** *f*, brass taping;
— **-gehäuſe** *n*, brass case;
— **-guß** *m*, cast brass;
— **-klemme** *f*, brass terminal;
— **-preßſtück** *n*, brass die pressing;
— **-ſchraube** *f*, brass screw;
— **-ſtöpſel** *n*, brass peg.

Meſſung *f*, testing; test, measur-
falſche —, mis-test; [ement;
regelmäßige —, routine test;
Abnahme- —, acceptance test, factory test;
Entdämpfungs-Frequenz- —, gain-frequency test;
Fabrik- —, factory test;
Fehlerorts- —, fault location test;
Früh- —, morning test;
— - —, regelmäßige, morning routine test;
Gleichſtrom- —, direct current measurement;
Schleifen- —, (**Erdfehler-**), loop test;
Spannungs- —, voltage measurement;
Strom- —, current measurement;
Überwachungs- —, maintenance test;
Wechſelſtrom- —, alternating current test, a. c. test.

Metall *n*, metal;
ganz aus —, all-metal;
edles —, nobler metal;
unedles —, baser metal;
Edel- —, nobler metal;
Lager- —, anti-friction metal;
— **-auflage** *f*, metal coating;
— **-band** *n*, metallic ribbon, metal tape;
— **-einlage** *f*, — **-einſatz** *m*, metal insert;
— **-faden** *m*, metallic filament;
— - — **-lampe** *f*, tungsten lamp;
— **-gehäuſe** *n*, metal(lic) case;
   im — - —, metal-cased;
— **-kappe** *f*, metal cover;
— **-kugel** *f*, metal ball;
— **-ſchlauch** *m*, metallic hose;
— **-ſchraube** *f*, metal screw;
— **-ſockel** *m*, metal base;
— **-überzug** *m*, metal coating;
metalliſch, metallic(al);
— **klingend**, metallic.

metallographiſch, metallographical.

Meter *n*, metre (*ab*: m = 3.2809 ft.);
Kubik- —, cubic metre (*ab*: cbm, m³; 1 cbm = 35.317 cub. ft.);
Quadrat- —, square metre (*ab*: qm, m²; 1 qm = 10.764 squ. ft.);
— **-ampere** *n*, meter/ampere;
— - — **-zahl** *f*, radiation constant R;
— - — - — geteilt durch die **Wellenlänge**, radiation factor R.

Methode *f*, method.
metriſch, metrical.
Mezzoſopran *m*, mezzo-soprano.
mieten, to lease.
Mietleitung *f*, leased wire, rented wire.
Mikanit *n*, micanite, built-up
Mikarta *f*, micarta. [mica.
Mikro-ampere *n*, micro ampere;
— **-analyſe** *f*, micro-analysis;
— **-coulomb** *n*, micro coulomb;
— **-farad** *n*, microfarad, mf(d);
mikrographiſch, micrographic(al)
Mikro-henry *n*, microhenry;
— **-meter** *n*, micrometer;
Funken- — - —, micrometric spark discharger, spark micrometer.

Mikro=meter
— = — =schraube *f*, micrometric screw, micrometer gauge, micrometer screw;
— =ohm *n*, microhm;
Mikrophon *n*, transmitter, microphone;
— = ..., microphonical;
— mit fester Rückwand, solid back transmitter;
Aufnahme= —, pick-up transmitter *R*;
Brust= —, breastplate transmitter;
Doppel= —, double button transmitter;
Druck=Zug= —, push-pull transmitter; double button transmitter;
Flammen= —, flame transmitter;
Flüssigkeitsstrahl= —, liquid jet transmitter;
Kapsel= —, inset transmitter, button transmitter;
— = —, Doppel=, double button transmitter;
Kohle= —, carbon transmitter;
Kohlenbeutel= —, carbon bag transmitter;
Kohlenkörner= —, carbon granule transmitter;
Kohlenpulver= —, carbon powder transmitter;
Kohlenstaub= —, carbon dust transmitter;
Kondensator= —, condenser transmitter;
Kontakt= —, contact microphone;
Körner= —, granular transmitter;
Lausch= —, pick-up transmitter;
Starkstrom= —, (high) power transmitter;

Unterwasser= —, hydrophone;
Walzen= —, pencil transmitter;
— =arm *m*, transmitter arm;
— =batterie *f*, transmitter battery, speaking battery;
— =einsatz *m*, transmitter inset;
— =geräusch *n*, side tone;
Schutzeinrichtung gegen — = —, anti-side tone device;
— =kapsel *f*, microphone button, transmitter inset, resistance cell;
— =kohle *f*, microphonic carbon;
— =normal *n*, transmitter standard;
— =relais *n*, receiver-transmitter amplifier;
— =speisung *f*, speaking current supply;
— =strom *m*, transmitter current;
— = — =kreis *m*, transmitter circuit;
— =summer *m*, microphone hummer; howler;
— =träger *m*, transmitter arm;
— =verstärker *m*, microphone amplifier.
mikrophonisch, microphonic(al).
Mikrophotographie *f*, photomicrograph(y).
mikrophotographisch, photomicrographical.
Mikro=siemens *n*, micromho;
— =skop *n*, microscope;
mikroskopisch, microscopic(al);
— =telephon *n*, microtelephone, telephone handset, combination;
— =volt *n*, microvolt;
— =watt *n*, microwatt.
Milchglas *n*, frosted glass.
milchig, milky.
Milchsaft *m* (der Gummipflanzen), rubber latex.
Milli=ampere *n*, milliampere;
— = — =meter *n*, milliammeter;

Milli
— -henry *n*, millihenry;
— -meter *n*, millimetre (*ab*: mm; 1 mm = 0.03937 in.).
Millionensystem *n*, million system *A*.
Milli-volt *n*, millivolt;
— -watt *n*, milliwatt.
Mindest-ansprechstrom *m*, minimum operating current.
— -maß *n*, minimum; auf das — = — zurückführen, to minimise;
— -strom *m*, minimum current;
— -wert *m*, minimum value.
Mineral *n*, mineral.
Minimal...., minimum...;
— -amplitude *f*, minimum amplitude;
— -ausschalter *m*, minimum cut-
Minuendus *m*, minuend. [out.
Minuszeichen *n*, minus sign.
Minute *f*, minute.
Minutenzeiger *m*, minute hand.
mischen, to mix, to blend, to compound; Verkehr: to merge *A*.
Mischen *n* des Verkehrs, merging of traffic *A*.
Mischwähler *m*, load distributing switch *A*.
Miß-klang *m*, discord;
— -ton *m*, jar;
— -verhältnis *n*, asymmetry, disproportion;
— -weisung *f*, deviation from the true bearing.
Mithöreinrichtung *f*, monitoring device.
mithören, to overhear; to monitor *F*; to tap, to listen in *R*.
Mithör-klinke *f*, monitor(ing) jack;
— -taste *f*, monitoring key;
— -übertrager *m*, monitoring coil.

mitklingen, to resonate.
Mitlaufwähler *m*, simultaneous movement selector, companion work switch, *A*.
Mitlauter *m*, consonant.
Mitlese-apparat *m*, control instrument, leak instrument, *T*;
— -drucker *m*, control printer;
— -streifen *m*, home record;
— -stromkreis *m*, leak circuit.
Mitnehmer *m*, striker;
— -scheibe *f*, driving disc;
— -schiene *f*, striker bar.
mitschwingen, to resonate;
— d, resonant;
nicht — d, non-resonant.
Mitsprechen *n*, overhearing *F K*;
Mitsprech-kopplung *f*, crosstalk path, phantom-to-side unbalance, *F K*.
Mitte *f*, centre, centre point; der Differentialspule: split point, centre tapping point; auf die — einstellen, to centre.
mitteilen, to communicate; einen Zustand: to impart (to).
Mittel *n*, medium (*pl* media); Hilfsmittel: means *pl*, Gerät: appliance; agent;
arithmetisches —, arithmetic mean;
geometrisches —, geometric
mittelbar, indirect; [mean;
Mittelgang *m*, gangway, main aisle;
mittelgroß, medium-sized;
Mittel-lage *f*, mid-position;
— -linie *f*, axis, centre line;
mittelmäßig, medium;
Mittel-punkt *m*, centre, centre point;
Zonen- — = —, zone centre *F*;
— -reihe *f*, central row;
— -schiene *f* der Taste, centre of key;
— -stellung *f*, mid-position;

Mittel
— -**teil** *m*, mid-portion;
— -**wert** *m*, mean value, average (value);
  auf einen — - — **bringen,** to equate;
  um einen — - — **schwingen,** to oscillate about an average value;
  quadratischer — - —, virtual value, r. m. s. value, root mean square value.
mittlere(r), mean, average; medium, mid-....
M. M. K. = **magnetomotorische Kraft,** magnetomotive force, m. m. f.
Modell *n*, model, pattern;
  Arbeits- —, working model.
Modul *m*, modulus.
Modulation *f*, modulation;
  — **der Sprache,** inflection of the voice;
  doppelte —, double modulation;
  gegenseitige —, intermodulation;
  prozentuale —, percentage of modulation;
  übertriebene —, overmodulation;
  Einseitenband- —, single side band modulation;
  Gitter- —, grid modulation;
  Parallelröhren- —, Heising modulation, choke control modulation, constant current modulation;
  Reihenröhren- —, constant potential modulation;
  — **durch Änderung der Anodenspannung,** plate modulation;
  — — — — **Gitterspannung,** grid modulation;
  — **in Röhrenabsorptionsschaltung,** valve absorption modulation;

Modulations-frequenz *f*, modulating frequency;
  — -**grad** *m*, amount of modulation, degree of modulation;
  — -**röhre** *f*, modulating valve;
  — -**spannung** *f*, modulating voltage;
  — -**strom** *m*, modulating current;
  — -**welle** *f*, wave of modulation.
Modulator *m*, modulator;
  besprochener —, voice-actuated modulator;
  Röhren- —, valve modulator;
  Zweiröhren- — **in Gegenschaltung,** balanced modulator;
  — -**röhre** *f*, modulating valve;
  — -**schaltung** *f*, modulating circuit, translating circuit.
modulieren, to modulate;
  über- —, to overmodulate.
moduliert, modulated;
  sprach- —, speech-modulated;
  — - —**e ungedämpfte Wellen** *pl*, speech-modulated continuous waves, type A 3 waves, *pl*.
Molekel *f*, **Molekül** *n*, molecule.
molekular, molecular.
Molekularmagnet *m*, molecular magnet.
Molybdän *n*, molybdenum (Mo).
Moment *m*, moment, instant;
  — *n* momentum;
  mechanisches —, momentum.
Momentanwert *m*, instantaneous value.
Momentschalter *m*, quickaction switch, quick break switch.
Monotelephon *n*, monotelephone.
Montage *f*, erection, fitting, mounting;
  — -**platte** *f*, mounting plate.
Monteur *m*, fitter, assembler.
montieren, to mount, to erect, to fit, to assemble.
Moorboden *m*, moor(land), marshy soil.

Morse=alphabet n, Morse code;
amerikanisches = —, American Morse code;
internationales — = —, Continental Morse code, land line Morse code;
Kabel= — = —, cable Morse code;
— =farbschreiber m, (Morse) inker
— =lampe f, flash lamp;
— =punkt m, dot;
breite — = —e pl, lengthened dots pl;
spitze — = —e pl, clipped dots pl;
— =streifen m, Morse slip;
— =strich m, dash;
gebrochene — = —e pl, split dashes pl;
— =system n, Hand=, key Morse system;
— = —, Schnell=, automatic Morse system;
— =taste f, Morse key;
— = — mit selbsttätiger Punktgebung, vibroplex key;
— =zahlen pl, abgekürzte, contracted Morse figure signals pl;
— =zeichen n, mark, Morse signal;
die — = — brechen, the marks split.

Mörtel m, plaster, mortar;
Zement= —, cement mortar.

Motor m, motor, mover, engine;
Asynchron= —, asynchronous motor;
Benzin= —, petrol engine;
Feder(kraft)= —, spring motor;
Gas= —, gas engine;
Gegenverbund= —, differential compound wound motor;
Gewichts= —, weight drive;
Gleichstrom= —, d. c. motor;
Hauptschluß= —, series (-wound) motor;
Induktions= —, induction motor;

Kommutator= —, commutator motor;
Nebenschluß=—, shunt (-wound) motor;
Reihenschluß= —, series (-wound) motor;
Synchron= —, synchronous motor;
Verbrennungs= —, (internal) combustion engine;
Verbund= —, compound (-wound) motor;
— =anlasser m, starter;
— =anlaßschalter m, motor starting switch;
— =antrieb m, motor drive;
mit — = —, motor driven;
— =generator m, motorgenerator;
Eingehäuse= — = —, dynamotor;
— =rad n, motor bicycle;
— =wagen m, motor car;
— =winde f, motor winch, power-driven winch.

Muffe f, socket, sleeve, bushing
Abschluß= —, cable head, (cable) pothead;
Abzweig= —, parallel jointing sleeve;
Blei= —, lead sleeve;
End= —, pothead terminal;
Kabel= —, sleeve, joint box;
Kondensatoren= —, condenser sleeve K;
Reduktions= —, reducing bush, reducing socket;
Röhren= —, pipe socket;
Verzweigungs= —, cable distribution plug.

Muffelofen m, muffle furnace.
Muffen=ende n eines Rohres, socket end of a tube;
— =rohr n, socket tube;
— = —Spitzende n, spigot end of a tube;

Muffen-rohr
— ⸗ — ⸗**verbindung** *f,* spigot (and socket) joint;
— ⸗**verbindung** *f,* sleeve joint.
Mulde *f,* tray.
Multiplikatorrahmen *m,* multiplier coil.
multiplizieren, to multiply.
Multizellular⸗ . . . ., multicellular.
Mundstück *n,* mouthpiece.
Münzbehälter *m,* coin receptacle, coin (collecting) box,
Münze *f,* coin. [cash box.
Münz-einwurf *m,* coin slot;
— ⸗**fernsprecher** *m,* coin collector telephone station, coin box call-office, unattended call office, pay station (*am.*), public call office;
— ⸗ — **für verschiedene Geldsorten,** multi-coin box call station;

— ⸗**sprecherleitung** *f,* coin box circuit.
Muschel *f,* cap;
 Hör⸗ —, earpiece, ear cap.
muscheliger Bruch *m,* conchoidal fracture.
Musik *f,* music;
— ⸗**instrument** *n,* musical instrument.
musikalischer Ton *m,* musical note, musical tone.
Muster *n,* sample, pattern, specimen.
Mutter *f,* nut;
 Flügel⸗ —, wing(ed) nut, butterfly nut;
 Gegen⸗ —, lock nut, clamping screw;
 Kordel⸗ —, milled nut;
 Schrauben⸗ —, screwed nut;
— ⸗**schlüssel** *m,* wrench, spanner, nut key;
— ⸗**schraube** *f,* bolt and nut.

N.

Nabe *f,* boss, hub.
Nachbar-schaft *f,* vicinity;
— ⸗**zone** *f,* adjacent zone.
nachbohren, to rebore.
nachbilden, to imitate, to simulate;
 genau —, to simulate closely.
Nachbilden *n,* simulation, balance.
Nachbildung *f,* simulation, imitation;
 Leitungs⸗ —, line balance; balancing network, artificial balancing line, *K*;
Nachbildungs-Frequenzbereich *m*- frequency range of simulation;
— ⸗**gestell** *n,* (balancing) network rack.

nacheichen, to recalibrate, to check (the calibration).
Nacheichung *f,* check(ing), check calibration, recalibration.
nacheilen, to go behind, to run slow; Strom: to lag (behind); um n^0 —, to lag by n^0.
nacheilend, behind; lagging (um, by).
Nacheilung *f,* lag, lagging;
 hysteretische —, hysteretic lag;
 zeitliche —, time lag.
nachfüllen, to fill up, to top up.
Nachfüllen *n,* topping-up, filling-up.
nachgeben, to yield.
nachgehen, to run slow, to go behind.
nachhallen, to reverberate.

Nachhallen n, reverberation.
Nachhallzeit f, time of reverberation, reverberating time.
Nachladung f, additional charge.
nachlassen, to slacken; to yield.
nachlaufen, to go behind, to lag.
nachprüfen, to check.
Nachprüfen n, Nachprüfung f, checking, check.
nachregeln, to readjust.
Nachregelung f, readjustment.
Nachricht f, communication, intelligence;
Nachrichten-technik f, communication art;
— -truppe f, signal corps;
— -übermittlung f, — -übertragung f, transmission of intelligence.
nachsuchen, to apply (bei, to; um, for).
Nachsuchende(r) m, applicant.
Nacht-belastung f, night load;
— -dienst m, night service;
— - -beamtin f, night operator;
— - -platz m, night service position;
— -gebühr f, night rate;
— -reichweite f, night range R;
— -wecker m, night bell, night alarm;
— -zentralschalter m, night concentrator.
Nachübertrager m, (repeater) output transformer, outlet transformer.
Nachwirkung f, after-effect, hysteresis;
dielektrische —, dielectric fatigue, dielectric hysteresis;
magnetische —, magnetic fatigue, magnetic after-effect;
Nachwirkungsverluste pl, hysteretic losses pl.
nachziehen, Schrauben, to screw up, to tighten screws.

Nadel f, needle;
Abfühl- —, pecker, selecting needle, T;
Magnet- —, magnetic needle;
Wähler- —, selecting needle, pecker, T;
— -ausschlag m, needle throw;
— -paar n, astatisches, astatic couple, two compound needles;
— -telegraph m, needle telegraph;
Doppel- — - —, double needle telegraph;
Ein- — - —, single needle (telegraph);
Mehr- — - —, multi-needle telegraph.
Nagel m, nail.
Nah- ...., short-range ....;
— -fernverkehr m, short-haul toll traffic (am.);
— -verkehrsamt n, toll exchange F.
Nähe f, proximity.
nähern (sich), to approach; to approximate M.
Näherungs-formel f, approximate formula;
— -größe f, approximate quantity;
— -wert m, approximate value, approximation.
Naht f, seam;
Löt- —, soldered seam;
Schweiß- —, welded seam.
— -schweißung f, seam welding.
Näpfchen n, cup.
Nasalkonsonant m, nasal consonant.
Nase f, nose, snug, Klaue: catch.
naß, wet;
nasses Element n, wet cell.
Natrium n, sodium (Na);
kohlensaures —, soda (Na_2CO_3);
— -karbonat n, carbonate of soda (Na_2CO_3);

Natrium
— **-thiosulfat** *n*, thiosulphate of sodium ($Na_2S_2O_3$).
natürliche Induktivität *f*, natural inductance.
Nebel *m*, fog.
nebelig, foggy.
Nebenamt *n*, minor office; minor exchange *F*.
Nebenanschluß *m*, extension (station) *F*;
— **-leitung** *f*, extension (circuit).
nebeneinander schalten, to join in parallel, to connect in parallel.
Nebeneinanderschaltung *f*, parallel connection, multiple connection.
Nebenentladung *f*, lateral discharge.
Nebenkopplungen *pl*, **kapazitive**, stray capacity.
Nebenfernamt *n*, sub-zone toll office;
Nebenlinie *f*, side line, secondary line, spur (from main) line.
Nebenschließung *f*, leak, leakage;
— **-weg** *m*, leakage path;
— **-widerstand** *m*, leak resistance, shunt resistance.
Nebenschluß *m*, shunt, sink; leak, leakage;
im — zu, in shunt with, shunted across;
in den — legen zu, to (put in) shunt to;
mit einem — behaftet, leaky;
mit einem — versehen, to shunt;
einen — bilden zu, to shunt;
ohne —, unshunted;
Ayrtonscher —, compensating resistance, universal shunt box;
induktiver —, inductive shunt;
magnetischer —, magnetic shunt, (electro)magnetic leak, inductive shunt;
Gitter- —, grid leak (resistance);
Leitungs- —, leakage;
Resonanz- —, resonant shunt;
Wetter- —, weather leakage;
— **-dynamo** *f*, shunt (-wound) dynamo;
— **-erregung** *f*, shunt excitation;
— **-motor** *m*, shunt (-wound) motor;
— **-regler** *m*, shunt regulator;
— **-resonanz** *f*, parallel resonance;
— **-widerstand** *m*, leak resistance.
Nebensprechen *n*, crosstalk *K*;
Gegen- —, far end crosstalk;
— **am Anfang**, near end crosstalk.
Nebensprechdämpfung *f*, crosstalk transmission equivalent.
nebensprechfrei, crosstalk-proof;
Nebensprech-kopplung *f*, crosstalk circuit, crosstalk path, unbalance;
Messung *f*, **der — - —en**, unbalance test;
— **-messer** *m*, crosstalk meter, crosstalk measuring set;
— **-messung** *f*, crosstalk measurement;
— **-strom** *m*, unbalance current, crosstalk current;
— **-weg** *m*, crosstalk path.
Nebenstelle *f*, extension station;
Amts- —, exchange extension set;
Außen- —, external extension;
Fernsprech- —, extension set;
Post- —, exchange extension set;
Privat- —, private telephone;
Nebenstellen-klappe *f*, extension indicator;
— **-klinke** *f*, extension line jack;

Nebenstellen
— -leitung f, extension line;
— -umschalter m, substation switchboard;
— -zentrale f, private branch exchange, ab: p. b. x.;
  Selbstanschluß- — - —, private automatic branch exchange, ab: p. a. b. x.
Nebenuhr f, auxiliary clock.
Nebenweg m, by-path.
Nebenwinkel m, adjacent angle, adjoining angle.
negativ, negative (gegen, with respect to);
— -er Widerstand m, negative resistance, third-class resistance;
— -e Zahl f, negative number.
neigen, (sich), to incline, to tend; kippen: to tilt, sich senken: to slope.
Neigung f, inclination, tendency; obliquity, slope, dip;
— der Kurve, slope of a curve;
Neigungs-grad m, obliquity;
— -winkel m, angle of slope.
Nenner m, denominator;
  gemeinsamer —, common denominator.
Nenn-reichweite f, nominal range;
— -wert m, nominal value.
Neon n, neon (Ne);
— -lampe f, neon lamp.
— -röhre f, neon tube.
Nernst-brenner m, Nernst needle;
— -lampe f, Nernst lamp.
Netz n, net, netting; network, external plant, FT;
  Draht- —, wire netting.
  (Erd)draht- —, earthed netting B; ground mat R;
  Fernkabel- —, long-distance cable system, toll cable system; [system];
  Fernleitungs- —, toll line

Fernsprech- —, telephone network;
Funk- —, radio system, radio network;
Gleichstrom- —, direct current supply;
Schnellverkehrs- —, no-delay telephone network; no-delay traffic area;
Schutz- —, protecting network;
Starkstrom- —, public supply, public mains pl;
Verbindungsleitungs- —, junction network;
— -anschluß m, public supply, commercial current supply;
— -gestaltung f, network layout;
— -gruppe f, subzone network A;
— -plan m, network plan; network map;
— -spinne f, junction network;
— -überwachung f, transmission maintenance work.
Netzwerk n, network (v. Kettenleiter).
Neubildung f, recreation, regeneration.
neu einregeln, to re-regulate.
Neueinregelung f, re-regulation.
neu einstellen, to readjust.
Neueinstellung f, readjustment;
neu stanzen, to repunch T.
Neusilber n, German silver (4 Cu 2 Ni 1 Zn), argentan.
neutral, neutral; non-polarized;
  ein Relais — einstellen, to set neutral a relay;
— eingestellt, neutrally adjusted;
— es Relais n, non-polarized relay.
neutralisieren, to neutralize, to compensate (for), to balance out.
Neutralisierung f, neutralization, balancing-out.

Neutralstellung *f*, neutrality; neutral adjustment.
Neutralität *f*, neutrality.
Neutrodyn-...., neutrodyne.
neuzeitlich, modern, up-to-date.
Neuzündung *f*, re-ignition.
Nf. = **Niederfrequenz** *f*, low frequency, *ab*: l. f.;
— **-Sperrkreis** *m*, high-pass selective circuit.
nichtig, void.
Nichtleiter *m*, insulator.
nichtrostend, rust-proof, rust-free.
Nickelin *n*, nickelin.
niederbrechen, to break down.
— **-drücken** *n*, to depress;
 eine Taste — - —, to depress, to strike, to touch a key.
Niederdrücken *n*, depression.
niederfrequent, low-frequent, low-frequency....
Niederfrequenz *f*, low frequency, *ab*: l. f.;
— **-drossel** *f*, low-frequency choke;
— **-siebgebilde** *n*, low-pass selective circuit;
— **-sperrkreis** *m*, high-pass filter (circuit);
— **-verstärker** *m*, low-frequency amplifier, note amplifier;
— **-verstärkung** *f*, l. f. amplification.
niederholen, to lower.
niederohmig, low-resistance....
niederperiodig, low-frequent.
Niederschlag *m*, deposit, deposition;
 galvanischer —, electro-deposition.
niederschlagen, to deposit, to precipitate;
 galvanisch, to electro-deposit.
Nieder-schlagung *f*, deposition, precipitation.

— **-spannung** *f*, low pressure, low tension;
— - —s -seite *f*, low-tension side, l. t. side.
niedrig, low.
Niet *m*, **Niete** *f*, rivet.
nieten, to rivet.
Niet-kopf *m*, rivet head.
— **-verbindung** *f*, rivet joint.
Niobium *n*, niobium (Nb).
Nische *f*, niche.
Niveau *n*, level;
 Energie- —, power level;
 Übertragungs- —, transmission level;
— **-karte** *f*, level diagram, level chart;
— **-linie** *f*, level;
— - —n -diagramm *n*, level diagram.
nivellieren, to level.
Nivellierschraube *f*, levelling screw.
Nocke *f*, **Nocken** *m*, cam;
Nocken-kontakt *m*, cam contact, cam springs *pl*;
— **-satz** *m*, battery of cams;
— **-welle** *f*, cam spindle.
Nomogramm *n*, straight-line chart, self-computing chart.
Nonius *m*, vernier.
Nordlicht *n*, aurora borealis.
nordmagnetisch, north-magnetic.
nördlich, northern;
— er Breite, n^0, latitude n^0 N.
Nordpol *m*, north pole.
nordsuchend, north-seeking.
normal, normal, standard;
— schalten, to set at normal.
Normal *n*, standard;
 Widerstands- —, resistance standard;
— **-belastung** *f*, normal load.
Normale *f*, normal (zu, to) *M*.
Normal-einstellung *f*, normal adjustment;
— **-element** *n*, standard cell;

Normal
— =fernhörer m, standard receiver;
— =Fernkabel n, standard long-distance telephone cable.
Normalien pl, standards pl.
Normal=instrument n, standard (reference) instrument;
— =ohm n, standard ohm;
— =spurbahn f, standard gauge railway, normal gauge railway;
— =stellung f, normal position;
— =tastenfeld n, universal keyboard;
— =wellenmesser m, standard wavemeter;
— =widerstand m, standard resistance.
normalisieren, to standardize.
Normalisierung f, standardization.
normen, to standardize.
Normung f, standardization.
Not=amt n, temporary exchange;
— =anlage f, provisional plant; emergency plant;
— =apparat m, emergency instrument;
— =ausgang m, emergency exit;
— =ausrüstung f, emergency outfit;
— =batterie f, emergency battery;
— =einrichtung f, emergency set, emergency apparatus;
— =fall m, emergency;
— =kabel n, interruption cable;
— =ruf m auf See, s o s call, distress call;
— =sender m, emergency transmitter;
— =zeichen n, distress signal.
Null, zero;
— werden, to approach zero.
Null=achse f, zero axis;
— =auslösung f, no-load release;
— =ausschalter m, zero cut-out, no-load cut-out, no-voltage circuit breaker;
— =hebel m, zero adjusting lever, unison lever;
— =leiter m, neutral wire;
geerdeter — = —, earthed neutral conductor, power ground wire;
— =linie f, zero line, zero axis;
wandernde — = —, shifting zero;
— =methode f, null method, zero method;
— =potential n, zero potential;
— =punkt m, zero point, null point; Drehstrom: neutral point; Koordinaten: origin; Teilung: zero (degree mark); falscher= — = —, false zero;
geerdeter — = —, grounded neutral point;
mittlerer — = —, centre zero (einer Teilung, of a scale);
wandernder — = —, wandering or fluctuating or shifting zero;
— = — =abweichung f, zero error;
— =spannung f, zero voltage;
— = —s =auslösung f, no-volt release;
— = —s =ausschalter m, no-voltage circuit breaker;
— =stellung f, home position;
— =strom m, zero current;
— = — =anzeiger m, zero current indicator;
— = — =auslösung f, no-load release.
numerieren, to number.
Numerierung f, numbering.
numerisch, numeric(al).
Nummer f, number;
unbenutzte —, dead number F;
unzugeteilte —, unallotted number F;

Nummer
Amts= —, exchange number;
Nummern=folge f, (in der), (in) consecutive order;
— =geber m, number indicating system, call sender, A;
Tasten= — = —, key set call sender, key sender A;
— =gebung f, numbering;
— =speicher m, (numerical) register A;
— =schalter m, dial switch A;
— =scheibe f, dial, dial switch, A; number plate, dial plate A;
die — = — ablaufen lassen, to release the dial;
die — = — aufziehen, to wind up, to pull round the dial;
Ablaufen der — = —, returning of the dial;
Aufziehen der — = —, winding-up of the dial;
— =stempel m, numbering machine;
— =wahl f, impulse action, impulse stepping;
— =wähler m (Gegensatz zu Amtsnamenwähler) numerical switch A.

Nußbaum(holz n) m, walnut.
Nute f, slot, notch, groove;
mit —n versehen, slotted, notched, grooved;
Kreis= —, recess;
Öl= —, Schmier= —, oil groove, oil way;
nuten, to groove, to slot, to notch.
Nuten=wellen pl, slot ripple;
— = =frequenz f, slot ripple frequency;
— =wicklung f, slot winding; mit — = — versehen, slot wound.
Nutzarbeit f, useful work.
nutzbar, useful;
— machen, to utilize, to employ; to economize.
Nutz=dämpfung f, useful resistance;
— =effekt m, useful effect, efficiency;
— =leistung f, useful output;
— =spannung f, useful voltage;
— =widerstand m, useful resistance;
— =wirkung f, useful effect.

O.

O. B. = Ortsbatterie f, local battery, ab: l. b., magneto...;
— =Amt n, l. b. exchange, magneto exchange;
— =Fernhörer m, l. b. receiver;
— =Klappenschrank m, magneto or l. b. (switch)board;
— =Mikrophon n, l. b. transmitter;
— =Schrank m, magneto board;
— = —für einen Arbeitsplatz, single position magneto board.
obere(r), upper, top;

— =Drahtlager n, top groove B;
— Reihe f, top row.
Oberbund m, top binding B;
Oberfläche f, surface, skin;
wirksame — = —, active surface.
Oberflächen=ableitung f, surface leakage;
— =entladung f, surface discharge;
— =spannung f, surface constraint, surface tension;

Oberflächen
— -ſtrom m, superficial current;
— -widerſtand m, surface resistance;
— -wirkung f, skin effect; in Kriſtallen: surface work.
oberflächlich, superficial.
Oberharmoniſche f, harmonic, — pl. upper harmonics, higher harmonics pl;
Erzeugung f, von — n, production of harmonics;
in einer — n ſchwingen, to vibrate to a harmonic;
dreifache —, triple harmonics pl;
fünffache —, quintuple harmonics pl;
ſiebenfache —, septuple harmonics pl;
gerade —, even higher harmonics pl;
ungerade —, odd harmonics pl.
oberirdiſch, overhead;
eine Leitung — führen, to run a line overhead;
— e Linie f, overhead line, open (wire) line.
Oberlicht n, skylight.
Oberſchwingung f, harmonic vibration, overtone.
Oberſeite f, top side.
Oberteil n, top.
Oberton m, overtone.
Oberwelle f, overtone, harmonic vibration.
Ocelitſtab m, ocelit rod.
Ofen m, furnace, stove;
Muffel- —, muffle furnace;
Trocken- —, drying stove.
offen, open;
— e Sprache f, open language;
— er Stromkreis m, open circuit.
offenkundig, öffentlich, public.
Öffentlichkeit f, public; publicity.

öffnen, to open;
einen Stromkreis —, to break, to open a circuit.
Öffnen n, opening.
Öffnung f, aperture, opening, hole;
Luft- —, vent, ventilation hole;
Schlitz- —, slotted hole;
Öffnungsfunke m, spark at break;
— -impuls m, break impulse.
Ohm n, ohm;
absolutes —, absolute ohm;
British Association- —, B. A. (standard) ohm (= 0,9866 int. ohm);
internationales —, international ohm, standard ohm (= 1,00052 abs. ohm);
Normal- —, standard ohm;
1 — -cm = 0,3937 ohm-inch;
1 — (Meter-Gramm) = 5710 ohm (mile-pound);
auf n — gewickelt, wound to a resistance of n ohms;
Normal- —, standard ohm;
— -meter n, ohmmeter, megger;
— -ſches Geſetz n, Ohm's law;
— -ſcher Widerſtand m, ohmic resistance, steady current resistance;
reiner — - — —, non-reactive resistance, dissipative resistance.
ohmiſch, ohmic.
Ohr n, ear.
Ohren- . . . ., aural.
okkludieren, to occlude.
okkludierte Gaſe pl, occluded gases pl.
Okonit n, okonite.
ökonomiſch, economical.
Oktave f, octave.
Öl n, oil;
Baumwoll(samen)- —, cotton seed oil;

Öl

Öl
- Erd- —, mineral oil;
- Harz- —, resin oil;
- Knochen- —, bone oil;
- Lein- —, linseed oil;
- — - —firnis m, boiled oil;
- Mineral- —, mineral oil;
- Paraffin- —, paraffin oil;
- Rizinus- —, castor oil;
- Roh- —, crude naphta;
- Schmier- —, lubricating oil;
- Schwer- —, heavy oil;
- Teer- —, (coal) tar oil;
- Terpentin- —, spirit of turpentine;
- Vaselin- —, vaseline oil.

ölgetränkt, oiled.

Öl-kanne f, oil can;
- -kondensator m, oil (dielectric) condenser;
- -leinen n, oil cloth, varnished cambric;
- -loch n, oil hole;
- -motor m, oil engine;
- -nute f, oil groove, oil run, oil way;
- -papier n, oiled paper;
- -schalter m, oil switch;
- -seide f, oiled silk;
- -stein m, oil-stone;
- -transformator m, oil (-cooled) transformer;
- -trennschalter m, oil-break switch;
- -tuch n, oil cloth, varnished cloth.

ölen, to oil, to lubricate, to grease.

Öler m, lubricator;
- Docht- —, wick lubricator.

ölig, oily.

Ölung f, lubrication;
- Docht- —, wick lubrication;
- Druck- —, forced oil feed.

Omnibusleitung f, omnibus circuit, way circuit (am.), YQ-circuit (engl.).

Operator m, operator M.
Opposition f, opposition.
Optik f, optics pl.
optisch, optical;
- -er Telegraph m, optical telegraph; semaphore.

orangefarben, orange.
Ordinate f, ordinate.
ordnen, to arrange, to range; to rearrange, to collect terms, M.

Ordnung f, order; Rang: rank;
n-ter —, of the nth order.

Organisation f, organization.
organisieren, to organize;
um- —, to reorganize.

Oersted n, oersted (Einheit der Reluktanz, unit of reluctance).

Ort m, place; locus pl loci M.
örtlich, local.
Orts- . . . ., local;
- -amt n, local exchange;
vereinigtes Fern- — und — - —, combined toll and local exchange;

ortsbesetzt, local busy, engaged on local call, F;

Orts-besetztsein n, local busy condition, engagement on local call;
- -bestimmung f, localization; position finding R;
Fehler- — - —, fault location;
- -bezirk m, local area;
- -fernsprech-anlage f, local telephone plant;
- — - — mit mehreren Ämtern, multi-office exchange (am.), multi-exchange system (engl.);
- — - -verkehr m, local telephone traffic;
- -gebiet n, local area;
- -gespräch n, local call, city conversation;
- -linie f, local line;

Orts
— ⸗postbezirk m, town postal district;
— ⸗relais n, local relay;
— ⸗ring m, local ring T;
— ⸗strom m, local current;
— ⸗ — ⸗kreis m, local circuit;
— ⸗verbindung f, local conversation;
— ⸗ — s⸗leitung f, junction, trunk;
— ⸗verkehr m, local traffic.

Öse f, eye, eyelet, loop;
Aufhänge⸗ —, suspension loop, suspension eye;
Löt⸗ —, (flat type) soldering tab, tag;
Metall⸗ —, metallic eyelet;
Ösenbolzen m, eye bolt.
Osmium n, osmium (Os).
Osmose f, osmose, osmosis.
osmotisch, osmotic(al).
Oszillator m, oscillator;
gerader —, dipole, straight oscillator;
Hertzscher —, Hertzian doublet;
offener —, open oscillator.
oszillieren, to oscillate, mechanisch auch: to rock.
oszillierend, oscillating, oscillatory.
Oszillograph m, oscillograph;
Kathodenstrahlen⸗ —, Braun tube oscillograph;
Saiten⸗ —, string oscillograph;
— en⸗aufnahme f, oscillograph record;
— en⸗schleife f, oscillograph loop, oscillograph vibrator;
oszillographisch, oscillographic(al);
Oszillogramm n, oscillogram, oscillograph curve, film.
Oxyd n, oxide;
wasserhaltiges —, hydroxide;
Hydr⸗ —, hydroxide;
— ⸗faden m, oxide-coated filament;
— ⸗häutchen n, film of oxide;
— ⸗kathode f, oxide cathode, Wehnelt cathode;
— ⸗ — n ⸗röhre f, oxide-coated filament vacuum tube.
— ⸗schicht f, film of oxide, oxide coating.
Oxydation f, oxidation.
oxydierbar, oxidizable;
nicht —, inoxidizable.
Oxydierbarkeit f, oxidizability.
oxydieren, to oxidize.
Ozeankabel n, ocean cable, submarine cable.
Ozokerit n, ozokerite.
Ozon n, ozone (O_3).
ozonisieren, to ozonize.

P.

Paar n, pair, couple;
Adern⸗ —, pair.
pachten, to lease.
Pack n, pack.
packen, to pack.
Packrolle f, guide pulley B.
Packung f, packing, serving;
Jute⸗ —, jute packing.
paginieren, to page.

Paket n, pack; set;
Federn⸗ —, set of springs, spring bank, spring assembly.
Palladium n, palladium (Pd).
Paneel n, panel.
Pantelephon n, pantelephone.
Panzer m, shield; — ⸗...,
shielded, ironclad;

Panzer
— -transformator m, ironclad transformer.
Papier n, paper;
geöltes —, oiled paper;
lichtempfindliches —, sensitized paper;
paraffiniertes —, paraffined paper;
Glas- —, glass paper;
Grund- —, body paper;
Holz- —, wood-pulp paper;
Japan- —, Japanese paper;
Kohle- —, carbon paper;
Koordinaten- —, (logarithmisches), (logarithmic) coordinate paper;
Manila- —, Manil(l)a paper;
Millimeter- —, cross-section paper, squared paper;
Öl- —, oiled paper;
Paus- —, tracing paper;
Polreagens- —, pole finding paper;
Reagens- —, test paper;
Sand- —, sand paper;
mit — - — abreiben, to sandpaper;
Schmirgel- —, emery paper;
Stanniol- —, tinfoil paper;
Träger- —, body paper;
Wachs- —, waxed paper;
Zeichen- —, drawing paper;
— -band n, paper tape, breites: web of paper;
— -blatt n, paper blank;
— - — -stärke f, thickness of paper;
doppelte — - — -, two thicknesses of paper;
— -führung f, paper guide T;
— - —-s -hebel m, paper feeding lever T;
— -Hohlraumkabel n, dry core cable, air space paper core cable, ab: a. s. p. c.
— -hülse f, paper sleeve B;

— -kabel n, paper cable, dry-core cable;
— -röhrchen n, paper sleeve, paper jointing tube;
— -rolle f, web of paper, paper roll;
— - —n -halter m, tape roll holder T;
— - —n -träger m, tape wheel T;
— -schlitten m, paper carriage T;
— - — -rückführung f, carriage return T;
— -streifen m, paper tape, paper slip, strip of paper;
— -umhüllung f, wrapping of paper;
— -vorschub m, paper feed(ing);
— -wagen m, carriage T.
Pappe f, cardboard, board, millboard.
Pappelholz n, poplar.
Parabel f, parabola.
Parabol- ...., parabolisch, parabolic;
— -spiegel m, parabolic mirror.
Paraffin n, paraffin(e);
festes —, paraffin wax;
Weich- —, soft paraffin;
— -öl n, paraffin oil.
paraffiniert, paraffined.
parallaktisch, parallactic(al).
Parallaxe f, parallax.
parallel, parallel, equidistant, paralleling;
gleichsinnig —, parallel aiding;
— geschaltet, in parallel (zu, with), paralleled, shunted (zu, across);
— schalten, to join in parallel, to connect in parallel, to shunt (zu, across), to tee together;
Parallelbetrieb m, operation in parallel, parallel operation (von Röhren, of valves).
Parallele f, parallel. [ance;
Parallelimpedanz f, leak imped-

Parallel
— -klinke f, branching jack, parallel jack; duplicate jack;
Vielfach- = -, parallel multiple jack;
Vielfachfeld n mit — = — en, parallel multiple;
— -kondensator m, shunting condenser;
Parallelogramm n, parallelogram.
Parallel-ohmmethode f, shunted-telephone method, parallel ohm method;
— -resonanz f, parallel resonance;
— = — -kreis m, parallel or multiple or branched resonant circuit;
— -schaltung f, parallel connection, multiple connection;
Arbeiten n in — = —, leak working;
— -verlauf m, parallelism;
— -weg m, by-path;
— -widerstand m, parallel resistance.
paramagnetisch, paramagnetical.
Paramagnetismus m, paramagnetism.
Pardune f, span rope, guy wire, backstay;
Stahldraht- —, steel span rope.
passen, to fit;
schlecht —, to misfit.
passiv, passive.
Paste f, paste;
Erreger- —, exciting paste;
Füll- —, filling paste.
Patent n, patent;
ein — anmelden ober beantragen, to apply for a patent;
ein — nehmen, to take out a patent (auf, for);
abgelaufenes —, expired patent;
angemeldetes —, pending patent;
schwebendes —, pending patent;
verfallenes —, void patent;
Haupt- —, parent specification;
Pionier- —, pioneer patent;
Zusatz- —, additional patent;
— -amt n, patent office;
— -anmeldung f, patent application;
Aktenzeichen n der — = —, file number;
eine- — = — einreichen, to file a patent application;
— -anspruch m, claim;
— -anwalt m, patent attorney;
— -beschreibung f, patent specification;
— -brief m, letters patent;
— -erteilung f, patent grant;
patentfähig, patentable;
Patent-gebühr f, patent fee;
— -gesetz n, patent law;
— -gesuch n, patent application;
patentierbar, patentable;
patentieren, to patent;
patentiert, im In- und Auslande patented at home and abroad;
Patent-inhaber m, patentee;
— -jahresgebühr f, patent renewal fee;
— -recht n, patent right;
— -register n, — -rolle f, patent rolls pl;
— -urkunde f, patent letter;
— -verlängerung f, renewal of a patent;
— = —s -gebühr f, patent renewal fee.
Patrone f, cartridge;
Feinsicherungs- —, heat coil;
Grobsicherungs- —, glass tube fuse;
Sicherungs- —, cartridge;

Patronen-sicherung f, cartridge fuse.
Pauke f, bass drum;
 Kessel- —, tymbal, kettle drum.
Pauschgebühr f, flat rate;
 — **en-tarif** m, flat rate tariff, bulk tariff;
 —**en-teilnehmer** m flat rate subscriber.
Pauschtarif m, flat rate tariff.
Pause f, tracing, copy;
 Blau- —, blue print.
Paus-leinen n, tracing cloth;
 — **-papier** n, tracing paper.
Pech n, pitch;
 Erd- —, mineral pitch, bitumen.
Peilantenne f, direction finder aerial.
Peilfunk-anlage f, direction finding plant;
 — **-einrichtung** f, radio direction finder;
 — **-empfänger** m, direction finder, direction finding receiver;
 — **-sender** m, wireless direction finding transmitter, radio beacon;
 — **-stelle** f, direction finding station.
peilen, to bear.
Peilung f, bearing(s pl).
Pendel n, pendulum;
 — **-gleichrichter** m, vibrating rectifier;
 — **-kontakt** m, pendulum contact;
 — **-linse** f, bob, ball;
 — **-regler** m, (conical) pendulum governor;
 — **-Selbstunterbrecher** m, pendulum self-interrupter;
 — **-telegraph** m, pendulum start-stop telegraph, Pendel telegraph;
 — **-umformer** m, vibrating recti-
pendeln, to swing. [fier.

Pendeln n (des Rotors), phase swinging, hunting.
Pendelung f, hunting.
per/sec, periods per second, p. p. s., cycles per second, c. p. s.
Pergament n, parchment.
Perikondetektor m, perikon detector.
Periode f, cycle, period;
 1000 —n, kilocycle,
 Eigen- —, natural period (of oscillation);
 Halb- —, half cycle, half period, semi-oscillation, semiperiod, alternation;
 Schwebungs- —, beat cycle;
 Viertel- —, quarter period;
Periodenzahl f, periodicity, number of periods.
 .... **-periodig**, ... cycle;
 600**-** —**er Ton** m, 600-cycle note;
 500**-** —**er Wechselstromgenerator** m, 500-cycle alternator.
periodisch, periodic(al), cyclical; recurrent.
Periodizität f, periodicity.
Peripherie f, circumference, periphery.
peripherisch, peripheral, peripheric.
Permalloy, permalloy (78,5%Ni 21,5% Fe).
permanent, permanent.
permeabel, permeable.
Permeabilität f, permeability;
 differentielle —, differential permeability;
 magnetische —, magnetic permeability;
 reversible —, **umkehrbare** —, reversible permeability;
 zusätzliche —, incremental permeability;
 reziproker Wert m **der** —, reluctivity;

Permeabilität
  Anfangs- —, initial permeability;
  — eins, unity permeability;
  — bei kleinen Feldstärken, permeability at low magnetizing forces.

Permeameter n, permeameter;
  — für Messungen bei erhöhter Temperatur, hot permeameter.

Permutation f, permutation.

permutieren, to permute.

Personal n, staff, personnel;
  — -ersparnis f, staff economies pl.

Perspektive f, perspective.

perspektivisch, perspective.

Petroleum n, petroleum;
  — -motor m, petroleum engine.

Pfahl m, post, peg, stake;
  Anker- —, stay block;
  Markier- —, marking post, peg;
  Prell- —, fender.

Pfeife f, pipe;
  Baß- —, bassoon;
  Zungen- —, reed pipe.

pfeifen, to sing, to squeal, to howl, V.

Pfeifen n (der Verstärker), singing, squealing, howling, self-oscillation, V;
  — durch gegenseitige Beeinflussung mehrerer Verstärker, end-to-end singing KV;
  — durch Selbsterregen eines Verstärkers, local singing KV;
  — -neigung f, near-singing condition KV;
  — -punkt m, singing point KV.

Pfeil m, arrow;
  — -spitze f, arrowhead;
  — -verzahnung, Getriebe n mit, herring-bone gear(ing).

Pfeiler m, pillar.

Pferdekraft f, horse power, h. p.

Pflaster n, pavement, paving;
  — -arbeiten pl, pavement work.

pflastern, to pave.

Pflege f, maintenance (work), routine repair work.

pflegen, to attend (to), to maintain.

Pflichtenblatt n, specification.

Pflichtwert m, specification value, contract value.

Pflock m, peg, stake, trenail, jack.

Pfropfen m, plug.

Pfund n, pound, lb. pl lbs. (= 453,59 g).

Phänomen n, phenomenon.

Phantomkreis m, phantom circuit, combined circuit;
  zum — schalten, to phantom;
  — -leitung f, phantom circuit;
  — -schaltung f, phantom connection.

Phase f, phase, epoch;
  außer — bringen, to dephase;
  in gleicher — mit, in phase with, cophasal to;
  gleich belastete — n pl, balanced phases pl;
  verschobene —, displaced phase;

Phasen-änderung f, phase change;
  — -anzeiger m, phase indicator;
  — -beziehung f, phase relation;
  — -bilanz f, phase balance;
  — -differenz f, difference of phase;
  — -entzerrung f, correction of phase;
  — -gleichheit f, phase coincidence;
  — -indikator m, phase indicator;
  — -schieber m, phase shifter, phase changer;
  — - -transformator m, phase shifting transformer;
  — -spannung f, phase voltage;
  — -spaltung f, phase-splitting;
  — -sprung m, shift in phase;

Phasen
— -stellung f, entgegengesetzte, opposition of phase;
— -teiler m, phase-splitting device;
— -teilung f, phase splitting;
— -umkehr f, phase reversal;
— -unterschied m, phase difference;
— - — zwischen Geber und Empfänger, orientation T;
— -verzerrung f, phase distortion;
— -verschiebung f, phase shift, phase displacement;
— - — um 90°, phase quadrature;

phasenverschoben, dephased, out of phase; um 90° —, in phase quadrature;
— er Strom m, out-of-phase current;
— e Welle f, out-of-phase wave.

phasenverspätet, phasenverzögert lagging.

Phasen-verzögerung f, lagging of phase, phase retardation;
— -voltmeter n, phase voltmeter;
— -voreilung f, leading of phase;
— -winkel m, phase angle; impedance angle;
negativer (positiver) — - —, negative (positive) impedance angle.

Phenolfiber f, phenol fibre.
phonisch, phonic;
— es Rad n, phonic wheel.

Phosphor m, phosphorus (P);
— -bronze f, phosphor bronze;
— - — -draht m, phosphor bronze wire;

phosphorhaltig, phosphorous.
photoelektrisch, photoelectric.
Photogramm n, photo, photograph, photoprint copy;
Photographie f, photo; photography.

photographisch, photographic(al);
— ein Zeichendruck, Telegraph m mit, photo-printing telegraph.

Phototelegraph m, telephotograph.
phototelegraphisch, telephotographic(al).
Physik f, physics pl.
physikalisch, physical.
Physiker m, physicist.
Picke f, pick.
piezoelektrisch, piezo-electric(al);
— -er Kristall m, piezo-electric crystal.

Piezoelektrizität f, piezoelectricity.
Pilzisolator m, mushroom insulator, umbrella insulator.
Pimpel m, nipple; stud, plunger (der Klinkenfeder, of jack spring);
Hartgummi- —, ebonite stud.
Pinsel m, brush;
— -detektor m, catwhisker detector.
— -elektrode f, catwhisker R.

Pionierpatent n, pioneer patent.
Pipette f, pipette.
Piston n, cornet (à pistons).
Pitchpineholz n, pitchpine.
Plan m, plan, scheme, lay-out, contrivance; map;
Lage- —, general plan;
Leitungs- —, **Netz-** —, map of network.

planen, to plan, to design, to contrive.
Planetengetriebe n, epicyclic (train of) gear.
Planimeter n, planimeter.
planparallel, plane parallel.
Planung f, design, planning.
plastisch, plastic(al).
Platin n, platinum (Pt).
Platine f. side plate.

Platinoid *n*, platinoid (W, Ni, Cu, Zn).

Platte *f*, plate, slab, sheet; plateau;
formierte —, formed plate;
Deck= —, cover, top-plate;
Erd= —, earth plate;
Gitter= —, grid plate, lattice plate;
Hartgummi= —, ebonite plate;
Kasten= —, box plate;
Kern= —, core plate;
Klang= —, sounding plate;
Klemm= —, Spann= —, clamping plate;
Trag= —, mounting plate;

Platten-blitzableiter *m*, plate lightning arrester;
— =kondensator *m*, disc condenser, plate condenser.

Platz *m*, place, room position, Grundstück: site;
besetzter, occupied position;
A= —, A-position;
Abfrage= —, home position, answering position;
Arbeits= —, operator's position;
Aufsichts= —, supervisor's position;
Auskunfts= —, information desk, enquiry position;
B= —, B-position;
— = —, halbautomatischer, semi-B-position;
Durchgangs= —, through-position;
Fern= —, trunk position;
Kontroll= —, monitor's position; supervisor's position;
Melde= —, record position;
Melde-Spitzen= —, record transfer position;
Nachtdienst= —, night (service) position;
Prüf= —, test(ing) position;

Schrank= —, operator's position;
Teilnehmer= —, A-position, home position, answering position;
Verbindungsleitungs= —, junction board;
Vorort= —, suburban position;
Zuleitungs= —, B-position;
— =ausrüstung *f*, position equipment;
— =lampe *f*, pilot lamp, pilot indicator;
— = —n =relais *n*, pilot relay;
— =schaltung *f*, operator's speaking circuit;
— =schnur *f*, switchboard cord;
— =umschalter *m*, position switching key, coupling key;
— =wechsel *m*, transposition, crossing, B.
— =zähler *m*, position meter.

Pliodynatron *n*, pliodynatron.
Pliotron *n*, pliotron.
Plombe *f*, seal; plumber's wiped joint *B*.
plombieren, to seal, to lead.
Pluszeichen *n*, plus sign.
pneumatisch, pneumatic(al).
Pockholz *n*, lignum vitae.
Pol *m*, pole; terminal.
entgegengesetzte —e *pl*, opposite or unlike poles *pl*;
gleichnamige —e *pl*, like poles, similar poles *pl*;
induzierender —, inducing pole;
induzierter —, induced pole;
nordsuchender (südsuchender) —, north-seeking (south-seeking) pole;
ungleichnamige —e *pl* opposite poles, unlike poles *pl*;
Einheits= —, magnetischer, unit magnetic pole;
Folge= —e *pl*, consequent poles *pl*;

Pol
Hilfs- —, auxiliary pole;
Hörner- —, horn-shaped pole;
Kohle- —, carbon terminal, carbon pole;
Kupfer- —, copper pole, copper terminal;
Nord- —, north pole;
Süd- —, south pole;
Wende- —, reversing pole;
Zink- —, zinc terminal, zinc pole.

Pol-abstand m, pole distance; freier — —, pole clearance;
polar, Polar- . . . ., polar.
Polar-charakteristik f, polar characteristic;
— -diagramm n, polar diagram.
Polarisation f, polarization; dielektrische —, dielectric polarization;
Polarisations-magnet m, polarizing magnet;
— -spannung f, electromotive force of polarization;
— -strom, m, polarizing current, polarization current;
— -zelle f, electrolytic valve, polarization cell.
polarisierbar, polarizable.
polarisieren, to polarize.
polarisiert, polarized, polar;
— es Relais n, polarized relay.
Polarität f, polarity;
Umkehr f der —, reversal of polarity;
Polaritätswechsel m, alternation of polarity.
Polarlicht n, aurora pl aurorae.
Pol-bogen m, pole arc;
— -draht m, connection (am Element, of cells);
polen, to pole.
Polfläche f, polar surface;
wirksame — - —, active polar surface;
polieren, to polish, to burnish.

poliert, bright, polished;
hochglanz- —, highly polished.
Politur f, polish.
Pol-klemme f, pole terminal;
— -lücke f, pole clearance;
— -paar n, pair of poles;
— -rad n, cog wheel, magnet wheel;
— -rand m, pole tip;
— -reagenspapier n, pole finding paper, pole test paper;
— -schuh m, pole piece;
— — -rand m, pole tip;
abgeschrägte — - -ränder pl, skewed pole tips pl;
— - — -spitze f, pole horn;
— -stärke f, strength of poles;
magnetische — - —, magnetic pole strength;
Polster n, pad.
Polsterung f, padding.
Pol-teilung f, pole pitch;
— -wechsler m, pole changer, ringing vibrator;
— - — -feder f, pole changing spring;
Polygon n, polygon.
polygonal, polygonal.
Pol-zahl f, number of poles;
— -zahn m, pole tooth, spoke;
— -zwischenraum m, pole clearance.
Pore f, pore.
porös, porous;
— -e Zelle f, porous pot.
Porosität f, porosity.
Portlandzement n, Portland cement.
Porto n, postage.
portofrei, frank.
Porzellan n, porcelain, china;
Hart- —, hard porcelain;
Weich- —, soft porcelain;
— -isolator m, porcelain insulator. [trombone stop.
Posaune f, trombone; Orgel-

positiv, positive.
Posten *m*, item.
Post-anweisung *f*, telegraphische, money telegram.
— -nebenstelle *f*, exchange extension set;
— -relais *n*, Post Office standard relay (*engl.*).
Potential *n*, potential;
elektrisches —, electric potential;
Entladungs- —, discharge potential;
Erd- —, earth potential;
Funken- —, spark potential;
Null- —, zero potential;
— -differenz *f*, potential difference, *ab*: p. d.;
— -einheit *f*, unit potential;
— -fläche *f*, potential surface;
Äqui- — - —, equipotential surface.
potentiell, potential;
— -e Energie *f*, potential energy.
Potentiometer *n*, potentiometer.
Potenz *f*, power;
in die *n*te — erheben, to raise to the *n*th power.
potenzieren, to involve. to raise to a higher power.
Potenzierung *f*, involution.
Pottasche *f*, potassium carbonate (K_2CO_3).
prägen, to stamp.
praktische Einheit *f*, practical unit.
Praxis *f*, practice.
Präzision *f*, precision;
— -s -instrument *n*, precision instrument.
prellen, to chatter.
Prellen *n* der Kontakte, contact chatter.
Prell-pfahl *m*, fender;
— -stein *m*, curbstone.
Presse *f*, press;
hydraulische —, Wasserdruck- —, hydraulic press;

— -telegramm *n*, news message, press message.
pressen, to press, compress; to stamp; in Formen: to mould.
Preß(-guß)-stück *n*, die pressing;
— -material *n*, (Isolier-), moulded insulation.
— -schraube *f*, clamp(ing) screw;
— -span *m*, pressboard, strawboard, fullerboard, press-spa(h)n;
— -stück *n*, pressing.
Pressung *f*, compression.
primär, Primär- . . . ., primary;
Primär-batterie *f*, primary battery;
— -element *n*, primary cell, voltaic cell;
— -empfänger *m*, single circuit receiving set, primary receiver;
— -strom *m*, primary current;
— -wicklung *f*, primary (winding).
Priorität *f*, priority.
Prisma *n*, prism.
Privat- . . . ., private;
— -fernsprechanlage *f*, private telephone plant, house telephone plant;
— -fernsprechleitung *f*, private (telephone) wire;
— -fernsprechzentrale *f*, private exchange, *ab*: p. x.;
selbsttätige — - —, private automatic exchange, *ab*: p. a. x.;
— -telegramm *n*, private message;
— -telegraphenanlage *f*, private telegraph plant;
— -telegraphenleitung *f*, private telegraph wire.
Probe *f*, sample, specimen;
Prüfung: trial, test;
— -länge *f*, sample (length).
Produkt *n*, product;

Sattelberg, Wörterbuch: Deutsch-Englisch. 13

Produkt
— aus Ampere und Volt, product of amperes by volts.

Profil n, profile.

Profil-...., profiliert, profile..., profilated, shaped;
— -eisen n, profile iron.

Progression n, progression.

progressiv, progressive.

Projekt n, project, scheme.

Projektion f, projection;
— s-apparat m, projector.

Proportion f, proportion.

proportional, proportional, proportionate;
umgekehrt —, inversely proportional.

Proportionale f, proportional.

Proportionalität f, proportionality;
— s-konstante f, constant of proportionality.

provisorisch, provisional.

Prozent n, percent;
Gewichts- —, percent by weight;
Raum- —, Volum- —, percent by volume;
— -satz m, percentage.

prozentual, percentage....;
— e Änderung f, percentage change;
— e Zunahme f, percentage increase.

Prozeß m, process; gerichtlich: suit.

Prüf-arm m, private wiper A;
— -batterie f, testing battery; Besetzt: engaged test battery F;
— -beamter m, testing officer, test clerk; checker T, am Wheatstone: key clerk T;
— -draht m, testing wire, pilot wire;
— -einrichtung f, testing set;
— -kasten m, test box;

— -klemme f, test clip;
— -klinke f, test jack;
— -leitung f, testing circuit, pilot wire; Besetzt: testing or holding or third wire F;
— -methode f, method of test;
— -platz m, testing (operator's) position, test position, monitor's position;
— -pult n, test desk;
— -raum m, testing room;
— -relais n, testing relay;
— -schaltung f, monitoring circuit;
— -schrank m, test box, test board;
kleiner — - —, test case;
Fern- — - —, trunk test board, toll test board;
— -spannung f, testing voltage;
— -stand m, testing shop;
— -stelle f, testing position;
— -stöpsel m, test plug;
— -tisch m, test desk;
— -verfahren n, testing method;
— -vorrichtung f, tester;
— -wert m, test value.

prüfen, to test, to verify, to examine, to inspect;
Telegramme: to check (messages); überholen: to overhaul;
auf Erdschluß (Berührung, Kurzschluß) —, to test for earth (contact, short-circuit).

Prüfen n, testing; checking.

Prüfer m, tester; checker;
Erdschluß- —, ground detector;
Ton- —, tone tester R. [or;

Prüfung f, test, testing, verification, examination, inspection; overhauling; checking;
— auf Betriebsfähigkeit, clear test;
in der — begriffen, under test;
Abnahme- —, acceptance test;
Frei- —, disengaged test, meist: engaged test;

Prüfungs-beamter *m*, test clerk; checker *T*.

Puffer *m*, pad, buffer:
— -**batterie** *f*, buffer battery, floated battery;
— -**dynamo** *f*, buffer dynamo;
— -**feder** *f*, buffer spring;
— -**wirkung** *f*, buffer action.

puffern, to float (**eine Batterie**, a battery).

Pulsation *f*, pulsation, beat.

pulsieren, to pulsate.

Pulsieren *n*, pulsating, pulsation.

Pult *n*, desk;
Schreib- —, writing shelf;
— -**gestell** *n*, desk stand.

Pulver *n*, powder;
Kohle- —, carbon powder;
— -**fritter** *m*, powder coherer;
— -**kern** *m*, powder core;
gepreßter Eisen- — - —, compressed iron powder core;
— -**mikrophon** *n*, powder transmitter.

pulverisieren, to powder, to pulverize.

Pumpe *f*, pump;
Luft- —, air pump;
— - —, **Quecksilber**-, mercurial air pump.

pumpen, to pump.

Punkt *m*, dot *T*, point;
spitze —e *pl*, sharp dots, clipped dots *pl*, *T*;
Anfangs- —, origin;
Fest- —, section point *K B*;
Kehr- —, cusp;
Knoten- —, centre;
Morse- —, dot;
Null- —, zero (point), null point;
— - —, **falscher**, false zero;
— - —, **wandernder**, wandering zero, fluctuating zero;
Wende- —, cusp.

punktförmig, in lumps;
— **verteilt**, lumped, in lumps; concentrated;

— **verteilte Induktivität** *f*, lumped inductance;
— **verteilte Ladung** *f*, lumped load(ing).

punktgeschweißt, spot-welded.

punktieren, to dot;
strich- —, to chain-dot, to dash-dot.

Punktschweißung *f*, spot-welding.

Pupinisation *f*, lump-loading, coil loading, pupinization.

pupinisieren, to coil-load, to lump-load, to pupinize.

pupinisiert, coil-loaded, lump loaded;
besonders leicht —, extra light loaded, *ab*: X. L. L.;
mittelstark —, medium heavy loaded, *ab*: M. H. L.;

Pupinisierung *f*, coil loading, pupinization, series-loading;
starke —, heavy loading, *ab*: H. L.;;
mittelstarke —, medium heavy loading, *ab*: M. H. L.;
leichte —, light loading, *ab*: L. L.;
besonders leichte —, extra light loading, *ab*: X. L. L.;
Vierer- —, phantom loading, composite or superposed loading;
— **s -festpunkt** *m*, section point.

Pupin-kabel *n*, coil(-loaded) cable;
— -**leitung** *f*, lump-loaded circuit, coil-loaded circuit;
— -**spule** *f*, load(ing) coil, Pupin coil; [loading coil.
längliche — - —, elongated

purpurrot, purple.

Putz *m*, plaster.

Pyramide *f*, pyramid.

Pyrit *m*, pyrite.

pyroelektrisch, pyroelectric(al).

Pyroelektrizität *f*, pyroelectricity.

Q.

Quadrant *m*, quadrant;
— **en-elektrometer** *n*, quadrant electrometer.
Quadrat *n*, square.
quadratisch, square, quadratic;
— **e Gleichung** *f*, quadratic equation;
— **e Spule** *f*, square coil.
Quadrat-meter *n*, square metre;
— **-wurzel** *f*, square root;
— **-zentimeter** *n*, square centimetre (= 0,155 squ. in.).
quadrieren, to square.
Quadruplextelegraph *m*, quadruplex, quad.
quantitativ, quantitative.
— **e Messung** *f*, quantitative measurement.
Quarz *n*, quartz (SiO_2).
quasi-stationär, quasistationary;
— **-unendlich**, quasi-infinite;
— **- — lange Leitung** *f*, quasi-infinite line, semi-infinite line.
Quecksilber *n*, mercury, quicksilver (Hg);
— **-chlorid** *n*, corrosive sublimate ($HgCl_2$);
— **-dampf** *m*, mercury vapour;
— **- — -gleichrichter** *m*, mercury arc rectifier, mercury vapour rectifier;
— **- — -lampe** *f*, mercury vapour lamp;
— **-druck** *m*, pressure in (terms of) mm of mercury;
— **-luftpumpe** *f*, mercurial air pump;
— **-näpfchen** *n*, mercury cup;
— **-oxyd** *n*, oxide of mercury (HgO);
schwefelsaures — - —, mercurous sulphate (Hg_2SO_4);
-säule, **Druck** *m* **in** mm, pressure in terms of *mm* of mercury;

— **-strahlunterbrecher** *m*, mercury jet interrupter.
Quelle *f*, source;
Kraft- —, source of power;
Strom- —, source of current.
Quer-...., **quer**, transverse, skew, cross....
queren, to cross.
Quer-feld *n*, transverse (magnetic) field, cross field;
— **-glied** *n*, stem, shunt element, (einer Kunstleitung, of a network);
in $1/2$ — - — **endender Kettenleiter** *m*, network terminated at mid-shunt (position);
— - —, **Abschluß** *m* **durch** $1/2$, mid-shunt termination;
— **-haupt** *n*, crosshead;
— **-impedanz** *f*, leak impedance;
— - — (**-glied** *n*) *f*, shunt impedance element;
— - — **-entzerrer** *m*, shunt-admittance type equalizer;
— **-kondensator** *m*, bridging condenser, shunting condenser, by-pass condenser;
— **-lochstreifen** *m*, cross-perforated tape *T*;
quermagnetisiert, cross-magnetic, cross-magnetized;
Quer-magnetisierung *f*, cross-magnetization;
— **-riegel** *m*, transom, traverse;
— **-rolle** *f*, bridging coil;
— **-schiene** *f*, crossbar;
— **-schnitt** *m*, (transverse) section, cross-section;
rechteckiger — - —, rectangular cross-section;
runder — - —, circular cross-section;
Wicklungs-—-—, cross-sectional area of winding; [al area;
— - —**-s** (**-fläche** *f*), (cross-) section-

Quer
— -spule f, leak coil;
mit — - —n belastet, leak-loaded;
— - —n-belastung f, leak-loading;
— -stück n, crosspiece, crossbar;
— -träger m, arm, crossarm, traverse;
Ausrüstung f mit — - —n, arming;
mit — - —n versehen, armed;
— - -für 4 (6, 8) Leitungen, 4 (6, 8) wire arm;
— -verbindung f, cross-connection, cross; tie line F;
— - -feld n, cross-connecting block;
— - -s -kabel n, tie cable;
— -widerstand m, leak resistance; Kettenleiter: stem, shunt element;
— -zug m, transverse stress B.
Quetschfuß m, squash, press, (einer Röhre, of a valve).
Quotient m, quotient.

R.

Raa f, spacer, spreader.
Rachenlehre f, gap gauge.
Rad n, wheel;
 mit Rädern versehen, wheeled;
 phonisches —, phonic wheel;
 — mit scharfen Zähnen, star wheel;
 — mit stumpfen Zähnen, cog wheel;
 — mit Schrägverzahnung, helically toothed wheel;
Drei- —, tricycle;
Fahr- —, bicycle;
Friktions- —, friction wheel;
Hand- —, handwheel;
Hemm- —, escape(ment) wheel;
Kegel- —, bevel wheel, conical wheel, mitre wheel;
Korrektions- —, correcting wheel, correction wheel, corrector wheel;
Kron- —, crown wheel;
Magnet- —, magnet wheel;
Motor- —, motor bicycle;
Pol- —, magnet wheel, cog wheel;
Reib- —, friction wheel;
Schnecken- —, worm wheel;
Schrittschalt- —, step(-by-step) wheel;
Schwung- —, flywheel;
Sperr- —, ratchet wheel;
Steig- —, escape(ment) wheel, ratchet wheel;
Stern- —, star wheel, pin wheel; am Locher, Streifensender: pin feed wheel;
Stift- —, pin (feed) wheel;
Stirn- —, spur wheel;
Typen- —, type wheel;
Vorschub- —, feed wheel;
Zahn- —, tooth(ed) wheel;
— -buchse f, bushing;
— -kranz m, rim;
— -linie f, epicycloid.
Rädchen n, wheel, roller;
Farb- —, inking roller;
Reiter- —, jockey wheel, jockey roller.
Rädergetriebe n, gear, gearing;
Kegel- —, bevel gearing, mitre (wheel) gearing;
Zahn- —, (toothed wheel) gearing.
Räderübertragung f, gearing.

Räderwerk *n*, train of wheels.
radial, radial.
radieren, to erase *T*.
Radio=...., radio, wireless.
radioaktiv, radioactive.
Radio=aktivität *f*, radioactivity;
— =amateur *m*, radio amateur;
radioelektrisch, radio-electric.
Radio=frequenz *f*, radio frequency;
— =goniometer *n*, radiogoniometer;
— =goniometrie *f*, radiogoniometry;
radiogoniometrisch, radiogoniometric(al);
Radio=gramm *n*, wireless message, radiogram;
— =großstation *f*, long-distance radio station.
Radius *m*, radius;
— Vektor *m*, radius vector, *pl*, radii vectores.
radizieren, to extract the root of.
Radizierung *f*, evolution.
raffinieren, to refine.
Rahmen *m*, frame, framework, Spulen: former; coil, frame aerial, loop, *R*;
drehbarer —, Dreh= —, rotatable coil, moving frame, rotating loop, *R*;
Empfangs= —, receiving loop;
Kontakt= —, contact carriage *TA*;
— =antenne *f*, frame aerial, coil, mit 1 Windung: loop (aerial);
drehbare — = —, moving frame;
— =ebene *f*, plane of frame;
— =effekt *m*, closed loop effect;
— =empfänger *m*, loop receiver;
— =empfang *m*, loop reception;
— =werk *n*, framework, frame structure;
— =wirkung *f*, closed loop effect.
Ramme *f*, ram.

rammen, to ram, to tamp.
Rand *m*, edge; vorstehend:flange;
vorspringender —, shoulder, ledge;
Pol= —, pole tip.
Rang *m*, rank (der Wähler, of switches).
Raspe *f*, rasp.
raspeln, to rasp.
rasseln, to rattle.
Rasseln *n*, rattle, rattling.
Rast *f*, rest.
rational, rational.
rationalisieren, to rationalize.
Rätsche *f*, ratchet drill.
rauh, rough; unbearbeitet: raw;
— =werben, to roughen.
Rauheit *f*, roughness.
Rauh=reif *m*, rime, frozen fog;
— = — =bildung *f*, frozen fog formation;
— =werden *n*, roughening.
Raum *m*, room, space, place; volume;
— gewähren für, to accommodate;
abgeschirmter —, screened room, screened cabin *R*;
ausgelichteter —, clearance *B*;
luftleerer —, vacuous space;
Apparat= —, instrument room;
Dunkel= —, dark space *V*;
Schalt= —, switch room;
Wicklungs= —, winding space;
Zwischen= —, space, interstice;
— =einheit *f*, unit (of) volume;
— =ersparnis *f*, saving in space;
— =inhalt *m*, volume, cubic contents *pl*;
— =lade=gitter *n*, space charge grid, filament-screening grid;
— = — =strom *m*, space current;
— = — =wirkung *f*, space charge effect;
— =ladung *f*, space charge;
— = —s =effekt *m*, space charge effect;

Raum
- -mangel *m*, space restriction;
- -prozent *n*, percent by volume;
- -strahlantenne *f*, radiator;
- -welle *f*, spherical wave.
Rauschen *n*, noise *F*.
Reagenspapier *n*, test paper;
Pol- —, pole finding paper, pole test paper.
Reaktanz *f*, reactance;
mit — behaftet, reactive;
induktive —, inductance, positive reactance, inductance reactance;
kapazitive —, condensance, capacity reactance, condensive reactance, negative reactance;
negative —, negative reactance;
positive —, positive reactance;
Gitterkreis- —, input reactance.
Reaktion *f*, reaction.
reaktionslos, non-reactive.
Rechen *m*, comb;
Übersetzer- —, combiner comb;
- -maschine *f*, calculating apparatus;
Additions- — - —, adding machine;
Universal- — - —, calculating machine;
- -schieber *m*, slide rule;
- -tafel *f*, computation table.
rechnen, to calculate, to rate.
Rechnung *f*, calculation, rating; bill, account;
- -s-stelle *f*, account section *FT*.
Rechteck *n*, rectangle.
rechteckig, rectangular, rectangled.
rechtwinklig, rectangular; im rechten Winkel zu, at right angles to;

— es Dreieck *n*, right-angled triangle.
Rechtsdrall *m*, right-handed lay or twist.
rechtsdrehend, clockwise.
Rechtsdrehung *f*, clockwise rotation, rotation to the right.
rechtsgängig, clockwise, right-handed.
Rechtsgewinde *n*, right-handed thread.
Rechtsnachfolger *m*, assign.
recken, to rack, to stretch, (einen Draht, a wire) *B*;
Recken *n*, racking, stretching.
Reduktion *f*, reduction;
Reduktions-faktor *m*, reduction factor;
- -getriebe *n*, reduction gear;
- -muffe *f*, reducing socket, reducing bush;
- -transformator *m*, step-down transformer.
reduzieren, to reduce; to step down.
Reede *f*, road.
reell, real *M*.
reflektieren, to reflect.
reflektierte Welle *f*, reflected wave.
Reflektion *f*, reflection.
Reflektor *m*, reflector.
Reflex-empfang *m*, reflex reception, dual reception;
- -empfänger *m*, (regenerative) reflex receiver, dual receiver;
- -verstärkung *f*, dual amplification.
Regal *n*, shelf, partition.
Regel *f*, rule;
regel-mäßig, regular; wiederholt: routine;
—e Prüfung *f*, routine test;
Regelmäßigkeit *f*, regularity.
regeln, to regulate, to govern; steuern: to control; einstellen: to adjust;

regelrecht, normal, regular;
— schalten, to set at normal;
Regel-stellung f, normal position;
— strom m, normal operating current.
Regelung f, regulation, adjustment; control;
gleichförmige —, smooth regulation;
stufenweise —, regulation in steps.
Nach- —, re-adjustment.
regenerieren, to regenerate.
Regenerierung f, regeneration.
Register n, register auch A; impulse storing device A
Amtsnamen- —, office code register A;
Nummern- —, numerical register A.
registrieren, to record;
selbst- —d, self-recording;
—der Strommesser m, recording ammeter.
Registrierung f, record.
Regler m, regulator, governor, controller;
Feld- —, field regulator, field rheostat;
Fliehkraft- —, centrifugal governor;
Geschwindigkeits- —, speed governor; speed controlling device;
Haupt- —, master regulator;
Kopplungs- —, coupling control R;
Nebenschluß- —, shunt regulator;
Pendel- —, (conical) pendulum governor;
Spannungs- —, voltage regulator;
Umlauf- —, speed governor;
Windfang- —, fan governor;
Zentrifugal- —, centrifugal governor;

— hebel m, speed lever;
— leitung f, pilot wire K;
— widerstand m, rheostat.
Regulator m, regulator.
regulierbar, regulable.
regulieren, to regulate, to adjust.
Regulier-schraube f, adjusting screw;
— transformator m, regulating transformer.
Regulierung f, regulation.
reiben, to rub, to chafe;
sich —, to rasp, to chafe.
Reib-antrieb m, friction drive;
(nachgiebiger) — - —, slipping drive, yielding drive;
— kupplung f, friction clutch;
— rad n, friction wheel;
— scheibe f, friction disc;
— sitz, m, friction-tight.
Reibung f, friction;
magnetische —, magnetic friction;
Reibungs-elektrizität f, frictional electricity;
— last f, friction(al) load;
— verlust m, friction load, frictional loss.
Reichweite f, range, range of transmission.
eine — haben (von), to range;
Nacht- —, night range;
Nenn- —, nominal range;
Sende- —, transmission range;
Sprech- —, speaking range;
Tages- —, day(light) range.
Reif m, rime, frozen fog;
— bildung f, frozen fog formation.
Reifen m, hoop.
Reihe f, series, row, senkrechte: column; progression, succession;
in —, in cascade; in tandem;
in — geschaltet, serially connected;

Reihe
gegensinnig in —, series-opposing;
gleichsinnig in —, series-aiding;
in — schalten, to connect in series, to join in series;
in einer —, in zwei —n, in one row, in two rows;
arithmetische —, arithmetical series;
obere (untere) —, top (bottom) row;
senkrechte —, vertical row, column;
unendliche —, infinite series;
wagerechte —, horizontal row, level;
Fouriersche —, Fourier's series;
Impuls- —, series or succession or train of impulses;
Kosinus- —, cosine series;
Stromstoß- —, series or succession of impulses;
Tasten- —, row of keys, bank of keys;
Telegramm- —, batch (of messages);
— - —n von ...., in, in batches of ....;
Reihen- ...., serial;
— -anlage *f*, **Fernsprech-,** intercommunication telephone plant, house telephone system;
— -beförderung *f*, batch working *T*;
— -betrieb *m*, tandem operation (von Verstärkerämtern, of repeater stations);
— -folge *f*, succession, sequence, order of succession:
der — — nach, in sequence, in succession;
— -funkenstrecke *f*, multiple spark gap;
— -glied *n*, series element (einer Kette, of a network);

— -impedanz *f*, (line) series impedance;
— - — -entzerrer *m*, series-impedance type equalizer *K*;
— - — -glied *n*, series impedance element;
— -induktivität *f*, series inductance;
— -resonanz *f*, series resonance;
— - — -kreis *m*, series-resonant circuit;
— -schaltung *f*, series connection;
Arbeiten *n* **in — - —,** series working;
— -schlußmotor *m*, series motor;
— -spule *f*, series coil;
— -spulen, Belastung *f* **durch,** lumped series loading *K*;
— -telegraph *m*, automatic telegraph, high-speed, telegraph;
— -Verlustwiderstand *m* (eines Kondensators), equivalent series resistance (of a condenser);
— -widerstand *m*, series resistance.
rein, clean, pure, Leitung clear;
chemisch —, chemically pure;
— e Sprache *f*, clear voice;
— er Ton *m*, pure note.
Reinheit *f*, clearness, purity.
reinigen, to clean (Kontakte, contacts).
Reinigen *n*, **Reinigung** *f*, cleaning.
reißen, to break, to crack.
Reiß-feder *n*, drawing pen;
— -länge *f*, breaking length (eines Drahtes, of a wire).
Reiterrädchen *n*, **Reiterröllchen** *n*, jockey roller, jockey (wheel).
Reizschwelle *f*, threshold of sensation.
Rekorder *m*, recorder;
Syphon- —, syphon recorder.
Relais *n*, relay;

Relais
mit — übertragen, to relay;
— — ausgerüstet, relayed;
das — spricht an, zieht an, the relay pulls up, the relay is pulled up;
— — wird (ab)erregt, the relay is (de)energized;
— — fällt ab, the relay releases, the relay is released;
— — liegt auf Zeichen=(Trenn)= seite, the relay marks (spaces);
— — sendet Zeichen= (Trenn)= strom, the relay marks (spaces);
— — umlegen, to reverse the relay;
— — auf einen Grenzstrom von .... einstellen, to margin a relay to pull up at . . . .;
— — mit Bremszylinder, dashpot relay;
— — zwei Schließ=(Trenn=)kon= takten, double make (break) relay;
— — zwei Wechselkontakten, double break and make relay;
— — Schneidenlagerung, knife-edge relay;
einspuliges —, single-spool relay;
langsam abfallendes —, slow-to-release relay, slow releasing relay;
neutrales —, nicht polarisier= tes —, non-polarized relay;
polarisiertes —, polar(ized) relay;
— — mit mittlerer Ruhestellung des Ankers, neutral relay;
unpolarisiertes —, non-polarized relay;
zweispuliges —, double-spool relay;
Abtrenn= —, cut-off relay;
Auflöse= —, clear-out relay A;
Aufnahme= —, receiving relay;

Auslöse= —, tripping relay; releasing relay A;
r=—, third conductor relay;
Differential= —, differential relay;
Doppelschließ= —, double-make relay;
Doppeltrenn= —, double-break relay;
Dosen= —, box relay;
Drehspul= —, moving coil relay;
Druck= —, printing relay, printer relay;
Durchruf= —, through-ringing relay, signalling relay;
Einheits= —, universal relay F;
Einschalt= —, cut-in relay;
Elektronen= —, electron relay, valve relay, thermionic relay;
Empfangs= —, receiving relay;
Fernsprech= —, telephone relay, telephonic relay;
Flügelanker= —, vane armature relay;
Flüssigkeitsstrahl= —, (liquid) jet relay;
Gasentladungs= —, gas discharge relay;
Gegenstrom= —, reverse current relay;
Gleichlauf= —, correcting relay, corrector relay;
Grenzstrom= —, marginal operation relay;
Gulstab= —, vibrating relay, Gulstad relay;
Impuls= —, impulsing relay, impulse relay;
Kontroll= —, supervisory relay, supervision relay;
Kupfermantel= —, coppered relay, copper-jacketed relay;

Relais
Kupferring= —, copper collar relay;
Linien= —, main (line) relay, line relay;
Mikrophon= —, receiver-transmitter amplifier;
Melde= —, pilot relay, supervisory relay;
Orts= —, local relay;
Platzlampen= —, pilot relay F;
Post= —, englisches, P.O. standard relay;
Prüf= —, testing relay;
Resonanz= —, resonance relay;
Rückführ= —, clear-out relay;
Rückstrom= —, reverse current relay;
Ruf= —, calling relay, signalling relay;
Schalt= —, switching relay;
Schlußzeichen= —, clearing relay, supervisory relay;
— = — für den rufenden (verlangten) Teilnehmer, answering (calling) supervisory relay;
Schneiden(anker)= —, knife-edge relay;
Selbstunterbrecher= —, buzzer relay;
Sende= —, transmitting relay, signalling relay;
Spannungsregler= —, voltage control relay;
Spannungswechsler= —, incrementer (Quadruplexbetrieb, quadruplex) T;
Speicher= —, storing relay, storage relay;
Speise= —, supply relay;
Sperr= —, locking relay;
Stanz= —, punching relay;
Summer= —, buzzer relay;

Tast= —, relay key, keying relay, contactor;
Tauchkern= —, plunger relay;
Thermionen= —, thermionic relay;
Trenn= —, cut-off relay;
Überstrom= —, overload relay;
Übertragungs= —, repeating relay, translating relay;
Überwachungs= —, supervisory relay, pilot relay; monitoring relay;
Umschalte= —, switching relay, auto-switch;
Verzögerungs= —, slow-acting relay, time-delay relay, mit Kupfermantel: copper-jacketed relay, mit Kupferring: copper collar relay;
Vibrations= —, vibrating relay;
Wechselstrom= —, alternating-current relay;
Zähl= —, meter(ing) relay;
Zeit= —, time-delay relay, mit Bremszylinder: dashpot relay;
— =anker m, tongue, armature;
— =Anrufsucher m, relay finder A;
— =brett n, abgefedertes, spring tray T;
— =gestell n, relay rack;
— =kette f, relay chain;
— =kontakte pl, relay points pl;
— =satz m, relay set, relay unit;
— = — und Folgeschalter für Leitungswähler, final sequence switch and relay set, A;
— =station f, — =stelle f, relaying station;
— =system n, Selbstanschluß=, relay automatic telephone system;
— -übertragung f, relay repeater T;
— =vorwähler m, relay preselector.

Reliefschreiber m, embosser.
Reluktanz f, reluctance, magnetic resistance.
Reluktivität f, reluctivity.
remanent, residual, remanent;
— er Magnetismus m, residual magnetism.
Remanenz f, remanence.
rentabel, productive.
Rentabilität f, productiveness.
Reparatur f, repair.
reparieren, to repair.
reproduzierbar, reproducible.
reproduzieren, to reproduce.
Reserve f, reserve;
— -ader f, reserve wire; reserve pair;
— -anker m, spare armature;
— -apparat m, spare instrument;
— -nummer f, unallotted number F;
— -satz m, spare set, spare unit;
— -teile pl, spares, spare parts pl.
Resonanz f, resonance, syntony;
auf — abgestimmt, tuned to resonance;
außer — befindlich, non-resonant;
der — entgegenwirkend, anti-resonant;
in — befindlich, resonant, in resonance (mit, with), in tune (mit, to);
in — sein, to resonate;
Parallel- —, parallel resonance;
Reihen- —, series resonance;
— -anzeiger m, resonance indicator;
— -bedingung f, condition of resonance, resonant condition;
— -bereich m, resonant range;
— -erscheinung f, resonance phenomenon;
— -frequenz f, resonant frequency, resonance frequency;
— -gebilde n, resonant combination;
— -grundfrequenz f, first resonating frequency;
— -kreis m, (Reihen-), series-resonant circuit;
— - —, Parallel-, parallel or multiple resonant circuit;
— - —, Spannungs-, series-resonant circuit;
— - —, Strom-, parallel resonant circuit;
— - und Drosselkreis in Reihe, series-multiple resonant circuit;
— -kurve f, resonance curve;
— -lage f, resonant range;
— -methode f, resonating method;
— -nebenschluß m, resonant shunt;
— -relais n, resonance relay;
— -schärfe f, sharpness of resonance;
— -spitze f, resonance peak;
— -transformator m, resonance transformer;
— -verlauf m, resonance curve;
— -wirkung f, resonant effect, resonance effect;
— -zustand m, resonant condition, condition of resonance.
Resonator m, resonator.
resonieren, to resonate.
Rest m, rest, remainder;
— -dämpfung f, net transmission equivalent, overall transmission loss, net attenuation;
— - —s -messung f, overall (toll circuit) transmission test.
Resultante f, resultant.
resultieren, to result;
— d, resultant.
Retortenkohle f, retort carbon.

Retransmetteur *m*, retransmitter *T*.

Rettungsboot-Funkeinrichtung *f*, lifeboat wireless set.

reversibel, reversible.

Revolverdrehbank *f*, capstan lathe.

reziprok(er Wert *m*), reciprocal.

Rheostat *m*, rheostat, resistor.

Rheotan *n*, rheotan.

Rhodium *n*, rhodium (Rh).

Rhombus *m*, lozenge.

Rhythmus *m*, rhythm.

Richt- ...., directive, directional;

— -antenne *f*, directional aerial;

— -dorn *m*, mandrel, mandril, *B*;

— -empfang *m*, directional reception;

— -empfänger *m*, directional receiver;

richten, to direct, to position, to set; geradlinig stellen: to align; gerade —: to straighten.

richtfähig, (stark), (highly) directive;

Richt-fähigkeit *f*, directional property;

— -größe *f*, directional quantity.

Richtigkeit *f*, correctness; faithfulness (der Wiedergabe, of reproduction).

Richt-kraft *f*, directing force, controlling force;

— -latte *f*, level;

— -linie *f*, aim;

— -magnet *m*, setting magnet, controlling magnet;

— -maß *n*, rule;

— -scheit *n*, batten;

— -senden *n*, directional transmission;

— -sender *m*, directional transmitter;

— -spule *f*, control coil;

— -vermögen *n*, directivity;

— -weiser *m*, drahtloser, radio beacon;

— -wirkung *f*, directional effect, (uni)directional effect;

— - —, einseitige, unidirectional action.

richtwirkungsfrei, non-directional.

Richtung *f*, direction;

in einer —, one-way;

ausgepeilte —, bearing;

Dreh- —, direction of rotation;

Richtungs-bestimmung *f*, drahtlose, wireless direction finding;

— -finder *m*, direction finder.

Ricinusöl *n*, castor oil.

riefeln, to groove.

Riegel *m*, bolt, slip bolt, sliding bolt, fastening;

Quer- —, traverse, transom.

Riemen *m*, belt;

Treib- —, (driving) belt;

— -antrieb *m*, belt drive;

— -scheibe *f*, pulley.

Riffelblech *n*, channeled plate.

Rille *f*, groove;

Hals- —, neck groove *B*;

Kopf- —, top groove *B*;

— n-rolle *f*, grooved roller.

Ring *m*, ring, torus, collar, washer;

geteilter —, segmented ring;

ungeteilter —, solid ring *T*;

Brems- —, brake ring;

Eisen- —, ferrule;

Empfangs- —, receiving ring *T*;

Gewinde- —, threaded ring;

Gleichlauf- —, correcting ring *T*;

Klemm- —, clamping ring;

Kollektor- —, collecting ring, collector ring;

Ring
 Kupfer= —, copper collar, copper jacket (am Fernsprech=relais, of telephone relays);
 Orts= —, local ring T;
 Schleif= —, slip ring, collecting ring;
 Segment= —, segmented ring;
 Spann= —, clamping ring;
 Sprih= —, thrower;
 Sende= —, transmitting ring;
 Verteiler= —, distributor ring, crown, T;
— =anker m, ring armature;
— =bolzen m, eye bolt;
ringförmig, ring-shaped, ring-like, annular, toroidal;
Ring=magnet m, annular magnet;
— =schmierung f, ring lubrication;
— =spule f, toroidal coil;
— =übertrager m, toroidal repeating coil, ring transformer;
 Doppelsprech= — = —, phantom repeating coil.
Rinne f, groove, trough(ing), ditch, run;
 Eisen= —, iron troughing B;
 Holz= —, wood trough;
 Kabel= —, cable trough(ing), cable channel;
 U=förmige —, U-troughing.
Rippe f, rib;
 Kühl= —, cooling vane, cooling flange;
— u =gefäß n, ribbed tank.
Risiko n, risk.
Riß m, Zeichnung: plan, drawing; im Draht: flaw, im Isolator: crack.
Ritz m, scratch.
ritzen, to scratch.
roh, raw;
— e Stange f, untreated pole, plain pole, B.
Roh=baumwolle f, cotton wool;

— =gummi m (n), raw rubber;
— =haut f, raw hide;
— = — =aufhänger m, raw hide suspender B;
— =öl n, crude naphta;
— =stoff m, raw material.
Rohr n, tube, pipe, duct, conduit; valve, tube (am.), V (v. Röhre);
 irdenes —, earthenware pipe B;
 zweiteiliges —, split pipe B;
 Beton= —, concrete pipe;
 Durchführungs= —, wall tube;
 Eisen= —, iron tubing;
 Erd= —, soil pipe;
 Isolier= —, insulating tube;
 Kapillar= —, capillary tube;
 Sprach= —, speaking tube;
 Zement= —, concrete pipe;
— =biegezange f, pipe bending tongs pl;
— =bogen m, bend;
— =flansch m, pipe flange, pipe socket;
— =haken m, pipe hook, wall hook;
— =kanal m, pipe line B;
— =knie n, pipe elbow;
— =kratzer m, pipe scraper;
— =krümmer m, bend;
— =leitung f, conduit, tubing;
— =mast m, tubular pole, tube pole;
 Stahl= — = —, tubular steel mast;
— =netz n, piping;
— =post f, pneumatic (dispatch) tube;
 durch — = —, by tube;
 Haus= — = —, house tube(s pl);
 Zettel= — = —, pneumatic ticket carrier;
— = — =anlage f, pneumatic tube plant, pneumatic tubes pl;
— = — =büchse f, (pneumatic dispatch) carrier;

Rohrpost
— = —**empfangsstelle** *f*, pneumatic tube receiving station;
— = —**rohr** *n*, dispatch tube;
— = —**sendestelle** *f*, pneumatic tube dispatching station;
— = —**zettelverteiler** *m*, pneumatic ticket distribution position;
— **-schelle** *f*, wall hook, wall clamp;
— **-schutzkappe** *f*, valve protecting cap V.
— **-ständer** *m*, tube pole, tubular pole;
  nach oben verjüngter — = —, tapered tube pole;
  zylindrischer — = —, parallel tube pole;
— **-stutzen** *m*, nozzle;
— **-strang** *m*, pipe line conduit;
  einzügiger — = —, single-duct conduit;
  mehrzügiger — = —, multiple-duct conduit;
  Eisen- = —, iron pipe conduit;
  Fiber- = —, fibre duct, fibre conduit, B;
  Kabel- = —, cable duct;
— **-zange** *f*, pipe wrench.
Röhrchen *n*, tube;
  Haar- —, capillary tube;
  Papier- —, paper jointing tube B,
  Schreib- —, pen, syphon.
Röhre *f*, pipe, tube; valve, tube (am.), V;
  anodenlose —, plateless valve;
  Braunsche —, Braun tube;
  fremderregte —, separately excited tube;
  gasgefüllte —, gas content tube;
  harte —, hard valve;
  hochbeheizte —, bright (emitting) valve;
  rückgekoppelte —, self-excited valve;
  schwach beheizte —, dull (emitting) valve;
  selbsterregte —, **selbstgesteuerte** —, self-excited valve;
  weiche —, soft valve;
  Beeinflussungs- —, modulator tube;
  Detektor- —, detecting tube, detector valve, audion;
  Doppelgitter- —, double-grid valve, negatron;
  — = — **in Schutznetzschaltung**, pliodynatron;
  Dreielektroden- —, triode (valve), triple electrode valve, oscillion, audion;
  — = —, **Hochvakuum-**, pliotron;
  Elektronen- —, thermionic valve, (electron) valve, electronic tube, pliotron, audion;
  Eingitter- —, single grid valve;
  Entladungs- —, (electron) discharge tube;
  Erreger- —, exciter tube;
  Fünfelektroden- —, pentode;
  Gleichrichter- —, rectifier valve rectifying valve;
  — = —, **Empfangs-**, audion;
  — = —, **Hochleistungs-**, power rectifying valve;
  — = —, **Hochvakuum-(Glüh-kathoden-)**, kenotron;
  Glühkathoden- —, thermionic valve, ionic valve;
  Hochleistungs- —, power tube;
  Hochvakuum- —, high vacuum valve;
  Hochvakuumgitter- —, pliotron;
  Kathoden(strahlen)- —, thermionic valve;
  Kraft- —, tube of force; große Senderöhre: power tube;
  Mehrgitter- —, multiple grid valve;

Röhre
Meß=—, graduated tube;
Modulator= —, modulating valve;
Oxyd(kathoden)= —, oxide-coated filament electron tube, dull emitter valve;
Schwing= —, oscillator valve, generating tube;
— = —, fremderregte, separately excited oscillator valve;
— = —, selbsterregte, self-excited oscillator valve;
Sende= —, transmitter valve, transmitting tube, generator triode;
— = —n =gestell n, power tube rack;
Steuer= —, master oscillator, pilot oscillator, exciter tube, control valve;
Thorium= —, thoriatid filament valve;
Vakuum= —, vacuum valve;
Ventil= —, valve, diode;
Verstärker= —, amplifier valve, amplifying tube, amplifier triode, strengthening tube;
— = —, Fernsprech=, telephone repeater valve, telephone amplifying tube;
Vierelektroden= —, four-electrode valve, tetrode;
Wehnelt= —, Wehnelt valve;
Zweielektroden= —, two-electrode valve, diode, Fleming valve;
Zweigitter= —, double grid valve;
— mit dunkelrotglühendem Faden, dull emitter valve.
röhrenförmig, tubular;
Röhren=anlage f, tubing;
— =detektor m, (thermionic) valve detector, audion;
— =empfang m, valve reception;
— = — mit gleichzeitiger Hoch= und Niederfrequenzverstärkung, dual reception, reflex reception;
— =empfänger m, valve receiver;
— = — ohne Anodenbatterie, solodyne receiver;
— =fassung f, valve holder, valve socket;
— =generator m, valve generator, electron tube generator;
— =gleichrichter m, vacuum tube rectifier, valve rectifier;
— =kabel n, conduit cable;
— =kapazitäten pl, inter-electrode capacities pl;
— =kennlinie f, valve characteristic;
— =kopplung f, intervalve linkage, intervalve coupling;
— =modulator m, vacuum tube modulator;
— =muffe f, pipe socket;
— =oszillator m, valve oscillator;
— =schaltung f mit induktiver Rückkopplung, Hartley circuit;
— = — mit kapazitiver Rückkopplung, Colpitts circuit;
— = — mit magnetischer Rückkopplung, Meißner circuit;
— =sender m, valve transmitter;
— =sockel m, valve socket;
— =spannungsmesser m, amplifying voltmeter;
— =summer m, electron tube generator;
— =verstärker m, valve amplifier;
— =voltmeter n, amplifying voltmeter;
— =widerstand m, tube resistance;
— =Zwischenstecker m, valve adapter, valve adaptor.
Röllchen n, roller;
Farb= —, inking roller;

Röllchen
Reiter= —, jockey roller, jockey wheel.
Rolle f, roller, pulley; Flaschenzug: block;
Druck= —, impression roller; platen T;
Führungs= —, guide pulley, guide roller;
Kordel= —, milled roller;
Leit= —, Pack= —, guide roller, guide pulley;
Papier= —, web of paper;
Quer= —, bridging coil;
Rillen= —, Schnur= —, grooved roller;
Streifen= —, tape roll, roll of slip, T;
rollen, to roll.
Rollen-halter m, tape roll holder T;
— =gewicht n, pulley weight (der Stöpselschnüre, of plug cords);
— =lager n, roller bearing.
Roll-leiter f, rolling ladder.
— =schuhe pl, roller skates pl;
— =tisch m, wheeled stand;
— =wagen m, truck.
rosa, pink.
Rost m, rust; Gitter: grating, rack;
Kabel= —, cable rack, cable shelf.
rosten, ·to rust.
rost=frei, rustfree;
— =sicher, rust-proof.
rot, red;
hell= —, bright red;
kirsch= —, cherry red;
purpur= —, purple.
Rotation f, rotation.
Rotbuche f, red beech.
rotglühend, red hot;
dunkel= —, dull red hot;
hell= —, bright red hot.
Rotglut f, red heat; [heat;
Dunkel= —, dim or dull red

Hell= —, bright red heat.
Rotguß m, red brass.
rotieren, to rotate, to revolve;
— =d, rotary.
Rotor m, rotor, rotator;
— der Funkenstrecke, spark gap rotor.
rotwarm, red hot.
Rotzinkerz n, zincite, red oxide of zinc (ZnO).
Rubrik f, column.
Ruck m, jerk.
Rück-anschlag m, back stop;
— =anschluß m, back connections pl (der Schalttafel, of switchboard);
— =ansicht f, back view, rear view;
— =auslösung f, back release A;
Vor= und — = —, first party release A;
— =bewegung f, back stroke, unshift; retrogression.
rucken, to jerk.
Rück-entladung f, back discharge;
— =frage f, request, ab: RQ;
— = —halten, to request T;
rückfragen, te request.
Rückführdaumen m, resetting cam;
— =führen, to reset;
Rückführ-feder f, restoring spring, controlling spring, retracting spring;
— =relais n, clear-out relay.
Rückführung f, return;
— des Papierschlittens, carriage return T; [turn.
Rückgang m, retrogression; rückgekoppelt, back-coupled, retroactive.
Rück-hub m, back stroke, return stroke;
— =kehr f, return;
rückkoppeln, to couple back, to feed back;

Rückkopplung *f*, reaction, reaction coupling (*ab*: r. c.), feed(ing)-back, back-coupling;
 induktive —, inductive feed-back;
 kapazitive —, electrostatic feed-back;
 negative —, negative reaction;
 zu starke — im Empfänger, too much tickler;

Rückkopplungs-...., regenerative;
— -audion *n*, regenerative valve detector, retroactive audion;
— -empfang *m*, regenerative reception, retroactive reception;
— - — mit Hilfsfrequenz, super-regenerative reception;
— -empfänger *m*, regenerative receiver;
 Hilfsfrequenz- — —, super-regenerative receiver;
— -empfangsschaltung *f*, feed-back receiving circuit, retroactive receiving circuit;
— -faktor *m*, regenerative coefficient of coupling;

rückkopplungsfrei, non-regenerative;

Rückkopplungs-grad *m*, reaction coefficient;
— -kondensator *m*, reaction condenser;
— -methode *f*, reaction method, feed-back method;
— -prinzip *n*, reaction principle;
— -schaltung *f*, feed-back connection; [circuit;
 Meißnersche - — —, Meißner
— -spule *f*, reaction coil;
— - — des Schwingaudions, tickler coil;
— -transformator *m*, reaction transformer;
— -verstärkung *f*, regeneration, regenerative amplification.

Rücklauf *m*, return.
rücklaufen, to return (to normal).
Rückleiter *m*, return, return wire, return conductor.
Rückleitung *f*, return path, return;
 gemeinsame —, common return;
 magnetische —, magnetic return path;
 metallische —, metallic return, return wire;
 Erd- —, earth return, ground return, ground circuit;
— - —, Leitung *f* mit, earth return circuit;
 Schienen- —, rail return;
 See- —, sea return.

Rückruf *m*, recall, back ring;
— auf eigene Leitung, reverting call *A*;
— -taste *f*, ring back key;
— -wähler *m*, reverting call switch *A*.

Rück-schlag *m*, back kick, back stroke;
— -schub *m*, unshift *T*;
— -seite *f*, rear, back;
— -speisung *f*, feeding-back, retransfer;
— -stand *m*, residue, residuum;
 Gas- — — —, residual gas;
 Luft- — — —, residual air.

rückstellen, to reset, to replace.
Rückstell-klappe *f*, plug-restored shutter, self-restoring indicator;
— -magnet *m*, release magnet *A*; resetting magnet;
— -strom *m*, releasing current;
— -taste *f*, resetting key.

Rückstellung *f*, release, replacement.

Fallklappe *f* mit elektrischer (mechanischer) —, electric (mechanical) replacement indicator.

Rückstrom *m*, return current;
— =relais *n*, reverse current relay.

rückübertragen, to retransfer.

Rückübertragung *f*, retransfer (von Energie, of energy).

rückübersetzen, to retranslate.

Rückübersetzung *f*, retranslation.

rückverwandeln, to retranslate.

Rückweg *m*, return path.

ruckweise, by jerks.

Rückwerfer *m*, rejector.

rückwirken, to re(tro)act (auf, on);
— d, reactive.

Rückwirkung *f*, re(tro)action, reactive effect;
Anker= —, armature reaction.

Rückzug-feder *f*, retracting spring;
— =taste *f*, back-spacing key *T*.

Ruf *m*, ring, call; ringing;
abgestimmter —, harmonic selective ringing;
intermittierender —, interrupted ringing;
selbsttätiger —, keyless ringing;
selbsttätig wiederholter —, interrupted ringing;
unbeantworteter —, no-reply call;
wahlweiser — nach einem Rufschlüssel, code ringing;
— — mit abgestimmten Einrichtungen, harmonic selective ringing.

rufen, to ring, to call;
wahlweise —, to call selectively;
zurück= —, to recall, to ring back.

Rufen *n*, ringing, calling;
— mit Durchrufrelais, relayed ringing;
— in Schleifenschaltung, loop ringing;
— in Simultanschaltung, composite (through-) ringing;
— mit gleichstromüberlagertem Wechselstrom, superposed ringing;
— mit Maschinen= oder Polwechslerstrom, power ringing.

rufender Teilnehmer *m*, calling party.

Ruf-lampe *f*, calling lamp, calling pilot;
— =relais *n*, signalling relay, calling relay;
Durch= — = —, through-ringing relay;
— =schaltung *f*, ringing connection;
Durch= — = —, ringing-through scheme;
— =schlüssel *m*, ringing code, calling code; Schalter: ringing key;
— =störung *f*, ringing failure;
— =strom *m*, ringing current;
gleichstromüberlagerter — = —, superposed ringing current;
— = =anzeiger *m*, ringing vibrator, ringing current indicator;
— = =dynamo *f*, ringing dynamo, ringer;
— = (— =) maschine *f*, ringer, ringing machine;
aus der Z.B. gespeiste — = — =, battery ringer;
— = — =quelle *f*, ringing source;
— = — =Umkehrtaste *f*, ringing reversing key (für Gesellschafts-leitungen, for party lines);
— = — =zuführung *f*, ringing lead(s *pl*);
— =taste *f*, ringing key;
Rück= — = —, ring back key;
— =zeichen *n*, call letter, calling code, code letters *pl*.

Ruhe *f*, rest; Stille: silence;
in — befindlich, idle, non-operative, inoperative;

Ruhe
— **anschlag** m, spacing stop.
ruhend, static; inoperative;
— **er Transformator** m, static transformer.
Ruhe-kontakt m, rest(ing) contact, spacing contact T,
— **lage** f, resting position, position of rest;
 in der — - —, at rest, inoperative;
— **schiene** f, backstop, spacing stop, T;
— **stellung** f, unoperated position, resting position, home position, idle position;
 in der — - —, at rest, normal;
— **strom** m, **Anoden-**, feed current V;
— - — **betrieb** m, closed circuit working;
— - — **schaltung** f, closed circuit connection;
— **weg** m, path of rest T;
— **zustand** m, state of rest.
rund, round;
 kreis- —, circular.
Runddraht m, round wire;
— **bewehrung** f, round wire armouring.
Rundfeuer n, flash(ing)-over.

Rundfunk m, broadcasting;
 durch — verbreiten, to broadcast;
— **empfang** m, broadcast reception;
— **empfänger** m, broadcast receiver;
rundfunken, to broadcast;
Rundfunk-gerät n, broadcast apparatus;
— **sender** m, broadcast transmitter;
— **station** f, — **stelle** f, broadcasting station;
— **Zwischensender** m, remotely controlled broadcast transmitter.
Rund-holz n, round timber, spar;
— **kern** m, round core;
— **kopf** m, round head.
rundköpfig, round-headed.
Rundspruch m, broadcasting;
 Draht- —, electrophone (engl.), program transmission over wires (am.);
— **empfänger** m, broadcast receiver.
Rundzange f, round nose(d) pliers pl.

S.

Saal m, room;
 Apparat- —, instrument room, instrument gallery; switch room F, auto room A.
 Auslands- —, foreign gallery (engl.);
— **geräusch** n, crowd noise.
Sack m, sack.
sacken, to sag.
saftreich, sappy.
Säge f, saw;

 Hand- —, hand saw;
— **blatt** n, sawblade;
— **mehl** n, saw-dust.
sägen, to saw.
Sägenfeile f, saw file.
Saite f, string, cord, chord;
 schwingende —, vibrating cord;
Saiten-elektrometer n, string electrometer;
— **galvanometer** n, string galvanometer;

Saiten
— ≈instrument n, Musikinstr.: string(ed) musical instrument;
— ≈oszillograph m, string oscillograph;
— ≈summer m, chord buzzer;
— ≈unterbrecher m, vibrating wire interrupter.

Salmiak m, sal-ammoniac, ammonium chloride, (NH_4Cl);
— ≈element n, Leclanché cell.

Salpetersäure f, nitric acid (HNO_3).

Salz n, salt;
— ≈säure f, muriatic acid, hydrochloric acid, (HCl);
— ≈wasser n, brine, salt water.

Sammel=amt n, smaller centre T;
— ≈dienstleitung f, split order wire (circuit).

sammeln, to collect.

Sammel=ring m, collecting ring, collector ring;
— ≈schiene f, bus bar, collecting bar.
— ≈stelle f, collecting point.

Sammler m, storage cell, accumulator, secondary cell;
— nachfüllen, to top up, to fill up storage cells;
alkalischer —, Edison storage cell;
tragbarer —, portable storage cell;
Blei= —, lead storage cell, lead-sulphuric acid cell;
Bleistaub= — lead dust storage cell;
Chlorid= —, chloride storage cell;
Edison= —, Edison storage cell;
— ≈batterie f, storage battery, secondary battery;

— ≈ — von 2000 Ah bei zehn=stündigem Betrieb, storage battery of 2000 AH at the 10 hours rate;
— ≈gefäß n, accumulator box or jar;
— ≈platte f, accumulator plate;
— ≈ —n, Wachsen n der, fanning-out of storage cell plates;
— ≈raum m, battery room;
— ≈zelle f, storage cell;
gegengeschaltete — ≈ —, counter-cell.

Sammlung f, collection.

Sand m, sand;
— ≈bank f, sandbank;
— ≈papier n, sand paper;
— ≈strahlgebläse n, sand blast;
— ≈uhr f, sand glass.

Saubelholz n, redwood.

Saphir m, sapphire.

sättigen, to saturate.

Sättigung f, saturation.

Sättigungs=dichte f, saturation density;
— ≈grad m, degree of saturation;
— ≈grenze f, saturation point;
— ≈knie n, saturation bend;
— ≈punkt m, saturation point;
— ≈strom m, saturation current;
— ≈wert m, saturation value.

Satz m, Apparate: set, gang, assembly; Verhältnis: rate; Apparat= —, set of instruments;
Federn= —, spring assembly;
Stempel= —, gang of punches T;
— ≈zeichen n, punctuation mark.

sauber, clean.

säubern, to clean, to clean out.

Säuberung f, cleaning.

Sauerstoff m, oxygen (O).

saugen, to suck, elektr.: to surge;

Saug=luft f, vacuum;
— ≈spule f, sucking solenoid;

Saug
— -transformator m, booster transformer, suction transformer;
— -welle f, surge.
Säule f, pillar, column; Batterie: pile;
Volta'sche —, voltaic pile;
Zamboni'sche —, Zamboni (dry) pile;
Untersuchungs= —, pillar test box;
— n -tischfernsprecher m, desk stand telephone set.
Säure f, acid;
von — angegriffen, attacked by acids;
Akkumulator= —, accumulator acid;
Chlorwasserstoff= —, hydrochloric acid (HCl);
Füll= —, accumulator acid;
Salpeter= —, nitric acid (HNO_3);
Salz= —, muriatic acid, hydrochloric acid, (HCl);
Sammler= —, accumulator acid;
Schwefel= —, sulphuric acid (H_2SO_4);
— -anfressung f, acid corrosion;
säurebeständig, acid-proof;
Säure=dämpfe pl, acid fumes pl;
säure=fest, acid-resisting, acid-proof;
— -haltig, acid-laden;
Säure-ion n, acid ion;
— -messer m, hydrometer;
Heber= — = —, hydrometer syringe;
— -prüfung f, acid test.
schaben, to scrape.
Schaber m, scraper.
Schablone f, former.
Schacht m, pit, chute, run;
gemauerter —, brick |pit;
Kabel= —, cable chute.

Schaden m, damage;
— nehmen, — zufügen, to damage.
schadhaft, damaged.
Schaft m, shank.
Schäkel m, shackle.
schälen, to bark.
Schale f, dish, bowl; Wecker: dome, gong;
Glocken= —, [bell dome, |flach: bell gong.
schalenförmig, dished.
Schalenhalter m, gong support.
Schall m, sound;
— -dämpfer [m, |damper, sourdine.
schalldicht, sound-proof;
— e Zelle f, silence cabinet.
Schall=dose f, sound box;
— -empfänger m, sound receiver;
Unterwasser= — = —, subaqueous sound receiver;
— -empfindung f, auditory sensation;
— = —, Schwellenwert schmerzhafter, threshold of feeling.
schallen, to sound.
— der Schlag m, bang.
Schall-erzeuger m, sound generator:
— -frequenz f, acoustic frequency;
— -kammer f, sound box, des Klopfers: screen;
— -stärke f, sound intensity;
— -medium n, sound propagating medium;
— -meßeinrichtung f, sound measuring device;
— -platte f, disc record;
— -quelle f, sound generator;
— -schwingung f, sound vibration;
— -spektrum n, sound spectrum;

Schall
- =trichter *m*, mouthpiece *F*; Lautsprecher: horn, trumpet, großer: projector;
- =welle *f*, sound wave; sinusförmige - = -, sine wave of sound.

Schalmei=glocke *f*, sheep gong;
- =wecker *m*, gong bell.

Schaltader *f*, jumper wire, cross-connecting wire;
- = - =feld *n* des Verteilers, jumpering field of distributing frame;
- =arm *m*, (rotary) wiper *A*; die - = -e weiterdrehen auf, to step round the wipers to, *A*;
- =bild *n*, circuit diagram, wiring diagram; schematisches - = -, skeleton; Gesamt= -=-, full connection diagram;
- =draht *m*, jumper (wire), cross-connecting wire; mit - = - verbinden mit, to jumper to;
- =einrichtung *f*, switching device, switching equipment;
- =element *n*, circuit element.

schalten, to switch, to connect, to join; mechanisch: to step; in Brücke - zu, to bridge across, to place across, to tee across;
gemischt -, to connect in series-multiple, to connect in multiple arc;
hintereinander -, to join in series;
nebeneinander -, to connect in parallel;
normal -, to set at normal;
parallel -, to join in parallel;
in Reihe -, to join in series;
vielfach -, to connect in multiple.

Schalten *n*, switching, joining, connection; stepping.

Schalter *m*, switch, key;
einen - öffnen, to open a switch;
einen - schließen, to close a switch;
einen - umlegen, to throw a switch;
- mit zwei Stellungen, two-way switch, two-position switch;
- - fünf Ausgängen, five-point switch;
einpoliger -, single pole switch;
mehrteiliger -, multiple point switch;
sechsteiliger - mit zwei Stellungen, two position six point switch;
umlaufender -, revolving switch;
vielstufiger -, multi-point switch;
zweipoliger -, double pole switch, d. p. switch;
Anlaß= -, starting switch;
Annahme= -, counter *T*;
Dreh= -, spindle switch, revolving switch; rotary switch *A*;
Dreiwege= -, three-way switch;
Erdungs= -, earthing switch;
Fern= -, remote control switch, teleswitch;
Folge= -, sequence switch *A*;
Fuß= -, foot switch;
Gruppen= -, group switch;
Haupt= -, main switch, master switch;
Hebel= -, lever switch;
- = -, zweipoliger, double lever switch;
Hörner= -, horn-type switch;
Kipp= -, switching key;

Schalter
 Kurbel- —, lever switch, radial arm switch;
 Lade- —, battery charging switch;
 Mast- —, pole switch;
 Messer- —, knife (blade) switch;
 Moment- —, quickaction switch, quick break switch;
 Motoranlaß- —, motor-starting switch;
 Nacht-Zentral- —, night concentrator;
 Post- —, shutter, counter;
 Schnapp- —, tumbler switch;
 Schritt- —, step-by-step switch, stepping mechanism;
 Selbst- —, relay;
 Sitz- —, socket contact F;
 Stern-Dreieck- —, star-delta switch;
 Steuer- —, sequence switch, control(ling) switch, master switch, A;
 Stöpsel- —, plug switch;
 Trenn- —, disconnecting switch, circuit breaker;
 — - —, Öl-, oil break switch;
 Tret- —, foot switch;
 Tür- —, door switch;
 Umkehr- —, reversing switch;
 Verstimmungs- —, wave changing switch;
 Wahl- —, selector switch;
 Walzen- —, barrel switch;
 Wechsel- —, double throw switch;
 Wende- —, reversing key;
 Zeit- —, time switch;
 Zentral- —, concentrator;
 — - —, Nacht-, night concentrator;
 Zug- —, pull switch;
 — -feder f, switch spring;
 — -gestell n, switch frame;
 — -reihe f, bank or row of keys;
 — -sockel m, switch base;
 — -tisch m, switching desk; Telegrammannahme: counter;
 Schalt-magnet m, switching magnet; driving magnet, stepping magnet, A;
 — -mittel n, switching device;
 — -plan m, wiring scheme, wiring diagram;
 — -pult n, switch desk;
 — -raum m, switch room;
 — -relais n, switching relay;
 — -satz m, End-, terminal repeater (am.);
 — -schema n, connections pl, wiring scheme;
 — -schlüssel m, switching key, throw key, push key;
 — -schrank m, switching cabinet;
 — -stange f, switching bar;
 — -stufe f, switch step;
 — -tafel f, switchboard, board;
  Kraft- — - —, power (switch-) board, power panel;
  Lade- — - —, charging switchboard;
  Wand- — - —, wall pattern switchboard.
Schaltung f, connection;
 gemischte —, series-multiple connection, multiple arc connection, parallel-series connection;
 Dreieck- —, delta connection, mesh connection;
 End- —, terminal circuit;
 Endkunst- —, terminal circuit, terminal network;
 Entzerrer- —, compensating network, correcting network;
 Fern- —, remote control;
 Kompensations- —, compensating circuit;
 Nebeneinander- —, Parallel- —, parallel connection;

Schaltung
 Reihen= —, series connection;
 Stern= —, Y-connection, star connection;
 Zweig= —, leak connection;
Schalt-verbindung f, circuit connection;
 — =vorgang m, switching operation;
 — =vorrichtung f, switching device, switch gear;
 — =welle f, wiper shaft A;
 — =werk n, switch gear;
 Fort= — = —, stepping mechanism;
 Schritt= — = —, step-by-step switch; stepping mechanism.
scharf, sharp.
Schärfe f, sharpness; harshness;
 Abstimm=—, sharpness of tune, sharpness of resonance.
schärfen, to sharpen.
Scharnier n, hinge;
 mit —en versehen, hinged;
 — =rahmen m, hinged frame.
Schatten m, shadow, cloud;
 — =stelle f, Empfangsloch: blind spot R.
schattieren, to shade, to hatch.
Schattierung f, shading, hatching.
schätzen, to rate, to estimate.
Schätzung f, estimate, rating.
Schaufel f, shovel.
schaufeln, to shovel.
schaukeln, to rock.
Schau-linie f, graph, curve;
 — =loch, n, inspection hole, sight opening;
 — =zeichen n indicator, visual signal;
 Gitter= — = —, grid indicator;
 Stern= — = —, white star indicator.

Scheibe f, disc; wheel; Verteiler: face, disc, plateau, plate, T;
 bewegliche —, movable disc T;
 feste —, fixed plate T;
 hintere —, rear plate T;
 stroboskopische —, stroboscopic disc;
 vordere —, front plate T;
 Empfangs= —, receiving disc T;
 Fall= —, drop shutter;
 Finger= —, finger disc;
 Friktions= —, friction disc;
 Kern= —, core disc;
 Kombinator= —, combiner disc, combiner wheel, T;
 Kupplungs= —, clutch disc;
 Mitnehmer= —, driving disc;
 Reib= —, friction disc;
 Spulen= —, spool flange, spool head;
 Übersetzer= —, combiner disc T;
Scheiben-anker m, disc armature;
 — =elektrometer n, disc electrometer;
 — =funkenstrecke f, disc discharger;
 glatte — = —, smooth disc discharger;
 Asynchron- — = —, asynchronous disc discharger;
 Synchron- — = —, synchronous disc discharger;
 Zahn= — = —, studded disc discharger.
Scheider m, separator;
 Glasrohr= —, glass tube separator;
 Holz= —, wood separator.
Scheidewand f, separator; diaphragm; Schott: bulkhead.
Schein= ...., apparent; dummy.

scheinbar, apparent;
— er Widerstand m, apparent resistance.

Schein-dämpfung f, apparent attenuation;
— -komponente f, apparent component;
— -leistung f, apparent power;
— -leitwert m, admittance, einer Leitung gegen Erde oder zwischen den Drähten einer Doppelleitung: line shunt admittance;
— -widerstand m, impedance, apparent impedance;
Blindkomponente f, des — - —s, reactive impedance;
Wirkkomponente f des — - —s, dissipative impedance;
— - — -meßbrücke f, impedance bridge.

Scheitel f, der Brücke: bridge apex, der Differentialspule: split point; einer Welle: wave crest, peak;
— -faktor m, amplitude factor;
— -messer m, crest meter;
— -punkt m, b. Brücke: apex;
— -spannung f, peak voltage;
— -wert m, peak, peak value, amplitude, crest, maximum value.

Schellack m, shellac.
schellackieren, to shellac.
Schelle f, clamp, clip, clamping ring;
Erd- —, earth clip, ground clamp.

Schema n, scheme, diagram, schedule;
Arbeits- —, schedule of operation.
schematisch, schematic;
— e Zeichnung f, skeleton sketch.
Schenkel m, limb, leg, foot;
Achs- —, journal, axle journal;
Magnet- —, magnet limb.

. . . -schenklig, . . . -legged;
zwei- —, two-legged.

Schere f, shears pl;
Blech- —, cutting shears pl.
scheren, to shear.
Scherfestigkeit f, shearing strength.

Scheuer-bock m, chafe rod B;
— -leiste f, skirting;
— -pfahl m, chafe rod, stay guard, B.
scheuern, to chafe.

Schicht f, layer, coat, coating, film, lamina; Luft: stratum; obere —en pl, upper strata pl R;
Heaviside- —, Heaviside layer;
Luft- —, stratum of air.
schichten, to pack.
Schiebe-gestänge n, sweep's rods pl B.
—, -kontakt m, sliding contact;
— -leiter f, travelling ladder.
schieben, to push, to slide, to shift.
Schieber m, slide(r), cursor;
— -ventil n, slide valve.
Schiebe(r)widerstand m, slide rheostat.
Schiebetür f, sliding door.
schief, oblique, inclined; bias; skew;
—e Ebene f, inclined plane;
Schiefe f, obliquity, inclination.
Schiefer m, slate;
— -platte f, slate slab;
— -tafel f, slate slab, slate board.
schieferfarben, slate.
Schiene f, rail; bar;
Gleit- —, sliding bar;
Quer- —, crossbar;
Sammel- —, collecting bar, bus bar;
Spann- —n pl, slide rails pl;

Schienen-kontakt *m*, rail contact;
— -rückleitung *f*, rail return;
— -stoß *m*, rail bond.

Schiff- *n*, ship, vessel;
Verkehr zwischen — und Land, ship-to-shore traffic; |
auf —en, on board ship;
— -funkstelle *f*, ship radio station;
— -körper *m*, hull.

Schild *m*, shield; plate;
Email- —, enamelled plate.

Schirm *m*, screen, shield;
kapazitiver —, electrostatic shield;
magnetischer —, magnetic shield;
— -antenne *f*, umbrella aerial;
— -fläche *f*, shielding surface;
— -isolator *m*, umbrella type insulator;
— -leiter *m*, screened conductor;
— -wirkung '*f*, screening effect, shielding action.

schirmen, to shield (from), to screen, to shade.

Schirmung *f*, shielding;
magnetische —, magnetic shielding.

schlaff, slack.

Schlag *m*, beat; Hub: stroke;
Kabel: lay, turn;
dumpfer—, thump;
elektrischer —, electric shock;
einen — erhalten, to receive a shock;
schallender —, bang;
— -länge *f*, (length of) lay, length of twist;
— -lot *n*, spelter solder.

Schlamm *m*, sediment, mud,
Meer: ooze;
Element- —, battery mud;
— -boden *m*, muddy soil, boggy soil;

— -grund *m*, mud, ooze.
schlammig, oozy, muddy.
Schlämmkreide *f*, whiting.
Schlauch *m*, hose;
Feuer- —, fire hose;
Metall- —, metallic hose.
schlechter Kontakt *m*, poor contact.
Schleifdraht *m*, slide wire; der Meßbrücke: differential slide wire;
— -Meßbrücke *f*, slide wire measuring bridge.

Schleife *f*, im Draht: bight; Leitung: loop;
über die — hinweg, round the loop *F*;
zur — schalten, to loop (Leitungen, lines);
zur — verbunden, looped;
Draht- —, wire loop;
Leitungs- —, loop;
— - —, für ein km, per kilometre loop;
Oszillographen- —, oscillograph vibrator, oscillograph loop.

schleifen, to grind, to smooth.
Schleifen-antenne *f*, loop aerial;
— -berührung *f*, loop, constant or permanent loop, short circuit;
— -kapazität *f*, wire-to-wire capacity *F*;
Vierer- — —, phantom capacity, side-to-side capacity, *F*;
— -schaltung *f*, loop connection;
in — - —, on the loop;
— -system *n*, metallic or two-wire automatic telephone system *A*;
— -wert *m*, loop value (einer Doppelleitung, of a metallic circuit;
— -widerstand *m*, (conductor) loop resistance *F*.

Schleif-feder f, wiper;
— **-maschine** f, grinding machine;
— **-ring** m, slip ring, collector ring;
— **- — -anker** m, slip-ring rotor;
— **-stein** m, grind(ing) stone;
— **-vorrichtung** f, grinding device.

Schlepper m, tractor.

Schlepp-kontakt m, continuity-preserving contact, make-before-break contact;
— **-tau** n, tow.

Schleuse f, lock.

Schlick m, ooze.

schließen, (sich) to close, to shut; verschließen: to lock;
einen Stromkreis —, to close or make or complete a circuit;
einen Schalter —, to close or throw a switch;
einen Vertrag —, to contract, to agree.

Schließen n, make, closure;
— und Unterbrechen, make and break.

Schließkontakt m, make contact;
Doppel- —, double make contact.

Schließung f, make, closure;
— en und Unterbrechungen pl, makes and breaks pl.

Schließungs-funke m, spark at make;
— **-impuls** m, make impulse.

Schlinge f, loop, im Draht: curl.
— n bilden, to curl.

schlingen, to string.

Schlipf m, slip.

Schlitten m, carriage, am Hughes-app: chariot;
Greifer- —, pick-up carrier;
Kontakt- —, contact carriage;
Papier- —, paper carriage T.

Schlitz m, slit, slot, slotted hole;
schräger —, sloping slot;
Kurven- —, curved slot, Nocke: cam slot;
Schrauben- —, nick.

schlitzen, to slit, to slot, to split.

Schloß n, lock, fastening.

Schlüpfung f, slip.

Schlüssel m, key; Buchstaben: code;
Chiffern- —, cipher code;
Geheim- —, code;
Mutter- —, nut key, wrench;
Ruf- —, ringing key; Rufzeichen: code of rings, ringing code;
— **- —**, Rufen n nach einem, code ringing;
— **- —**, Sprech- und, speaking and ringing key;
Schalt- —, switching key, throw key;
Schrauben- — spanner, wrench, nut key.
— **- —**, englischer, coach wrench, monkey wrench.
Telegraphen- —, telegraph code;
Verwürfelungs- —, jumble code;
— **-brett** n, key shelf F;
— **-buchstabe** m, cipher;
— **-reihe** f, bank of keys F.

schlüsseln, to code, to (en)cipher.

Schlüsselung f, coding, enciphering.

Schluß-klappe f, ring-off indicator;
— **-lampe** f, clearing lamp, supervisory lamp;
— **- — am B-Platz**, junction clearing lamp;
— **-zeichen** n, clearing signal; clear; Klappe usw: clearing indicator;
Induktor- — **-** —, ring-off signal;
— **- — -gebung** f, clearing;

**Schluß-zeichen-gebung
selbsttätige** – – – –, automatic clearing, central battery signalling, *ab*: c. b. s.
– – -**relais** *n*, clearing relay, supervisory relay;
– – – – **für den rufenden (verlangten) Teilnehmer,** answering (calling) supervisory relay;
– – – -**strom** *m*, clearing current.
Schmalspurbahn *f*, narrow gauge railway.
schmelzbar, fusible;
schwer –, refractory.
Schmelzeinsatz *m*, fuse.
schmelzen, to melt, to fuse.
Schmelzen *n*, melting, fusing.
Schmelz-punkt *m*, melting point, fusing point;
– -**sicherung** *f*, fuse, fusible cutout;
– -**strom** *m*, fusing current;
– -**tiegel** *m*, crucible, f. Blei: melting pot, melting tank.
Schmerstein *m*, soapstone.
schmiedbar, malleable;
– **er Guß** *m*, malleable casting.
Schmiedeeisen *n*, wrought iron;
schmiedeeisern, wrought iron ....
schmieden, to forge.
Schmiere *f*, grease.
schmieren, to lubricate.
schmierig, smudgy.
Schmier-loch *n*, oil hole, oil run;
– -**mittel** *n*, lubricant;
– -**nute** *f*, oil groove, oil way;
– -**öl** *n*, lubricating oil;
– -**ring** *m*, oil ring.
Schmierung *f*, lubrication;
Docht- –, wick lubrication;
Druck- –, forced oil feed;
Fett- –, grease lubrication;
Ring- –, ring lubrication.
Schmirgel *m*, emery;
– -**leinen** *n*, emery cloth;
– -**papier** *n*, emery paper;
– -**pulver** *n*, emery powder;
– -**scheibe** *f*, emery wheel.
Schnabel *m*, nib, nose.
Schnappschalter *m*, tumbler switch.
schnarren, to jar, to buzz, to burr.
Schnarre *f*, buzzer.
Schnarrwecker *m*, buzzer.
Schnecke, worm.
schneckenförmig, snail formed, helical.
Schnecken-getriebe *n*, worm gear;
– -**rad** *n*, worm wheel.
Schnee-belastung *f*, snow load;
– -**last** *f*, snow load;
– -**sturm** *m*, snow storm;
– -**wehe** *f*, snowdrift.
Schneid-anker *m*, cutting grapnel (für Seekabel, for submarine cables);
– -**backen** *pl*, screw dies *pl*.
Schneide *f*, edge;
Messer- –, knife edge.
schneiden, to cut;
Kraftlinien –, to cut, to intersect magnetic lines of force.
Gewinde –, außen: to thread screws, innen: to tap.
Schneiden *n*, cutting;
– -**aufhängung** *f*, knife-edge suspension;
– -**blitzableiter** *m*, knife-shaped or wedge-shaped lightning arrester, spark gap;
– -**lagerung** *f*, knife-edge suspension;
Relais *n* **mit** – – –, knife-edge relay.
Schneidkluppe *f*, screw stock.
schnell ansprechend, quick operating.
schnellaufend, high speed ....
schnell lösend, quick release ...., quick releasing.

schnell wirkend, quick acting.
schneller werden, to accelerate, to speed up.
Schnelligkeit f, speed, rapidity.
Schnell=morsesystem n, automatic Morse system;
— =telegraph m, high-speed telegraph, automatic telegraph;
— =unterbrecher m, rapid interrupter; ticker R;
— =verkehr m, no-delay service F;
— = — s=amt n, no-delay telephone exchange;
— = — s=netz n, no-delay telephone network.
Schnepper m, latch, tumbler.
Schnitt m, section, intersection; — =...., sectional;
Längs= —, longitudinal section;
Quer= —, cross-section;
— = — =fläche f, cross-sectional area;
— =fläche f, sectional area;
— =punkt m, (point of) intersection;
— =zeichnung f, sectional drawing.
Schnur f, cord, lace, string;
— mit zwei Steckern, loose cord and plugs, double-plugged cord;
biegsame —, flexible cord;
einadrige —, single conductor cord;
fehlerhafte —, defective cord;
zweiadrige —, two-way cord, double conductor cord;
Abfrage= —, answering cord;
Fernhörer= —, phone cord;
Fernleitungs= —, trunk cord;
Lahnlitzen= —, tinsel cord;
Litzen= —, flexible cord;
Platz= —, switchboard cord;
Stecker= —, cord, plug-ended cord;
Stöpsel= —, plug-ended cord;

— = —, lose, loose cord and plugs.
Trag= —, strain cord;
Verbindungs= —, (connecting) cord;
schnüren, to lace.
Schnurklemme f, cord fastener;
schnurlos, cordless;
Schnur=paar n, pair of cords;
— =prüfklinke f, cord testing jack;
— =prüfung f, cord test;
— = — durch Schütteln, cord shake test;
— =rolle f, grooved roller;
— =schutz m, cord protecting means;
— = — =spirale f, cord protecting wire helix (des Steckers, of the plug);
— =stromkreis m, cord circuit;
— =verstärker m, cord circuit repeater;
— =werkstatt f, cord repairing centre.
schokoladenbraun, chocolate.
Schornstein m, smoke stack.
Schott n, bulkhead.
schraffieren, to shade, to hatch, kreuzweise schraffiert, cross-hatched.
Schraffierung f, hatching, shading.
schräg, skew, oblique, bevelled, geneigt: inclined, sloping.
Schräge f, slope, inclination, obliquity;
Schrägverzahnung, Getriebe n mit, helical gear;
Rad n mit, —, helical(ly) toothed wheel.
Schramme f, scratch.
Schrank m, cabinet, cupboard; switchboard, board; kleiner —, case;
A= —, A- (switch)board;
B= —, B- (switch)board;

Schrank
Batterie= —, battery cupboard;
Fern= —, toll switchboard, long-distance switchboard;
— = —, Durchgangs=, l. d. through switchboard;
Fernpruf= —, toll test board;
Glühlampen= —, lamp switchboard;
Klappen= —, board, switchboard;
— = —, schnurloser, cordless switchboard;
O.B.= —, l. b. board, magneto board;
Prüf= —, test board, test box;
— = —, kleiner, test case;
Schalt= —, switch cabinet;
Verbindungsleitungs= —, junction board;
Vorort= —, suburban switchboard;
Z.B.= —, c. b. board;
— =abteilung f, switchboard section;
— =beamter m, switch clerk;
— =kabel n, switchboard cable;
— =platz m, operator's position.
Schranke f, enclosure, barrier.
Schraub= ...., screwed;
— =deckel m, screwed-on cover.
Schraube f, screw, bolt;
eine — nachziehen, to tighten a screw, to screw up;
versenkte —, sunk screw;
Ankerspann= —, stay tightener B;
Befestigungs= —, tightening screw;
Einstell= —, adjusting screw;
Erd= —, earth screw B;
Feinmeß= —, micrometer gauge;
Feinstell= —, micrometer screw, vernier;
Flügel= —, wing screw;
Fundament= —, foundation screw;
Haken= —, hook screw;
Holz= —, wood screw;
Klemm= —, binding screw; clamping screw;
Knebelgriff= —, tommy screw;
Kordel(kopf)= —, thumb screw, knurled screw;
Kreuzloch= —, capstan (head) screw;
Messing= —, brass screw;
Metall= —, metal screw;
Mikrometer= —, micrometer screw;
Mutter= —, bolt and nut;
Nivellier= —, levelling screw;
Preß= —, clamping screw;
Regulier= —, adjusting screw;
Spann= —, tightening screw;
— = —, Feder=, spring tensioning screw;
Stein= —, wall screw, rag bolt;
Stell= —, adjusting screw;
Stellmacher= —, coach screw.
schrauben, to screw.
Schrauben=bolzen m, screw bolt;
— =feder f, helical spring;
schraubenförmig, helical, twisted
Schrauben=fuß m, earth screw (einer Stange, of a pole) B;
— =gewinde n, screw thread;
— =kopf m, screw head;
— =mutter f, screwed nut;
— =schlitz m, nick;
— =schlüssel m, spanner, wrench
englischer — = —, monkey wrench, coach wrench;
— =sicherung f, nut lock;
— =stütze f, insulator bolt B;
gerade — = —, straight bolt;
J=förmige — = —, J-bolt;
— =zange f, handvice;
— =zieher m, screwdriver.
Schraub=haken m, hook screw

Schraub
— -kappe f, screwed cap;
— -klemme f, screw terminal;
— -stock m, jaw vice, vise (am.);
Bank- — -, bench vice;
— -stöpsel m, screwed plug;
— -verbindung f, bolted connection;
— -zwinge f, screw clamp.

Schreib- ...., recording;
— -empfang m, visual reception.

schreiben, to write (up), to record;
Maschine —, to typewrite.

Schreiber m, recorder, writer;
Farb- —, writer, inker, T;
Licht- —, photographical recorder;
Maschine- —, typist;
Relief- —, embosser.

Schreib-feder f, stylus, (recording) pen;
— — (-spitze f), nib;
— -maschine f, typewriter;
— -pult n, writing shelf, writing desk;
— -röhrchen n, pen, syphon;
— -stift m, stylus, style;
— -strommesser m, recording ammeter.

Schrift f, record, signals pl, T;
richtige —, straight signals pl, T;
umgekehrte —, reversed signals pl T;
Block- —, block signals pl T;
Druck- —, printed characters pl T;
Kontroll- —, home record T;
Kurz- —, shorthand;
— -zeichen n, signal, character.

Schritt m, step;
Dreh- —, rotary step A;
Höhen- —, vertical step A;
Strom- —, pulse, unit, T;

— -schaltelektromagnet m, stepping electromagnet;
— -schalter m, step-by-step switch;
— -schalt-rad n, step wheel;
— — -Selbstanschlußsystem n, step-by-step automatic telephone system;
— — -telegraph m, step-by-step telegraph;
— — -wähler m, step-by-step selector;
— — -werk n, step-by-step switch; stepping mechanism.

schrumpfen, to shrink, to contract.

Schrumpfring m, shrunk-on ring.

Schrumpfung f, contraction, shrinkage.

Schub m, shift, push;
— -kasten m, — -lade f, drawer;
— -lehre f, slide gauge.

Schuh m, shoe, jaw;
mit einem — versehen, to shoe;
Kabel- —, lug, cable eye;
Pol- —, pole piece;
Stangen- —, shoe.

schütteln, to shake.

Schütz n, relay.

Schutz m, protection, shelter;
— -abdeckung f, protecting cover;
— -behandlung f, preservative treatment (der Stangen, of poles) B;
— -blech n, guard, protecting sheet, guard plate;
— -brille f, protective goggle;
— -deckel m, guard, cover;
— -draht m, armouring wire, sheathing wire (der Kabel, of cables);
geerdeter — — -, grounded guard wire B;
— -drossel f, protective choke (coil).

schützen, to protect (vor, from), to guard (vor, against).
schützend, protective, erhaltend: preservative, verhindernd: prohibitory.
Schutzerdung f, protective ground;
— ‑gestell n, barrier guard B;
— ‑gitter n, barrier guard B; Anoden‑ — ‑ —, anode-screening net V;
— ‑haube f, protecting cap;
— ‑kappe f, protecting cap; Draht‑ — ‑ —, wire cage; Lampen‑ — ‑ —, Rohr‑ — ‑ —, valve protecting cap V;
— ‑kasten m, (protecting) cover;
— ‑leiste f, guard strip;
— ‑maßnahme f, protective means;
— ‑mittel n, preservative;
— ‑netz n, protecting network, guard net B; geerdetes — ‑ —, earthed cradling B; U‑förmiges — ‑ —, cradle guarding B; Anoden‑ — ‑ —, anode-screening grid V;
— ‑platte f, guard plate;
— ‑ring m, guard ring;
— ‑spirale f, protecting wire helix;
— ‑vorrichtung f, protective device, guarding, safety device;
— ‑widerstand m, protective resistance.
schwach, weak, feeble; Draht: small gauge...., fine; Zeit —en Verkehrs, slack period;
— beheizte Röhre f, dull (emitting) valve V;
— ‑drähtig, small gauge or light gauge wire.... [ate L.
schwächen, to weaken; to attenu-

Schwachstrom m, (im Englischen unbekannt) weak current;
— ‑leitung f, signalling circuit;
— ‑technik f, communication art.
Schwächung f, weakening, attenuation;
— s‑anker m, adjusting slide T;
— s‑widerstand m, gain controller, gain regulator, potentiometer, (der Fernsprech‑ verstärker, of telephone repeaters).
schwalbenschwanzförmig, dovetailed.
Schwammgummi m, sponge rubber.
schwanken, to fluctuate, to vary, schnell: to flutter.
Schwanken n, variation, fluctuation; flutter(ing).
Schwankung f, variation, fluctuation;
— en pl im Wellenwiderstand, impedance irregularities pl K;
jährliche —en pl, annual variations pl;
tägliche —en pl, diurnal variations pl;
Strom‑ —, current variation, fluctuation of current.
Schwanz m, tail; Wellen‑ —, wave tail.
schwarz, black.
schwärzen, (sich), to blacken.
Schwebung f, beat, surge; mittels —en empfangen, to heterodyne R;
— en pl von Hörfrequenz, beats pl of audible frequency;
— en bilden mit, to beat with;
Schwebungs‑...., heterodyne, surging;
— ‑amplitude f, surging amplitude;

Schwebungs
— -empfang m, beat reception, heterodyne reception;
— - — mit besonderem Überlagerer, separate heterodyne reception;
— - — mit Selbsterregung, self-heterodyne reception;
— -empfänger m, (heterodyne) beat receiver, heterodyne receiver;
— -frequenz f, beat frequency, frequency of beats, combination frequency;
— - — null, zero beat frequency; auf — - — — eingestellt, set for zero beat;
— -periode f, beat cycle;
— -strom m, beating current;
— -ton m, beat note;
— - — -höhe f, pitch of the beat note;
— -verfahren n, beat method;
— -verstärker m, heterodyne amplifier;
— -vorgang m, beating effect.

Schwefel m, sulphur (S);
— -kies m, pyrite;
— -säure f, sulphuric acid, hydrogen sulphate (H_2SO_4); verdünnte — - —, diluted sulphuric acid.

Schweigen n, silence;
zum — bringen, to silence.

schweißen, to weld;
elektrisch —, to electro-weld.

Schweiß-naht f, welding seam, welded joint;
— -stelle f, welded joint; überlappte — - —, welded overlap joint;
Stumpf- — - —, welded butt joint.

Schweißung f, welding;
elektrische —, electro-welding;
Naht- —, seam welding;
Punkt- —, spot welding.

Schweißvorrichtung f, welder;
— für Kontakte, contact welder.

Schwelle f, threshold, beam, B;
Hör- —, threshold of audibility;
Reiz- —, sensation level.

schwellen, to swell.

schwemmen, to float.

Schwemmsand m, drifting sand.

Schwenkung f, turn (um, by).

schwer, heavy, ponderous.

Schwere f, heaviness, gravity.

Schwer-fälligkeit f, heaviness;
— -kraft f, gravitational force;
— -öl n, heavy oil;
— -punkt m, centre of gravity.

schwimmen, to float.

schwinden, to shrink; to fade (away) R.

Schwinden n, shrinkage; fading.

Schwinderscheinung f, fading effect R.

Schwing-audion n, oscillating detector, self-heterodyne amplifier;
— - — -empfang m, autodyne reception;
— - — -empfänger m, autodyne beat receiver;
— -bewegung f, vibratory movement.

schwingen, to vibrate, to oscillate, to undulate, to swing;
hin und her —, elektrisch: to surge back and forth, mechanisch: to swing, to rock;
in einer Oberharmonischen —, to vibrate to a harmonic;
um einen Mittelwert —, to oscillate about an average value.

Schwingen n, oscillation, vibration, undulation, swing(ing);
Selbst- —, self-oscillation, self-excitation.

schwingend, vibratory, oscillatory, undulatory;

schwingend
— **er Kontakt** *m*, vibrating contact;
— **e Zunge** *f*, vibrating reed.

Schwing-entladung *f*, oscillating discharge, oscillatory discharge.

Schwinger *m*, oscillator, vibrator;

Hertzscher —, Hertzian oscillator;

offener —, open oscillator.

schwingfähig, able to vibrate, capable of oscillation.

Schwingfähigkeit *f*, ability to vibrate.

Schwing-leistung *f*, oscillatory power;

— **-röhre** *f*, oscillating tube, oscillator valve, generator valve;

fremderregte — - —, separately excited oscillator valve;

selbsterregte — - —, self-excited oscillator valve;

Schwingung *f*, vibration, oscillation, undulation; swing;

in — versetzen, to throw into vibration, to set into oscillation;

ungedämpfte — en erzeugen, to set up continuous oscillations;

abklingende —, dying-out oscillation;

einfallende —, incoming oscillation;

erzwungene — en *pl*, forced vibrations, constrained oscillations *pl*;

freie — en, free oscillations;

gedämpfte — en, damped oscillations;

harmonische —, harmonic (oscillation);

hochfrequente — en, high-frequency oscillations;

ungedämpfte — en, undamped oscillations, sustained vibrations, persistent oscillations;

zusammengesetzte —, complex harmonic wave;

durch Nebenkopplungen hervorgerufene — en, spurious oscillations;

Eigen- —, natural period, natural vibration;

Grund- —, fundamental (oscillation), first harmonic (vibration), fundamental period;

Längs- — en, Longitudinal- — en, longitudinal vibrations;

Ober- —, (upper) harmonic vibration;

Schall- —, sound vibration;

Sinus- —, harmonic oscillation, sinusoidal vibration;

Stör- —, disturbing wave;

Transversal- — en, transverse vibrations;

Schwingungs-amplitude *f*, vibrational amplitude, amplitude of oscillation;

— **-anzeiger** *m*, oscillation detector;

— **-bauch** *m*, vibration loop, antinode;

— **-dauer** *f*, period of oscillation, time of vibration, time of swing, periodic time;

— **-energie** *f*, oscillation energy.

schwingungserzeugende Kraft *f*, vibromotive force.

Schwingungs-erzeuger *m*, endodyne, oscillation generator;

Lichtbogen — - —, arc generator;

Summer — - —, buzzer wave generator;

— **-erzeugung** *f*, generation or production of oscillations.

schwingungsfrei, non-oscillating;

— **er Zustand** *m*, non-oscillating condition.

15*

Schwingungs-knoten *m*, vibration node, interference point, null point, nodal point of vibration;
— **-kreis** *m*, oscillatory circuit; des Röhrensenders: tank circuit;
gekoppelte — - —**e** *pl*, coupled oscillatory circuits *pl*;
geschlossener — - —, closed oscillating circuit;
offener — - —, open oscillating circuit, open radiative circuit;
— **-transformator** *m*, oscillation transformer *R*;
— **-weite** *f*, amplitude;
— **-zahl** *f*, frequency, rate of vibration;
(**Eigen**- — - —, natural frequency.
Schwundperioden *pl*, fading periods *pl R*.
Schwung *m*, swing;
— **-rad** *n*, flywheel;
— - — **-kreis** *m*, flywheel circuit, parallel resonant circuit.
sechseckig, hexagonal.
sechsfach, **Sechsfach**- ..., sextuple.
Sechskantkopf *m*, hexagonal head.
See-höhe *f*, sea level;
— **-kabel** *n*, submarine cable, ocean cable;
— - — **-alphabet** *n*, cable (Morse) code;
— **-karte** *f*, chart;
— **-meile** *f*, nautical mile, *ab*: n. m. (= 1,854965 km = 2029 yards);
— **-rückleitung** *f*, sea return *L*;
— **-wasser** *n*, sea water.
Seele *f*, core;
Kabel- —, cable core.
Segment *n*, segment;
in —**e geteilt**, segmented;
eingefressenes —, **verbranntes**—, burnt segment *T*;
verkürzte —**e** *pl*, shortened segments *pl T*;
Aufnahme- —, **Empfangs**- —, receiving segment;
Gleichlauf- —, correcting segment;
Kreis- —, segment;
Verzögerungs- —**e** *pl*, idle segments, propagation segments *pl*, (am Baudotverteiler, of the Baudot distributor).
segmentförmig, segmental.
Segment-ring *m*, segmented ring;
— **-stück** *n*, segmental piece.
Sehne *f*, chord *M*.
Seide *f*, silk;
geklöppelte —, braided silk;
mit — **umsponnen**, (**zweifach**), (double) silk-covered;
Kunst- —, imitation silk;
Öl- —, oiled silk;
Trama- —, tram (silk);
Seiden-faden *m*, silk fibre;
— **-kabel** *n*, silk-covered cable;
Baumwoll- — - —, silk and cotton insulated cable.
Seifenwasser *n*, soapy water.
Seignettesalz *n*, Rochelle salt ($NaKC_4H_4O_6 + 4H_2O$).
Seil *n*, rope;
Abspann- —, span rope;
Draht- —, wire rope, wire cable;
Hanf- —, hemp rope, manila rope;
Trag- —, messenger, supporting strand;
— **-bahn** *f*, cord carrier; Lamson carrier;
— **-post** *f*, Lamson carrier (plant)
— - — **-wagen** *m*, Lamson carrier.
Seite *f*, side, flank; Papier: page.

Seiten=anfer *m*, lateral stay, side guy *B*;
— =anficht *f*, side elevation, profile, side view, end view;
— =band *n*, side band *R*;
oberes (unteres) — = —, upper (lower) side band;
Übertragung *f* eines — = —es (beider — =bänder), single (double) side band transmission *R*;
Unterdrückung eines — = —es, side band suppression *R*;
— = —frequenz *f*, obere (untere), upper (lower) side frequency *R*;
— =bezirk *m*, lateral area;
— =bund *m*, side binding *B*;
— =drucker *m*, page printer *T*;
— =drucktelegraph *m*, page printing telegraph;
— =induftion *f*, inductive interference;
— = —s =schutz *m*, anti-induction device;
— =schnitt *m*, sectional side elevation;
— =vorschub *m*, page feed *T*; den — = — ausführen, to page up *T*;
— =wand *f*, side wall;
— =zug *m*, transverse stress, lateral pull, *B*.
seitlich, lateral.
Sekante *f*, secant, *ab*: sec.
Sektor *m*, sector.
Sekundär=batterie *f*, secondary battery;
— =element *n*, secondary cell;
— =empfang *m*, secondary reception, double circuit reception;
— =empfänger *m*, double circuit or secondary or coupled type receiving set;
— =kreis *m*, secondary circuit;
— =strahlung *f*, secondary emission;
— =strom *m*, secondary current;
— =wicklung *f*, secondary (winding).
Sekunde *f*, second.
Selbstanlasser *m*, automatic starter;
Selbstanschluß=amt *n*, Fern= sprech=, automatic telephone exchange, mechanical telephone office, automatic central office (*am*.);
zwei= (drei=, vier=)stelliges — = —, two (three-, four-) digit automatic telephone exchange, two-(three-, four-) figure automatic telephone exchange;
— =Nebenstellenzentrale *f*, private automatic branch exchange, *ab*: p. a. b. x.;
— =Privatzentrale *f*, private automatic exchange, *ab*: p. a. x.;
— =system *n*, automatic telephone system, machine-switching telephone system, electro-mechanical switching system;
drei= (zwei=)drähtiges — = —, three- (two-) wire automatic telephone system;
Erdschleifen= — = —, earth return automatic telephone system;
Kreislauf= — = —, by-pass automatic telephone system;
Relais= — = —, relay automatic telephone system;
Schleifen= — = —, metallic automatic telephone system, two wire automatic telephone system;
Schrittschalt= — = —, step-by-step automatic telephone system;
Umgehungs= — = —, by-pass automatic telephone system;

Selbstanschluß-system
— = — mit unmittelbarer Stromstoßgebung, direct impulse automatic telephone system;
— = — mit Stromstoßempfängern, stored impulse automatic telephone system.

Selbstausschalter m, elektromagnetischer, electromagnetic cutout.

selbsteinstellend, self-adjusting.

Selbst-einstellung f, self-adjustment;
— -entladung f, self-discharge;
— -erhitzung f, self-heating;
selbst-erregend, self-exciting;
— -erregt, self-excited;
Selbst-erregung f, self-excitation;
— -induktion f, self-induction;
E. M. K. der — = —, e. m. f. of self-induction;
selbstinduktiv, self-inductive;
Selbst-induktivität f, self-inductance;
Antennen- — = —, aerial inductance;
— = —s -koeffizient m, coefficient of self-inductance, coefficient of self-induction;
— -laut m, vowel (sound);
— -schalter m, auto-switch;
— -schwingen n, self-excitation, self-oscillation, V;

selbsttätig, self-acting, automatic;

Selbsttönen n, squealing V;

selbsttragend, self-supporting (Mast, pole);

Selbst-überlagerer m, autodyne, auto-heterodyne;
— -überlagerungsempfang m, self-heterodyne reception, autodyne reception;
— -unterbrecher m, trembler (contact), self-interrupter, buzzer;

— = —, Pendel-, pendulum self-interrupter;
— = — -relais n, buzzer relay;
— -unterbrechungs- . . . , self-interrupting;
— = — -kontakt m, trembler contact, self-interrupter contact.

Selektion f, selection.

selektiv, selective (gegen, to);
nicht —, non-selective.

Selektivität f, selectivity.

Selektivkreis m, selecting circuit, selective circuit.

Semaphor m, semaphore.

Sende-amt n, transmitting station;
— -antenne f, transmitting aerial;
— -beamter m, transmitter operator, beim Wheatstone: key clerk;
— -hebel m, transmitting lever;
— -kondensatoren pl in den Brückenarmen, signalling condensers, sending condensers pl;
— -leistung f, sending power.

senden, to transmit. to send;
nochmals —, to re-run (einen Streifen, a slip) T.

Senden n, transmission;
rein ungedämpftes —, cut-in c. w. transmission;
Doppel- —, double radio transmission;
Mehrfach- —, multiple transmission;
Richt- —, directional transmission;
Ton- —, modulated c. w. transmission;
— durch Verstimmung, compensated c. w. transmission.

Sender m, sender, transmitter, emitter;
gedämpfter —, spark transmitter;

Sender
tönender —, musical spark transmitter;
ungedämpfter —, c. w. (=continuous wave) transmitter;
nkW= —, nKW transmitter;
Einstrahl= —, uni-directional transmitter, beam transmitter;
Funk= —, radio transmitter, wireless transmitter;
Funken= —, spark transmitter;
Hand= —, manual transmitter T;
Lichtbogen= —, arc, arc transmitter;
Lochstreifen= —, tape transmitter, auto transmitter, T;
Löschfunken= —, quenched spark transmitter;
Not= —, emergency transmitting set;
Richt= —, directive transmitter, directional transmitter;
Röhren= —, valve transmitter;
Rundfunk= —, broadcast transmitter;
Speicher= —, storage transmitter T;
Stimmgabel= —, vibrating reed transmitter T;
Streifen= —, tape transmitter T;
Telegraphie= —, (radio-) telegraphic transmitter;
Telephonie= —, radio-telephonic transmitter;
Tonfunken= —, musical spark transmitter;
Zwischen= —, remotely controlled transmitter, repeating station, R;
— =amt n, transmitting station;
— =dekrement n, transmitter decrement;

Sende=reichweite f, transmission range;
— =relais n, transmitting relay, signalling relay;
— =ring m, transmitting ring T;
— =röhre f, transmitter valve, generator triode, power tube;
— = —u=gestell n, power tube rack;
Senderfucher m, sender finder A;
Sende=segment n, transmitting segment T;
— =seite f, sender end, sending end, generator end;
— =stelle f, transmitter station, sending station;
Strahl= — = —, beam transmitter;
Zwischen= — = —, repeater station R;
— =streifen m, transmitting tape;
— =strom, m, sending current;
— =taste f, sending key;
— =verteiler m, sending or transmitting distributor T;
— =welle f, transmitted wave.
Sendung f, transmission;
Lochstreifen= —, perforated tape transmission;
Maschinen= —, auto-transmission T.
senken, to depress, to lower; sich —, to lower, Kurve: to slope; sacken: to sag.
senkrecht, vertical, perpendicular (auf, to);
— zueinander, in quadrature to each other;
räumlich — aufeinander, in space quadrature;
— er Schnitt m, vertical section.
Senkrechte f, vertical, normal, (zu, to).
Senkung f, depression, lowering. [meter.
Senkwage f, hydrometer, densi-

setzen, to put, to set; eine Stange: to erect.

Shapingmaschine *f*, shaping machine. [chine.

sicher, safe;
  betriebs= —, reliable in operation.

Sicherheit *f*, safety;
  Betriebs= —, reliability of operation;

Sicherheits=faktor *m*, factor of safety;
— **=funkenstrecke** *f*, safety (spark) gap;
— **=gürtel** *m*, safety belt *B*;
— **=ventil** *n*, safety valve;
— **=vorrichtung** *f*, safety device;
— **=vorschriften** *pl*, safety rules.

sichern, to guard, to protect, (vor, gegen, against, from); festlegen: to secure (in position).

Sicherung *f*, protection; elektrische —: safety fuse, cut-out, fuse;
  Abschmelz= —, (safety) fuse, safety cut-out;
  — = —, **durchgebrannte,** blown fuse;
  Blind= —, dummy fuse;
  Fein= —, heat coil, *ab*: h. c., *FT*;
  Glasrohr= —, glass tube fuse;
  Grob= —, fuse;
  Haupt= —, main fuse;
  Hochfrequenz= —, h. f. cut-out;
  Licht= —, lamp fuse;
  Mehrfach= — multiple fuse;
  Melde= —, alarm type fuse;
  Patronen= —, cartridge fuse;
  Schmelz= —, fuse, fusible cut-out;
  Spannungs= —, voltage cut-out;
  Stöpsel= —, plug fuse;
  Streifen= —, strip fuse;
  Überspannungs= —, excess voltage cut-out;

Sicherungs=brett *n*, fuse panel;
— **=draht** *m*, fuse wire;
— **=element** *n*, fuse element;
— **=gestell** *n*, fuse rack, fuse board;
— **=kästchen** *n*, (combined) protector, heat coil and fuse *F*;
— **=kasten** *m*, fuse box;
— **=leiste** *f*, fuse fitting, fuse strip; Feinsicherung und Blitzableiter: heat coil and protector strip;
— **=patrone** *f*, cartridge;
  Fein= — = —, heat coil;
— **=sockel** *m*, fuse mounting;
— **=stöpsel** *m*, safety plug, fuse plug;
— **=streifen** *m*, fuse strip; Feinsicherung und Blitzableiter: heat coil and protector strip;
— **=tafel** *f*, fuse panel, fuse board.

sichtbar, visual.

Sieb *n*, sieve; filter.

sieben, to sift, to select, to screen, to filter out.

Sieb=gebilde *n*, selective system, filter circuit;
— **=kreis** *m*, filter(ing) circuit, selective circuit;
— **=kette** *f*, band-pass filter, filter chain, wave-band filter, network;
  eingliedrige — = —, single-mesh filter;
  mehrgliedrige — = —, multi-mesh filter circuit;
  zweigliedrige — = —, two-section filter;
  Hochfrequenz= — = —, high-pass filter, high-pass selective circuit, lower limiting filter;
  Niederfrequenz= — = —, low-pass filter, low-pass selective circuit, higher limiting filter;
— = — **von großer Lochbreite** *f*, broad band filter;

Siebung *f*, filtration.

sieben, to boil.
Siedepunkt *m*, boiling point.
Siegellack *m*, sealing wax.
Siemens *n*, mho, mo;
  Mikro- —, micromho;
  —-einheit *f*, Siemens unit.
Signal *n*, signal; indicator;
  Block- —, block signal;
  Gruppenleit- —, pilot signal;
  Licht- —, luminous signal;
  —-lampe *f*, pilot lamp, supervisory lamp.
signalisieren, to signal.
Signalisierung *f*, signalling;
  — mit abgestimmten Einrichtungen, harmonic selective signalling.
Silber *n*, silver (Ag);
  —-bronze *f*, silver bronze;
  —-nitrat *n*, nitrate of silver (AgNO$_3$).
Silizium *n*, silicon (Si);
  —-bronze *f*, silicon bronze;
  — - —-draht *m*, silicon bronze wire;
  —-eisen *n*, silicon steel;
  —-karbid *n*, silicon carbide, carborundum, (SiC);
  —-stahl *m*, silicon steel.
simplex, simplex.
Simplex-betrieb *m*, simplex operation;
  —-leitung *f*, simplex circuit.
Sims *m*, ledge.
Simultan-betrieb *m*, composite working;
  —-einrichtung *f*, (telegraph) composite set *T*;
  — - — für Doppelleitung, metallic composite set;
  —-geräusch *n*, thump;
  —-leitung *f*, composite circuit, plus circuit, *T*;
  —-telegraph *m*, superposed telegraph;
  — - —-en-leitung *f*, telegraph superposed circuit;

  —-verbindung *f*, superposed circuit, composite circuit, phantom circuit;
  Bildung *f* **von — - —-en**, compositing, superimposing.
sinken, to fall off, to sink.
Sinken *n*, sinking, fall(ing-off), drop.
Sinkstelle *f*, sink.
sintern, to shrink.
Sintern *n*, shrinkage.
Sinus *m*, sine, *ab*: sin;
  hyperbolischer —, hyperbolic sine *ab*: sinh;
  —-form *f*, sine shape;
sinusförmig, sine-shaped, sinusoidal;
  rein —, simple harmonic;
  — —-e EMK, pure sine e. m. f., simple harmonic e. m. f.;
Sinus-gesetz *n*, sine law;
  —-linie *f*, sinuous line, sine curve;
sinusoidal, sinusoidal;
Sinus-reihe *f*, sine series;
  —-schwingung *f*, sinusoid, sine wave;
  gedämpfte — - —, damped sine wave;
  ungedämpfte — - —-en *pl*, sustained sinusoids *pl*;
  —-strom *m*, sine current;
  —-welle *f*, sine wave;
  zusammengesetzte — - —, complex sine wave.
  reine — - —, pure sine wave;
  — - —-n-erzeuger *m*, sine wave alternator, harmonic generator.
Sitz *m*, seat;
  —-(um)schalter *m*, socket contact *F*.
Skala *f*, **Skale** *f*, scale, dial;
  Kreis- —, dial, circular scale;
Skalen-ablesung *f*, **direkte**, direct scale reading;
  —-aräometer *n*, graduated hydrometer;
  —-teil *m*, scale division.

Skineffekt *m*, skin effect.
Skizze *f*, sketch, outline, drawing;
   Bleistift- —, pencil drawing.
skizzieren, to sketch, to outline.
Sockel *m*, socket, holder, base;
   Gußeisen- —, cast-iron base;
   Lampen- —, lamp holder, lamp socket;
   Metall- —, metal base;
   Röhren- —, valve holder, valve socket;
   Sicherungs- —, fuse mounting.
Soda *f*, soda (Na_2CO_3).
Sohle *f*, bottom *B*.
Solenoid *n*, solenoid, helix *pl* helices;
   einlagiges —, single-layer solenoid;
   eisenloses —, air core solenoid.
Solodynempfang *m*, solodyne reception *R*.
Sonnen-fleck *m*, sun spot;
— **-höhe** *f*, sun's altitude.
Sopran *m*, soprano.
sortieren, to sort (out); einreihen: to step in.
Sourdine *f*, sourdine.
Spalt *m*, gap;
   Luft- —, air gap;
— **-breite** *f*, gap separation, gap
Spalte *f*, column. [width.
spalten, to split, to slit.
Spaltung *f*, splitting;
   Phasen- —, phase-splitting.
Späne *pl*, chips *pl*;
   Dreh- —, turnings;
   Eisenfeil- —, iron filings.
Spanndraht *m*, span wire.
spannen, to tighten, to span,
   Federn: to tension, to bend;
   einspannen: to clutch, to grip;
   Draht —: to strain (wire), to rack.
Spanner *m*, rack, strainer, *B*.
   Draht- —, strainer, wirestretcher, *B*.

Spann-platte *f*, clamping plate;
— **-ring** *m*, clamping ring, locking ring;
— **-schienen** *pl*, slide rails *pl*;
— **-schloß** *n*, (stay) tightener, swivel, rod strainer, stretching screw, spanner, *B*;
— **-schraube** *f*, tightening screw;
   Feder- - —, spring tensioning screw;
— **-vorrichtung** *f*, tightening device;
— **-weite** *f*, (length of) span, span length, *B*.
Spannung *f*, strain, tension, pressure; elektr.: voltage, potential, tension, pressure;
   die — erhöhen (verringern), to increase (decrease) the power *T*;
   an — liegende Ader *f*, negative wire *A*;
   unter — befindlich, alive;
— **— setzen**, to make alive (eine Leitung, a circuit);
   effektive —, r. m. s. voltage, virtual voltage;
   elektrische —, voltage, electrical tension, electric pressure, (zwischen, across);
   flüchtige —, transient voltage;
   pulsierende —, pulsating voltage, pulsatory voltage;
   sinusförmige —, sine wave of voltage, harmonic voltage;
   Anfangs- —, initial voltage;
   Anoden- —, discharge voltage, plate voltage, anode potential;
   Ausgleichs- —, transient voltage;
   Besetzt- —, busying potential;
   Betriebs- —, operating voltage;
   Blind- —, reactance voltage;
   Bruch- —, breaking strain *B*;
   Dreieck- —, delta voltage;
   Effektiv- —, r. m. s. voltage, virtual voltage;

Spannung
End= —, final voltage;
Entladungs= —, discharge potential;
Erreger= —, exciting voltage;
Faden= —, filament voltage;
Feder= —, spring tension;
Gegen= —, counter e. m. f., counter voltage;
Gleich= —, d. c. potential, continuous e. m. f., continuous or direct voltage;
— = —, wellige, ripple voltage;
Heiz= —, filament voltage, filament volts *pl*;
Hoch= —, high tension, *ab*: h. t., high pressure;
Höchst= —, super tension;
Ionisations= —, wahre, true ionisation potential;
Klemmen= —, terminal voltage;
Kontakt= —, contact potential;
Lade= —, charging voltage;
Leerlauf= —, no-load voltage, open-circuit voltage;
Meß= —, measuring voltage;
Modulations= —, modulating voltage;
Nieder= —, low tension, *ab*: l. t.;
Null= —, zero potential;
Nutz= —, useful voltage;
Oberflächen= —, surface tension;
Phasen= —, phase voltage;
Polarisations= —, electromotive force of polarization;
Prüf= —, testing voltage;
Scheitel= —, voltage peak;
Sinus= —, sine wave of voltage, harmonic voltage;
Stern= —, Y-voltage, star voltage;
Steuer= —, control voltage;
Teil= —, component voltage;
Überschlag= —, spark-over voltage;

Über= —, excessive voltage, overtension;
Verkettungs= —, interlinked voltage;
Vor= —, biasing voltage, initial voltage;
— = —, Gitter=, biasing or initial grid voltage, grid bias;
Zünd= —, ignition voltage (des Lichtbogens, of the arc), breakdown voltage;
Zusatz= —, additional voltage, boosting voltage;
— am Anfang, initial voltage L;
— — Ende, final voltage L;
spannungführend, live, alive;
Spannungs=abfall *m*, drop of potential (in, zwischen, across);
ohmscher — = —, ohmic drop of voltage, resistance drop;
räumlicher — = —, potential fall;
— = — durch Blindwiderstand, reactance drop of voltage;
— =anstieg *m*, rise of potential;
— =bauch *m*, potential loop, potential antinode;
— =empfindlichkeit *f*, voltage sensitivity;
— =gradient *m*, potential gradient;
— =knoten *m*, potential node;
— =komponente *f*, component voltage;
Blind= — = —, wattless component of e. m. f.;
Gleich= — = —, direct component of voltage;
Wechsel= — = —, alternating component of voltage;
Wirk= — = —, energy component of voltage;
spannungslos, dead;
— machen, eine Leitung, to kill a (power) circuit;
Spannungs=messer *m*, voltmeter, detector;

Spannungsmesser
  kombinierter Strom- und — - —,
  combined volt-and ammeter;
  statischer — - —, static voltmeter;
  Hitzdraht- — - —, hot-wire voltmeter;
  Röhren- — - —, amplifying voltmeter;
— quelle f, source of e. m. f.;
  Gleich- — - —, constant potential supply;
— -regelung f, voltage control;
— -regler m, voltage regulator;
— - — -relais n, voltage control relay;
— -reihe f, contact series;
  Thermo- — - —, thermo-electric series;
— -resonanz f, series resonance;
— - — -kreis m, senries-resonant circuit;
— -schwankungen pl, voltage variations pl;
— -sicherung f, voltage cut-out;
  Über- — - —, excess voltage cut-out;
— -spule f, pressure coil (des Ohmmeters, of the megger);
— -teiler m, voltage divider, potentiometer (resistance);
  kapazitiver — - —, capacitive voltage divider;
— -unterschied m, potential difference, ab: p. d.;
— -verstärkung f, voltage amplification;
— -verteilung f, distribution of voltage;
— -welle f, voltage wave; Wanderwelle: voltage surge;
— -wellen pl, p. d. ripple.
sparen, to save; arbeitsparend, labour saving.
Sparren m, spar;
  Dach- —, rafter, spar.

Spartransformator m, autotransformer.
Spaten m, spade.
speckiger Bruch m, lardaceous fracture (des Porzellans, of porcelain).
Speckstein m, steatite, soapstone.
Speiche f, spoke.
Speicher m, store; register A;
  Impuls- —, Stromstoß- —, impulse storing device, register, A;
  Leitbuchstaben- —, office code register A;
  Nummern- —, numerical register A;
  Stromstoß- —, impulse storing device, digit storing register, A;
— -geber m, storage transmitter T;
— -kondensator m, reservoir condenser, tank condenser; storage condenser T;
— -relais n, storing relay; storage relay T A.
speichern, to store (up), to accumulate.
Speicherung f, storing, storage; accumulation.
Speise-drücke f, feed retardation coil, feeding circuit, F;
— -drossel f, feed coil;
— -leitung f, feeder, supply circuit;
  Einphasenbahn- — - —, single phase electric railway power circuit.
speisen, to supply (to), to feed (into).
Speise-punkt m, input terminals pl; distributing point;
— -relais n, supply relay F;
— -seite f, generator end;
— -strom m, supply current, feeding current, F;
— - — -kreis m, supply circuit.

Speisung *f*, supply, feed;
Anoden= —, plate supply;
Gleichstrom= —, d. c. supply, c. c. supply;
Heizfaden= —, filament supply;
Mikrophon= —, Sprechstrom= —, speaking current supply.

Spektrum *n*, spectrum;
Frequenz= —, frequency spectrum;
Schall= —, sound spectrum;
Ton= —, tonic spectrum.

Sperre *f*, block, locking device, lock;
Tastenfeld= —, keyboard lock *T*.

sperren, to block, to lock, to stopper; einen Wähler: to make busy, to busy, *A*; einen Platz: to guard a position, to make inaccessible a position.

Sperrer *m*, suppressor;
Echo= —, echo killer, echo suppressor.

Sperr=filter *n*, rejector circuit, stopper circuit, suppression filter;
— =haken *m*, click, latch, (locking) pawl;
— =kegel *m*, pawl, detent, click;
— =kette *f*, Hf.=, low-pass filter, higher limiting filter;
— =klinke *f*, (holding) pawl, lock pawl, stop pawl, detent, ratchet, click;
Doppel= — = (des Strowger=wählers), pair of pawls, double dog, double detent, (of the Strowger switch);
— =kreis *m*, stopper (circuit), rejector circuit, suppression filter, rejective circuit;
Hf.= — = —, low-pass selective circuit, higher limiting filter;
Nf.= — = —, high-pass selective circuit, lower limiting filter;
— =magnet *m*, locking magnet;
— =nute *f*, locking notch;
— =rad *n*, ratchet wheel;
— =relais *n*, locking relay;
— =vorrichtung *f*, locking device;
— =zahn *m*, detent, dog;
— = — =kranz *m*, ratchet drum.
spezialisieren, to specialize.
Spezialisierung *f* specialization.
spezifisch, specific(al);
— er Widerstand *m*, specific resistance, resistivity.
spezifizieren, to specify.
sphärisch, spherical.
Spiegel *m*, mirror; reflector;
rotierender —, revolving mirror;
Parabol= — parabolic mirror;
— =ablesung *f*, mirror reading;
— =galvanometer *n*, reflecting or mirror galvanometer.
spiegeln, to reflect.
Spiegelung *f*, reflection;
— s=verlust *m*, loss at a junction, reflection loss, *L*.

Spiel *n*, play; Ausschlag b. Nadel: throw; toter Gang: lost motion, der Zähne: backlash; Satz: gang, set;
Anker= —, play of tongue;
— =raum *m*, play; backlash; zulässiger Unterschied: margin.

Spill *n*, vertical capstan winch.
Spindel *f*, spindle, arbor, axle;
— =blitzableiter *m*, reel protector.
Spirale *f*, spiral; helix *pl* helices;
logarithmische —, logarithmic spiral, equiangular spiral;

Spirale
  Draht= —, wire spiral;
  Flach= —, flat spiral;
  Flachfeder= —, flat spiral spring;
Spiral-antenne f, flat coil aerial, flat spiral coil;
— =band n, helical tape;
— =bohrer m, twist drill;
— =feder f, helical spring; flach: coiled spring, spiral spring:
— =linie f, spiral curve;
— =vierer m, spiral quad.
spiralig, helical; spiral.
Spiritus m, spirit, alcohol;
— =lötlampe f, alcohol blow [torch.
spitz, pointed;
— zulaufend, taper.
Spitzbock m, A-pole B.
Spitze f, point, tip, Ecke: corner, Vorderteil: head; der Kurve, Belastung: peak.
  Verkehrs= —, traffic peak;
Spitzenaufhängung f, point suspension, pivot suspension;
— =belastung f, peak load;
— =blitzableiter m, point lightning arrester.
Spitzende n, spigot end (des Muffenrohrs, of socket tubes).
Spitzen-entladung f, point discharge;
spitzengelagert, suspended in points;
Spitzenlagerung f, point suspension;
— =leistung f, peak power;
— =platz m, transfer position F; Melde= — = —, record transfer position F;
— = — =beamtin f, transfer operator F;
— =verkehr m, peak traffic;
— =wirkung f, needle effect.
Spleißblock m, jointer's vice.
spleißen, to joint (to), to splice (to).

Spleiß-gerät n, splicing tool;
— =stelle f, splice, joint;
  Abzweig= — = —, Y-splice.
Spleißung f, splicing, jointing.
Sp-Leitung f, rural phonogram circuit.
Splint m, split pin;
— =holz n, sap wood.
Splisser m, splicer; Gerät: splicing tool.
Splitter m, chip.
Sporn m, spur.
Sprache f, speech, voice; language;
  deutliche —, articulated voice, clear voice;
  geheime —, secret language T;
  offene —, open language T;
  — = —, Telegramm n in, open language message;
  verschwommene —, blurred voice;
Sprach-energie f, voice power;
— =frequenz f, voice frequency;
  mittlere — = —, mean frequency of speech;
— = — =band n, speech band;
— =lautstärke f, speech volume;
— =messung f, voice test(ing);
— =modulator m, voice-actuated modulator;
sprachmoduliert, speech-modulated;
— = —e ungedämpfte Wellen pl, speech-modulated continuous waves, type A 3 waves pl;
Sprach-rohr n, speaking tube;
— =schwingungen pl, speech waves pl;
— =übertragung f, speech transmission (über eine Leitung, over a circuit);
  Wirkungsgrad m der — = —, speaking efficiency;

Sprach
— **verstärker** *m*, speech amplifier;
— **wellen** *pl*, speech waves *pl*.
Sprech-...., *cf.* **Sprach**-....;
— **apparat** *m*, speaking instrument, telephone station;
— **batterie** *f*, speaking battery;
— **einrichtung** *f*, speaking equipment; operator's (phone) set *F*.
sprechen, to speak, to talk.
Sprechen *n*, talking, speaking.
Sprechende(r) *m*, talker.
Sprecher *m*, talker; Ansager: speaker *R*.
Sprech-frequenzbereich *m*, speech frequency range;
— **frequenzen** *pl*, voice frequencies, telephonic frequencies *pl*;
— **garnitur** *f*, operator's set *F*;
— **hörer** *m*, telephone handset, micro-telephone, combination;
— **leitung** *f*, speaker wire, Dienstleitung: order wire *F*;
— **prüfung** *f*, voice testing;
— **reichweite** *f*, speaking range;
— **schlüssel** *m*, speaking key;
— **und Rufschlüssel** *m*, speaking and ringing key;
— **schwingungen** *pl*, speech waves *pl*;
— **stelle** *f*, telephone station;
Teilnehmer- — = —, subscriber's set, subscriber's station; subset (*am.*), substation (*am*);
— = —n**störung** *f*, subscriber's apparatus fault;
— **stellung** *f*, b. Schalters: speaking position; allgemein: talking condition;
— **strom** *m*, talking current;
— **ströme** *pl*, speech currents, speaking currents, voice currents;
— **strom-kreis** *m*, speaking circuit;
— = — **speisung** *f*, — = — = **zu**= **führung** *f*, speaking current supply;
— **taste** *f*, am **Sprechhörer**, talk-listen button;
— **trichter** *m*, mouthpiece;
— **verbindung** *f*, connection;
— **versuch** *m*, talking test, speech test;
vergleichender — = —, comparative speech test;
— **weg** *m*, speaking circuit, talking path, *F A*.
Spreize *f*, spreader, spacer, strut, outrigger.
springen, to jump.
spritzen, to spray.
Spritzguß-...., die cast;
— **kondensator** *m*, die-cast condenser;
— **metall** *n*, die-cast metal;
— **stück** *n*, die-casting.
Spritzring *m*, thrower.
spröde, brittle.
sprühen, to spray.
Sprühen *n*, am Luftdraht usw.: corona.
Sprung *m*, crack, flaw.
Sp-Telegramm *n*, phonogram (message).
Spule *f*, spool, coil, bobbin;
drehbare —, rotatable coil, revolving coil;
einlagige —, single-layer coil;
feste —, fixed coil;
mit einer halben — **beginnend**, beginning at mid-load *K*;
ineinanderschiebbare —n *pl*, telescoping coils *pl*;
mehrlagige —, multi-layer coil;
quadratische —, square coil;
verschiebbare —, sliding coil;
— **mit zwei Gleitkontakten**, double slider coil;

Spule

Abstimm- —, syntonising coil, tuning coil, tuner;
Abzweig- —, bridging coil F;
Anker- —, armature coil;
Anzapf- —, tapped coil;
Auffang- —, search coil;
Aufsteck- —, plug-in coil;
Band- —, ribbon coil;
— - —, flach gewickelte, flatwise wound ribbon coil;
— - —, hochkant gewickelte, edgewise wound ribbon coil;
Belastungs- —, load(ing) coil, loading inductance;
Drahtkern- —, wire-core coil;
Dreh- —, moving coil, rotating coil;
Drossel- —, choking coil, choke (coil), impedance coil, reactive coil, reaction coil, retard(ation) coil, reactor;
— - —, veränderliche, reactance regulator;
Eisenblätterkern- —, laminated iron core coil;
Eisenstaubkern- —, iron dust core coil;
Feld- —, field coil, magnetizing coil;
Flach- —, flat coil, plane coil;
— - —, quadratische square plane coil, pancake coil, flat square coil;
Halte- —, holding coil;
Induktanz- —, inductance coil, reactor, inductor, retard(ation) coil, graduator;
— - —, eisengeschlossene, ferric inductance coil;
Induktions- —, induction coil F;
Käfig- —, cage coil;
Kopplungs- —, coupling coil, coupler; am Wellenmesser: exploring coil, search coil;
— - —, veränderliche, variocoupler;
Korb(boden)- —, basket type coil, basket-wound coil; spider web coil;
Leitungs- —, line coil (des Differentialrelais, of the differential relay);
Luft- —, air core coil, air core solenoid;
Luftdraht- —, antenna helix;
Luftdraht-Abstimm- —, aerial tuning inductance, a. t. i.;
Luftdraht-Verlängerungs- —, aerial loading inductance, antenna helix;
Magnet- —, magnet coil, solenoid;
Prüf- —, exploring coil, search coil;
Pupin- —, load(ing) coil, Pupin coil;
Quer- —, leak coil L;
— - —n, Belastung f mit, leak-loading L;
Reihen- —, series coil L;
— - —n, Belastung f mit, series loading L;
Richt- —, control coil;
Ring- —, toroidal coil;
Rückkopplungs- —, reaction coil, feed-back coil; beim Schwingaudion: tickler coil;
Saug- —, sucking solenoid;
Schiebe- —, sliding coil;
Schieber- —, slider coil;
Spannungs- —, pressure coil (des Ohmmeters, of the megger);
Stamm- —, side circuit coil K;
Steck- —, plug inductor, plug-in coil, R;
Tauch- —n pl, telescoping coils pl;
Tauchkern- —, sucking solenoid;
Übertrager- —, repeating coil;

Spule
Variometer= —, variometer coil;
— = —, **drehbare,** variometer rotor; [tor;
— = —, **feste,** variometer sta-
Verlängerungs= —, loading inductance R;
Vierer= —, phantom (circuit) coil;
Waben= —, honeycomb coil;
Widerstands= —, resistance coil;
— = —, **Einer=** (**Zehner=, Hunderter=, Tausender=**), units (tens-, hundreds-, thousands-) resistance coil;

Spulen=abschnitt m, (loading) coil section, loading section K;
— =**abstand** m, coil spacing K;
halber — = —, half a coil spacing;
— =**abzweig** m, coil tap;
— =**antenne** f, coil aerial;
— =**anzapfung** f, coil tap, tapping;
spulenbelastet, coil-loaded;
Spulen=belastung f, coil loading;
— =(**draht**)**enden** pl, lead-in wires, leads pl;
— =**entfernung** f, load-spacing, coil spacing, K;
— =**feld** n, loading (coil) section, pupinization section;
mit $7/10$ **des** — = — **es beginnend,** beginning at 0,7 section;
mit einem halben — = — **beginnend,** beginning at mid section;
— =**festpunkt** m, (loading) section point K;
— =**flansch** m, spool flange, spool head;
— =**gestell** n, coil rack;
— =**halter** m, coil holder;
— =**label** n, coil (-loaded) cable;
— =**kapazität** f, coil capacity;
— =**kasten** m, bobbin, spool; für **Pupinspulen:** loading-coil case or pot;
— =**kette** f, low-pass filter, ultra filter, upper limiting filter;
— =**körper** m, spool, bobbin;
— =**punkt** m, loading point;
erster — = —, first loading point;
— =**rahmen** m, coil former;
— =**satz** m, loading coil unit K;
Vierer= — = —, phantom coil set K;
— =**scheibe** f, spool flange, spool head, cheek;
— =**stück** n, coil piece (im **See= kabel,** of the submarine cable).

Spulmaschine f, winding machine.
Spur=kranz m, flange;
— =**lager** n, thrust bearing.

Staats=gespräch n, government call;
— =**telegramm** n, government message.

Stab m, bar, rod;
Anker= —, armature bar;
Kollektor= —, commutator segment, commutator bar;
stabförmig, bar-shaped.
Stabmagnet m, bar magnet.
stabil, stable, steady.
stabilisieren, to stabilize, to steady.
Stabilisierung f, stabilization, steadying.
Stabilit n, stabilit.
Stabilität f, stability, steadiness;
magnetische —, magnetic stability.
Stacheldraht m, barbed wire.
Stadtgebiet n, urban area, city area.
Staffel=betrieb m, echelon working T;
— =**gebühr** f, graduated rate;

Staffel
— **-gegensprechen** n, echelon duplexing T;
— **-leitung** f, echelon circuit, series circuit, T.

staffeln, to stagger, to grade A.

Staffeln n, staggering, grading, A;
— **von Gruppen**, staggering of groups A.

Staffel-tarif m, graduated tariff;
— **-telegraph** m, series multiplex telegraph.

Staffelung f, grading;
Bündel-—, group grading A;
— **S-plan** m, grading scheme.

Stahl m, steel;
— mit hohem (niedrigem) **Kohlegehalt**, high (low) carbon steel;
legierter —, alloy steel;
weicher —, mild steel;
Band- —, ribbon steel;
Bessemer- —, Bessemer steel;
Chrom- —, chromium steel;
Dreh- —, turning knife;
Kohlenstoff- —, (high) carbon steel;
Magnet- —, magnet steel;
Mangan- —, manganese steel;
Silizium- —, silicon steel;
Tiegel- —, crucible steel;
Wolfram- —, tungsten steel;
stahlband-armiert, — **-bewehrt**, steel tape armoured;

Stahl-drahtseil n, steel wire rope;
— **-guß** m, cast steel;
— **-platte** f, steel plate;
— **-rohr** n, steel tube;
— — **-mast** m, wrought steel pole, tubular steel mast;
— — **-ständer** m, tubular steel pole;
— **-spitze** f, steel tip;

mit einer — - — versehen, steel-tipped;
— **-stab** m, steel bar;
— **-turm, freitragender,** self-supporting steel tower;
— **-zunge** f, steel reed.

Stamm m, trunk; **Stammleitung**: side circuit;
— **-ende** n, butt (end), der Stange: pole butt B;
zubereitetes — - —, treated butt B;
— **-kreis** m, — **-leitung** f, side circuit, physical circuit, combining circuit, transformer circuit, component line;
— **-spule** f, side circuit coil.

stampfen, to tamp, to ram, B.
Stampfen n, tamping.
Stampfholz n, tamping bar.
Standardkabel n, standard cable;
n **Meilen** —, n miles of standard cable, ab: n m. s. c.;
— **-äquivalent** n, standard cable equivalent.

Ständer m, standard, pole, stand, pedestal; **Dynamo**: stator, field system;
Dach- —, roof standard, roof pole;
— - —,**Abspann-** —, roof end standard;
Gitter- —, lattice(d) pole;
Rohr- —, tubular pole, tube pole;
Roll- —, wheeled stand;
— **-fernsprecher** m, pedestal (desk) telephone station.

standfest, stable.
Standfestigkeit f, stability.
Standort m, position, stand.
Stange f, pole, post, rod;
— mit **zubereitetem Ende**, butt-treated pole;
eine — **anschuhen**, to shoe a pole;

Stange
eine — ausrichten, to straighten a pole;
eine — setzen, to set or erect a pole;
angeschuhte —, shoed pole;
getränkte —, treated pole;
— = —, mit Kreosot, creosoted pole;
mittelstarke —, medium pole;
rohe —, plain pole, untreated pole;
starke —, stout pole;
unzubereitete —, untreated pole, plain pole;
verankerte —, stayed pole, pole and stay;
verstrebte —, strutted pole;
zubereitete —, treated pole;
Abspann- —, stay pole, terminal pole;
Absteck- —, stake, peg;
Brech- —, crowbar;
Holz- —, wooden pole;
Kreuzungs- —, transposition pole B;
Kuppel- —, coupled poles pl B;
Kupplungs- —, coupling bar;
Schalt- —, switching bar;
Untersuchungs- —, pole test box;
Überführungs- —, distributing pole, terminal pole;
Verteilungs- —, distributing pole;
Winkel- —, angle pole;
Zahn- —, tooth rack;
Stangen-abdachung f, pole roof(ing);
— -abstand m, pole distance;
— -ausrüstung f, pole fittings pl;
— -bild n, pole diagram;
— -blitzableiter m, pole lightning arrester;
— -ende n, pole butt;
zubereitetes — = —, treated butt;
mit zubereitetem — = —, butt-treated;
— -fuß m, pole pedestal, pole footing;
— -holz n, pole timber;
— -linie f, pole line;
an einer — = — geführt, carried on a pole line;
— -loch n, post hole, pole hole;
— = — -bohrmaschine f, post-hole drilling machine;
— -schuh m, pole shoe, Dach-ständer: chair (of a roof pole);
— -spitze f, pole top end;
— -untersuchungskasten m, pole test box;
— -wähler m, panel switch, panel type selector.
Stanniol n, tin-foil;
— -papier n, tinfoil paper.
Stanz-beamtin f, perforator operator T;
— -block m, die block, punch block, T.
Stanze f, punch T; stamp.
stanzen, to punch, to perforate, T; aus Blech usw.: to stamp, to blank out (aus Blech, from sheet);
nochmals —, to repunch T.
Stanzen n, punching;
blindes —, touch-typing T.
Stanz-loch n, signal hole (des Lochstreifens, of the perforated tape);
— -magnet m, punch(ing) magnet;
— -matrize f, (cutting) die plate, punching die;
— -relais n, punching relay;
— -stempel m, punch, die;
— = — -führung f, die block;
— -stück n, stamping, blank.
Stapel m, pile, file.

ſtapeln, to pile.
ſtark, heavy, strong, intense; Magnet: powerful;
— er Verkehr m, heavy traffic;
— drähtig, heavy gauge wire....
Stärke f, intensity, strength; Dicke: thickness; Mehl: starch; Papierblatt- —, doppelte, two thicknesses of papers.
Starkſtrom m, power current, heavy current;
— -anlage f, power plant;
— -kreuzung f, power line crossing;
— -leitung f, power (transmission) line, power circuit;
— -mikrophon n, high power transmitter;
— -netz n, öffentliches, public supply, public mains pl;
— -ſtörung f, power failure; interference from power systems;
— -technik f, heavy current engineering;
— -wecker m, power bell;
— -zuführung f, power lead.
Stärkung f, strengthening.
ſtarr, rigid.
Starrheit f, rigidity.
ſtationär, stationary;
quaſi- —, quasi-stationary;
— bleiben, to stay;
— er Zuſtand m, steady state.
Statik f, statics pl.
ſtatiſch, static(al).
Statiſtik f, statistics pl.
ſtatiſtiſch, statistical.
Stativ n, stand.
Stator m, stator.
Staub m, dust.
ſtaubdicht, dust-proof;
— -e Kappe f, dust-proof cover.
Staub-(ſchutz)abdeckung f, dust (-proof) cover;
— -kappe f, dust cover;
— -kern m, dust core;

— - — -ſpule f, iron dust core coil;
— -ſauger m, vacuum cleaner.
ſtaubſicher, dust-proof.
Steatit n, steatite.
Steck-buchſe f, plug socket, connector socket;
— -doſe f, wall socket, wall plug;
unverwechſelbare — - —, non-interchangeable wall socket;
Licht- — - —, light socket;
— -faſſung f, lamp jack (für Ruflampen, for calling lamps) F; (plug-in) socket;
— -lampe f, jack(s) lamp F;
— -ſpule f, plug-in coil, plug inductor, R.
Stecker m, plug, connector;
dreiteiliger —, three-point plug, three-way plug, triplug;
unverwechſelbarer —, non-interchangeable plug;
Doppel- —, double plug, biplug, two-pin plug;
Dreifach- —, Drilling- —, three-pin plug, triplug;
Zwillings- —, biplug, pair of plugs;
Zwiſchen- —, socket adapter (or adaptor);
— -buchſe f, plug socket, connector socket;
— -ſchnur f, cord and plug.
Steg m, bridge, strap;
durch —e verbinden, to strap together.
ſtehen bleiben, to run down, to stop.
ſtehende Wellen pl, stationary waves, standing waves, pl.
Stehlager n, footstep bearing,
ſteif, stiff. [vertical bearing.
Steife f, strut, prop, brace.
Steigeiſen n, climbers pl, climbing iron; Stufe: pole step.

steigen, to rise.
Steigrad *n*, ratchet wheel, escapement wheel.
Steigung *f*, gradient, rise; Kurve: slope; Windung, Gewinde: pitch;
— s-änderung *f*, variation in slope;
— s-winkel *m*, angle of slope.
steil, steep, abrupt.
Steilheit *f*, steepness; slope *V*.
Stein *m*, stone; Achat usw=: jewel;
Kabel= —, cable tile;
Lager= —, jewel cup;
— =bohrer *m*, wall chisel, stone drill;
— =gut *n*, stone ware, pottery;
— = — =formstück *n*, tile;
— =schotter *m*, macadam; mit — = — belegen, to macadamize;
— =schraube *f*, wall screw, rag bolt.
Stellenzahl *f*, (number of) digits *A*.
Stell=macherschraube *f*, coach screw;
— =ring *m*, cursor, collar;
— =schraube *f*, adjusting screw, set screw;
— =stift *m*, capstan spike, tommy.
Stellung *f*, position;
in — bringen, to position;
in richtiger —, in register;
Arbeits= —, operative position;
— = —, in der, off-normal;
Ausgangs= —, Grund= —, home position, normal position;
Normal= —, normal position, idle position;
Regel= —, normal position;
Ruhe= —, unoperated position, resting position, home position;

— = —, in der, normal, idle, inoperative.
Stemmeisen *n*, chisel.
stemmen, to chisel.
Stempel *m*, stamp; Lampe, Röhre: press, squash; Stanze: punch;
Laufnummern= —, numbering machine;
Loch= —, die;
Nummern= —, numbering machine;
Stanz= —, punch, die;
Zeit= —, time stamp;
— =kissen *n*, stamp pad;
— =matrize *f*, cutting die-plate;
— =satz *m*, gang of punches (im Locher, of the perforator).
stempeln, to stamp.
Stenographie *f*, shorthand.
stenographieren, to write (in) shorthand.
stereophonisch, stereophonical.
Stern *m*, star;
— =Dreieckschalter *m*, star-delta switch;
sterngeschaltet, Y-connected, star-connected;
Stern=glied *n*, star circuit, T-mesh;
— = —er-kette *f*, T-mesh network;
— =rad *n*, star wheel; pin feed wheel *T*;
— =schaltung *f*, Y-connection, star connection; star circuit;
— =schauzeichen *n*, white star indicator;
— =spannung *f*, Y-voltage, star voltage;
— =vierer *m*, spiral quad, spiral four;
— = — =kabel *n*, spiral(led) four cable, spiral quad cable.
stetig, uniform.
Stetigkeit *f*, uniformity. [*A*;
Steuer=bürste *f*, private wiper

Steuer
— -elektrode f, control electrode;
— -gitter n, control grid.
steuern, to control.
Steuer-röhre f, control valve; pilot oscillator, master oscillator, exciter tube;
— -schalter m, master switch, control switch; sequence switch A;
— -spannung f, control voltage.
Steuerung f, control;
  Erregerkreis- —, control of excitation;
  Fern- —, remote control, distant control;
  Gitter- —, grid control.
Stichel m, style.
Stich-kabel n, branch cable;
— -leitung f, tie line.
Stickstoff m, nitrogen (N);
— -oxyd, nitric oxide (NO).
Stiel m, handle, stem, shank.
Stift m, pin, spike; Lötstift: tag, tab;
  konischer —, taper pin;
  Abstand- —, distance piece;
  Draht- —, wire nail;
  Feder- —, spring pin;
  Gewinde- —, grub screw, headless screw;
  Kleb- —, distance piece, stop;
  Löt- —, (wire type) soldering tab, tag;
— -büchse f, pin barrel T;
  Deckplatte f der - -, pin plate T;
— -lager n, hinge pin bearing (der Fernsprechrelais, of telephone relays);
— -lagerung f, pin suspension;
— -rad n, pin wheel.
Still-setzung f, arresting, stopping;
— -stand m, stop, cessation;
  zum - - bringen, to arrest;

zum - - kommen, to come to rest.
Stimme f, voice;
  Modulation f, der —, inflection of the voice.
Stimm-gabel f, forked reed, tuning fork;
  Geschwindigkeitsregelung f mittels - - —, reed control of speed;
— - -sender m, vibrating reed transmitter T;
— - -unterbrecher m, tuning fork circuit breaker, vibrator.
Stirn f, front, head;
  Wellen- —, wave front L;
  Zeichen- —, signal head L;
— -fläche f, face;
— -rad n, spur wheel;
— -rädergetriebe n, spur gearing.
Stockwerk n, floor;
  im ersten —, on the first floor.
Stoff m, material; substance, matter.
Stopfen m, plug, stopper;
  Gummi- —, rubber plug;
stoppen, to stopper.
Stopper m, stopper.
Stoppuhr f, stop watch.
Stöpsel m, plug, peg;
  einen — einsetzen, to insert a plug, to plug in;
  einen — herausziehen, to withdraw a plug;
  mit —n einschalten, to plug in, to plug up (to);
  dreiteiliger —, three-point plug, three-way plug;
  geschlitzter —, split plug;
  Abfrage- —, answering plug;
  Blind- —, dummy plug;
  Doppel- —, double plug;
  Hinweisungs- —, (indication) peg F;
  Isolier- —, insulating plug, infinity plug;

Stöpsel
Kontakt= —, contact plug;
Messing= —, brass peg;
Prüf= —, test plug;
Schraub= —, screwed plug;
Sicherungs= —, safety plug, fuse plug;
Trenn= —, infinity plug;
U= —, U-link plug;
Verbindungs= —, connecting plug; calling plug, ringing plug, F;
Verbindungsleitungs= —, junction plug;
— =brett n, plug shelf;
— =brücke f, plug bridge, contact plate of plug;
— =griff m, plug handle;
— =hals m, plug sleeve, plug neck;
— =hülse f, plug cover;
— = — n=gewinde n, plug cover thread;
— =körper m, plug body;
— =linienwähler m, plug selector;
— =loch n, plug hole.
— =ring m, (isolierter), (dead) ring of the plug.
stöpseln, to plug in; zustöpseln: to plug up.
Stöpsel=ring m, ring of the plug;
— = — zuführung f, ring wire, R-wire, F;
— =schalter m, plug switch;
— =schnur f, plug-ended cord; lose = — —, loose cord and plugs;
— =sicherung f, cartridge fuse, plug fuse;
— =spitze f, plug tip;
— = — n=zuführung f, tip wire, T-wire, F;
— =umschalter m, plug switch, plug commutator.
Stör=befreiung f, elimination of interference, elimination of jamming R; Luftstörungen:

X. stopping, elimination of strays, R;
Einrichtung f zur — = —, antiparasitic system, X. stopper.
stören, to disturb, to trouble, to interfere (with), to jam R;
Störer m, disturbing station, jamming station; Luft= —: X.'s, strays, pl;
durch — verdeckte Zeichen pl, swamped signals pl.
Störfaktor m, interference factor F;
störfrei, immune against interference, undisturbed;
Stör=freiheit f, immunity from interference;
— =frequenz f, interfering frequency;
— =geräusch n, interfering noise;
— =schwingung f, disturbing wave;
— =sender m, interfering transmitter;
— =strom m, disturbance current;
— = — infolge schlechter Ausgleichung, unbalance current TK;
— =ton m, interfering tone.
Störung f, disturbance, trouble, interference (aus Nachbarleitungen, from adjacent lines, der Nachbarleitungen, into adjacent lines); der Leitung usw.: fault, breakdown; durch fremde Sender: jamming, interference, R;
luftelektrische — en pl, statics, X.'s, atmospherics, pl;
magnetische —, magnetic perturbation;
starke —, heavy interference;
Erd= —, earth disturbance;
Fernleitungs= —, toll line fault;
Induktions= —, inductive interference;

Störung
Leitungs= —, line fault;
Luft= —, atmospheric disturbance, X.'s, strays, statics, *pl*;
— = — en, Beseitigung von, elimination of statics;
Ruf= —, ringing failure;
Starkstrom= —, interference from power system; der Starkstromanlage: power failure;
Verbindungsleitungs= —, junction fault;
Zünd= —, ignition interference (durch Explosionsmotoren, from internal combustion engines) *R*;

Störungs=aufsicht *f*, fault clerk;
— =beamter *m*, test clerk;
— =beseitigung *f*, fault clearance;
— =faktor *m*, interference factor;
— = — =messer *m*, interference factor set;
— =meldung *f*, fault docket;
— =personal *n*, fault staff;
— = — für Amtsstörungen, exchange fault staff, internal fault staff;
— = — für Außenstörungen, external fault staff;
— =platz *m*, trouble desk;
— =stelle *f*, fault section, trouble desk;
— =sucher *m*, faultsman, troubleman;
— = — im Außen=(Innen)dienst, external (internal) faultsman;
— =trupp *m*, repair gang;
— =verhinderung *f*, interference prevention.

Stoß *m*, pulse, impulse, impact, shock, thrust; Ruck: jerk; Stapel: pile;
Strom= —, current (im)pulse;
— =bohrer *m*, thrust borer *B*.

Stößel *m*, striker.
stoßen, to push, to thrust rucken: to jerk.
Stößer *m*, push rod; Hughes: rejector.
Stoß=erregung *f*, shock excitation, excitation by impact;
stoß=fest, shock-proof;
— =frei, smooth;
Stoß=ionisation *f*, ionisation by collision;
— =klinke *f*, driving pawl, thrust pawl, propelling pawl;
— =kreis *m*, impulsing circuit;
— =stange *f*, push rod;
— =stelle *f*, joint.
straff, tight, rigid.
Strahl *m*, ray, beam, jet;
Flüssigkeits= —, liquid jet;
Kathoden= —en *pl*, cathode rays *pl*;
Licht= —, luminous ray, light ray;
— =empfänger *m*, uni-directional receiver *R*;
— =sender *m*, — =sendestelle *f*, beam transmitter, uni-directional transmitter, *R*.
strahlen, to radiate, to emit.
Strahler *m*, radiating system, emitter, radiator;
— =gebilde *n*, radiating system, emitter;
— =kreis *m*, radiating circuit.
Strahlung *f*, radiation, emission;
elektromagnetische —, electromagnetic radiation;
Sekundär= —, secondary emission;
Strahlungs= ...., radiative;
— =energie *f*, radiated energy;
— =feld *n*, radiation field;
magnetisches — = —, radiation magnetic field;
— =fläche *f*, emitting surface, emitting area, (des Heizfadens, of the filament);

Strahlungs
- -höhe f, radiation hight;
- -leistung f, radiated power;
- -ökonomie f, efficiency of radiation;
- -verluste pl, corona(l) losses pl;
- -vermögen n, radiating capacity;
- -widerstand m, radiation resistance, characteristic impedance;
- -wirkungsgrad m, efficiency of radiation;
.... -strähnig, ... strand; sieben- -, seven strand ....
Strand m, beach, shore.
Straße f, way, road;
öffentliche -, public road;
Fahr- -, road, carriageway;
Land- -, road;
Straßen-arbeiten pl, road work;
- -bahn f, tram, tramway;
- - -schiene f, tramway rail;
- -kiosk m, street kiosk;
- -planum n, street level.
Strebe f, strut(ting), brace, prop;
Diagonal- -, diagonal strut.
Streckbalken m, balk.
Streckbarkeit f, ductility.
Strecke f, stretch, Abschnitt: section, Baustrecke: field;
Bau- -, field;
Block- -, block section;
Fehler- -, faulty section;
Gas- -, gas(eous) path.
strecken, to stretch, to rack.
Strecken n, racking;
- -dämpfungs-messer m, transmission efficiency measuring set, transmission measuring set for straightaway tests, K F;
- - -messung f, transmission efficiency test;
- -fernsprecher m, portable telephone station;
- -versuch m, field test.
Streckgrenze f, yield point.
streichen, to paint; bestreichen: to wipe, to sweep (over);
eine Gesprächsanmeldung - (lassen), to cancel a call.
Streichung f, cancellation.
Streifen m, strip, strap, Papier usw: ribbon, tape, slip, Ansatzlappen: tab;
- herstellen, to prepare the tape T;
- zu 10 Lampen, strip of ten lamps F;
unpassender -, misfitting slip T ;
Abschmelz- -, fuse strip;
Bezeichnungs- -, designation strip;
Blitzableiter- -, protector strip;
Klappen- -, strip of indicators;
Klinken- -, jack strip;
Lampen- -, lamp strip;
Loch- -, perforated slip T;
- - -, Längs-, lengthwise perforated tape;
- - -, Quer-, cross-perforated tape;
Lötösen- -, terminal strip, strip of tags;
Messing- -, brass strap;
Mithöre- -, home record;
Morse- -, Morse slip;
Papier- -, paper tape, paper slip;
Sende- -, transmitting tape;
Verbindungs- -, connection strip, strap;
Wheatstone- -, Wheatstone tape;
- -bahn f, tape race, tape platform, T;
- -druck m, tape printing T;

Streifen
— -drucker m, tape printer T;
— -geber m, (perforated) tape transmitter;
— -lade f, slip drawer, paper drawer;
— -rolle f, tape roll;
— -schublade f, slip drawer;
— -sender m, (perforated) tape transmitter;
— -sendung f, tape transmission;
— -sicherung f, strip fuse;
— -vorschub m, paper feeding;
— - -daumen m, paper feeding cam;
— - -einrichtung f, paper-feeding device.

streng, rigorous;
— -flüssig, refractory.
streuen, to stray, to leak; zerstreuen: to disperse.
Streu=feld n, stray field, leakage field, extraneous field;
— -fluß m, stray flux, leakage flux, cross-flux;
— -impedanz f, leakage impedance;
— -induktivität f, stray inductance;
— -kapazität f, stray capacity, spurious capacities pl;
— -ströme pl, stray currents, eddy currents, pl.
Streuung f, stray(ing), leakage, dispersion;
elektrostatische —, stray capacity;
magnetische —, magnetic dispersion, magnetic cross-flux, magnetic leakage;
Flanken= —, side leakage.
Strich m, line, dash;
Kompaß= —, rhumb;
Morse= —, (Morse) dash.
strichpunktiert, dash-dotted, chain-dotted.
Strick m, rope.

Stroboskop n, stroboscope.
stroboskopisch, stroboscopic(al);
— e Scheibe f, stroboscopic disc.
Strom m, current, Fluß: stream;
— aufnehmen, to take current;
— führen, to carry a current;
— senden, to feed current (in den Luftdraht, into the aerial);
abgehender —, outgoing current, sending current;
abfallender —, decreasing current;
abklingender —, decaying current;
ankommender —, incoming current, receiving current;
effektiver —, r. m. s. current;
eingeschwungener —, steady state current;
gleichförmiger —, steady current;
entgegengesetzte Ströme pl, inverse currents, opposed currents, pl;
induzierender Strom, inducing current;
induzierter —, induced current;
kommutierter —, commutated current;
oszillierender —, oscillatory current, oscillating current;
phasenverschobener —, out-of-phase current;
pulsierender —, pulsating current;
resultierender —, resultant current;
schwacher —, weak current, feeble current;
schwingender —, oscillatory current;
sinusförmiger —, harmonic current;
stationärer —, steady state current;   [current;
thermoelektrischer —, thermo-

Strom
überlagerter —, superposed current;
undulierender —, undulating current;
vagabondierende Ströme *pl*, leakage currents, stray currents, vagabond currents, *pl*;
wattloser Strom *m*, wattless current, reactive current;
welliger —, ripple current;
Ableitungs= —, leakage current, leak(ance) current, stray current;
Abschmelz= —, fusing current;
Absorptions= —, absorption current;
Anker= —, armature current;
Anoden= —, plate current, discharge current, anode current, space current, *V*;
Anodenruhe= —, feed current *V*;
Anspruch= —, (Mindest=), (minimum) operating current;
Antennen= —, aerial current;
Ausgleichs= —, compensating current; flüchtiger: transient current;
Auslöse= —, releasing current *A*, Einschalten: tripping current, starting impulse;
Ausschwing= —, decaying current;
Außen= —, foreign current;
Betriebs= —, (normal) operating current;
Blind= —, reactance current, wattless current;
Dauer= —, permanent current, continuous current, steady current;
Doppel= —, double current *T*;
Dreh= —, threephase current, triphase current;
Druckluft= —, forced draught;

Echo= —, echo current(s *pl*);
Effektiv= —, effective current;
Einfach= —, single current *T*;
Einphasen= —, single-phase current, monophase current;
Einrück= —, tripping current;
Einschwing= —, building-up current;
Ein= und Ausschwing= —, transient current;
Einzel= —, single current *T*;
Elektronen= —, electron current, ionic current; stream of electrons;
Emissions= —, emission current;
Entlade= —, discharging current;
Erd= —, earth current;
Erreger= —, exciting current;
Faden= —, filament current;
Foucault=ströme *pl*, eddy currents, Foucault currents *pl*;
Fremd=strom, foreign current;
Gesamt= —, total current;
Gitter= —, grid current;
— = —, negativer, reverse grid current;
Gleich= —, direct current (*ab*: d. c.), continuous current (*ab*: c. c.), steady current;
Glimm= —, glow current;
Halte= —, holding current, retaining current;
Haupt= —, main current;
Heiz= —, heating current, filament current *V*;
Induktions= —, induced current;
Jonen= —, stream of ions;
Jonisations= —, ionisation current;
Kompensations= —, compensating current;
Konvektions= —, convection current;
Kreis= —, circular current;

Strom
Kriech= —, surface leakage current;
Kurzschluß= —, short-circuit current;
Lade= —, charging current,
Kabel auch: surge current;
Leck= —, leak(age) current;
Leerlauf= —, no-load current;
Linien= —, line current;
Luftdraht= —, aerial current;
Magnetisierungs= —, magnetizing current;
Meß= —, testing current;
Modulations= —, modulating current;
Nebensprech= —, unbalance current KF;
Oberflächen= —, superficial current;
Polarisations= —, polarizing current;
Primär= —, primary current;
Prüf= —, testing current;
Raumlade= —, space current;
Regel= —, normal operating current;
Rück= —, reverse current;
Rückstell= —, releasing current;
Ruf= —, ringing current;
— = —, überlagerter, superposed ringing current;
— = —, gleichstromüberlagerter, ringing current superposed on d. c.;
Sättigungs= —, saturation current;
Schlußzeichen= —, clearing current;
Schwebungs= —, beating current;
Sekundär= —, secondary current;
Sende= —, sending current;
Sinus= —, sine current;

Speise= —, feeding current, supply current;
Sprech= —, speaking current, talking current; speech currents pl, voice currents pl;
Stark= —, power current;
Stör= —, disturbance current;
Streu=ströme pl, stray currents;
Teil=strom, partial current, component current;
Thermionen= —, thermionic current;
Träger= —, carrier (current);
Trenn= —, spacing current T;
Über= —, excess current;
Verschiebungs= —, displacement current;
Vormagnetisierungs= —, biasing current;
Wechsel= —, alternating current, ab: a. c.;
— = —, schneller, oscillatory current, undulating current;
Weck= —, ringing current;
Wirbel=ströme pl, eddy currents, Foucault currents;
Wirk=strom, energy current, active current;
Zeichen= —, marking current T;
Zweig= —, branch current;
— =abnahme f, decrease or fall of current;
— =amplitude f, current amplitude;
— =änderung f, current variation;
— =anzeiger m, current indicator, detector, galvanometer;
— =art f, kind of current;
— =bahn f, current path;
— =bauch m, current loop, current antinode;
— =begrenzer m, current limiter;
— =belastung f, current load;

Strom
— **-dichte** *f*, current density;
stromdurchflossen, current carrying;
Strom-einheit *f*, unit (of) current;
— **-empfindlichkeit** *f*, current sensitivity;
strömen, to flow, to stream.
Strom-erzeuger *m*, (current) generator;
— **-erzeugung** *f*, generation of current;
— **- —s-anlage** *f*, generating plant;
— **-faden** *m*, current path, stream line, current tube;
stromfähig, clear *F T*;
Stromfähigkeitsprüfung *f*, continuity test *K*;
stromführend, current-carrying, current-conveying, live, alive;
Strom-impuls *m*, current impulse;
— **-knoten** *m*, current node;
— **-komponente** *f*, current component;
Blind- — — —, wattless component of current;
Gleich- — — —, direct current component, d. c. component;
Wechsel- — — —, alternating current component, a. c. component;
Wirk- — — —, energy component of current;
Stromkreis *m*, circuit;
einen — öffnen, to open, to break a circuit;
einen — schließen, to make, to complete, to close a circuit;
mit Blindwiderstand behafteter —, reactive circuit;
aus ohmschen Widerständen bestehender —, resistive circuit;

äquivalenter —, equivalent circuit;
geschlossener —, closed circuit;
offener —, open circuit;
überlagerter —, superposed circuit;
Absorptions- —, absorbing circuit;
Anoden- —, plate circuit, plate-to-filament circuit;
Ausgangs- —, output circuit;
bei Röhren auch: plate-to-filament circuit;
Ausgleichs- —, compensation circuit;
Bezugs- —, reference circuit;
Eingangs- —, input circuit;
Entlade- —, discharging circuit;
Entnahme- —, load circuit, receiver circuit;
Erreger- —, exciting circuit;
Halte- —, holding circuit, retaining circuit;
Haupt- —, main circuit;
Kontroll- —, checking circuit;
Lade- —, charging circuit;
Meß- —, measuring circuit;
Mikrophon- —, transmitter circuit;
Mitlese- —, leak circuit *T*;
Schnur- —, cord circuit;
Speise- —, supply circuit;
Verbraucher- —, load circuit;
Wecker- —, bell circuit;
Zweig- —, branch circuit.
Strom-kurve *f*, current curve;
— **-lauf** *m*, circuit diagram;
Teil- — — —, circuit detail;
— — **-skizze** *f*, circuit diagram;
— **-linie** *f*, stream line;
stromlos, currentless, dead;
Stromlosigkeit *f*, absence of current;
Strommesser *m*, ammeter;
registrierender —, recording ammeter;

**Strommesser
vereinigter Spannungs-
und —,** combined volt- and
ammeter;
Drehspul- —, moving coil
ammeter;
Gleich- —, d. c. ammeter;
Hitzband- —, hot band ammeter;
Hitzdraht- —, hot wire ammeter;
Schreib- —, recording ammeter;
Thermo- —, thermo-ammeter;
Wechsel- —, a. c. ammeter;
Weicheisen- —, moving iron
ammeter, soft iron vane
ammeter.

Strom-messung *f*, current measurement;
Gleich- — —, d. c. measurement;
Wechsel- — —, a. c. measurement;
— **-pulsationen** *pl*, pulsations *pl*
of current;
— **-quelle** *f*, current source;
Gleich- — - —, d. c. source;
Wechsel- — - —, a. c. source;
— **-resonanz** *f*, parallel resonance;
— **- — -kreis** *m*, parallel resonant
circuit;
— **-richtung** *f*, direction of current;
— **- —s-anzeiger** *m*, polarity indicator;
— **-schlußhebel** *m*, circuit-closing
lever;
— **- — -taste** *f*, connection key;
— **-schritt** *m*, signal element,
unit, pulse, selten: signal, *T*;
— **-schwankungen** *pl*, current variation, fluctuation of current;
— **-schwebungen** *pl*, current
beats *pl*;

— **-Spannungskurve** *f*, current-
voltage characteristic;
— **-spule** *f*, current coil;
— **-stärke** *f*, current intensity;
Abschmelz- — - —, blowing
point (einer Sicherung, of a
fuse);
Ansprech- — - —, operating
current intensity, figure of
merit;
Heiz- —, heating current intensity;
— **-stoß** *m*, current impulse,
pulse;
flüchtiger — - —, transient impulse;
kürzester Telegraphier- - - —,
signal element, unit;
plötzlicher — - —, current rush;
Wähl- — - —, dialling impulse *A*;
— **- — -dauer** *f*, impulse period
A;
— **- — -empfänger** *m*, impulse
receiver, impulse storing device, *A*;
S A-System mit — - — - —u,
stored impulse automatic
telephone system;
— **- — -gabe** *f*, impulsing *A*;
rückwärtige — - — - —, reversed
impulsing;
unmittelbare — - — - —, direct
impulsing;
— **r — - — - —, S A-System** *n*
mit, direct impulse automatic telephone system;
— **- — -geschwindigkeit** *f*, impulse
frequency *A*;
— **- — -reihe** *f*, series or train or
succession of impulses *A*;
pulsations *pl*;
— **- — -speicher** *m*, (digit-storing)
register, impulse storing device;
— **- — -teilung** *f*, impulse ratio
A;

Stromstoß
— = — =**übertrager** m, impulse repeater A;
Wähler m mit — = — = —, selector repeater A;
— = — =**verhältnis** n, break-to-make ratio of an impulse A;
— =**übergang** m, leakage of current;
— =**vektor** m, current vector;
— =**verbrauch** m, current consumption;
— =**verbraucher** m, utilization device;
— =**verdrängung** f, skin effect;
— =**verlust** m, current loss;
— =**verstärkung** f, current amplification;
— =**verteilung** f, current distribution;
— =**wage** f, current balance;
— =**wandler** m, current transformer, series transformer;
— =**wechsel** pl, reversed currents, reversals, pl T;
— =**weg** m, current path;
— =**welle** f, current wave;
— = —**n** pl current ripple(s pl) (des kommutierten Stromes), of the commutated current);
— =**wender** m, commutator, reversing switch;
— =**wendung** f, commutation, reversal of current;
— =**zuführung** f, current supply (to);
— = — z=**bürste** f, current supply brush;
— =**zunahme** f, growth or increase or rise of current.
Strontium n, strontium (Sr).
Strowgerwähler m, Strowger switch, Strowger selector, A.
Struktur f, structure.
Stücklohn m, piece wages pl.
Studienplan m, plan of study.

Stufe f, step, Stadium: stage, Grad: grade; digit A; Leiter: step;
in —n von...., by gradations of....;
Einer= —, units digit A;
Zehner= —, tens digit A;
Hunderter= —, hundreds digit A;
Tausender= —, thousands digit A;
Wahl= —, digit A;
Wähler= —, rank of switches A;
Widerstands= —, resistance step;
—**n=folge** f, gradation;
stufenweise, gradual, in steps;
Stufen=wicklung f, bank(ed) winding R.
Stumpf m, stud;
— =**mast** m, stub mast;
— =**verbindung** f, butt joint.
Stunde f, hour.
Sturm m, gale, storm;
elektrischer —, electric storm;
magnetischer —, magnetic storm;
Hagel= —, sleet storm;
Schnee= —, snow storm;
— =**warnung** f, weather warnings pl.
Stütze f, rest, support, bracket B, crutch;
Haken= —, hook-shaped bracket;
Isolator= —, (insulator) bracket, gerade: insulator spindle, insulator pin;
Kabel= —, cable bearer, cable bracket;
Lager= —, bearing bracket;
Mauer= —, wall bracket;
Schrauben= —, bolt;
— = —, **gerade**, straight bolt, spindle;
— = —, **J=förmige**, J-bolt;

Stütze
Wand= —, wall bracket.
Stütz-balken m, brace;
— **-isolator** m, pin-type insulator;
— **-punkt** m, support; eines Hebels: fulcrum pl fulcra.
Stutzen m, stud;
Rohr= —, nozzle.
Sublimat n, sublimate.
Subskription f, subscription.
Substanz f, substance.
substituieren, to substitute.
Substitution f, substitution.
subtrahieren, to subtract.
Subtrahendus m, subtrahend.
Subtraktion f, subtraction, deduction.
Subvention f, subsidy.
subventionieren, to subsidize.
Suchanker m, grapnel.
suchen, to seek, Fehler: to trace; ein Seekabel: to drag (for); to select, to find (out), A;
frei —, to hunt (for) A;
Suchen n, tracing; finding, selecting auch A; hunting A;
Frei= —, hunting A.
Sucher m, selector, seeker; finder A;
Anruf= —, finder (switch) A;
Klinken= —, jack finder;
Sender= —, sender finder;
Verbindungsleitungs= —, junction finder;
— **-fuß** m, seeker toe T;
— **-hebel** m, seeker lever, selecting lever T.
Südlicht n, aurora australis.
südmagnetisch, south-magnetic.
Südpol m, south pole.
Sulfat n, sulphate.
sulfatieren, to sulphate.
Sulfatierung f, sulphation.
Sulfid n, sulphide.
Summe f, sum, total;
ganze —, grand total;

Vektor= —, vector sum.
summen, to hum, to buzz.
Summen n, hum(ming), buzz, buzzing.
Summer m, buzzer, vibrator, starker: howler;
durch — **erregter Kreis** m, buzzer-driven circuit;
Mikrophon= —, microphone hummer, howler;
Röhren= —, electron tube generator, valve generator;
Saiten= —, chord buzzer;
Zungen= —, reed hummer;
— **-erregung** f, buzzer excitation R;
— **-generator** m, buzzer generator;
— **-relais** n, buzzer relay;
— **-ton** m, buzzing sound, humming tone;
— **-unterbrecher** m, buzzer interrupter;
— **-zeichen** n, humming sound, tone F;
Amts= — = —, dialling tone A;
Frei= (— =) —, ringing tone A;
Gestört= — = —, out-of-order tone, ab: o o o tone, A;
— = —**zur Anzeige unbenutzter Leitung**, dead number tone A;
— = — — —**unausführbarer Verbindung**, number unobtainable tone. ab: n. u. tone, A;
summieren, to add, to sum (up).
Summierung f, summation.
Sumpf m, swamp.
sumpfig, swampy, marshy.
Superoxyd n, peroxide.
Superposition f, superposition.
Supererregenerativempfang m, super-regenerative reception.
Support m, slide;
Kreuz= —, cross slide.

Sufzeptanz f, susceptance;
induktive —, inductive susceptance;
kapazitive —, capacity susceptance.

Suszeptibilität f, susceptibility;
magnetische —, magnetic susceptibility.

Symbol n, symbol.

symbolisch, symbolic(al).

Symmetrie f, symmetry;
mangelnde —, lack of symmetry;
Induktivitäts- —, inductance balance (von Pupinspulen, of loading coils);
Kapazitäts- —, capacity balance;
Widerstands- —, resistance balance;
— -achse f, axis of symmetry;
— -bedingung f, condition for symmetry;
— -ebene f, plane of symmetry.

symmetrisch, symmetrical.

synchron, synchronous.

Synchron-Funkenstrecke f, synchronous rotating spark gap;
— -motor m, synchronous motor.

synchronisieren, to synchronize.

Synchronisierung f, synchronization.

Synchronismus m, synchronism.

Syphonrekorder m, syphon recorder.

System n, system;
bewegliches —, moving system;
— -kabel n, switchboard cable, multiple cable;
63-adriges — - —, 63 wire switchboard cable.

T.

Tabellarisieren, to tabulate.

Tabelle f, table, chart.

Tableau n, indicator board;
Licht- —, luminous indicator (board).

Tachometer n, tachometer, speedometer.

Tafel f, board, table; graph, chart; plate, panel; slab;
Fluchten- —, self-computing chart;
Kurven- —, curve sheet, graph;
Marmor- —, marble slab;
Rechen- —, computation table;
Umrechnungs- —, reduction table;
— -feld n, panel.

Tages-belastung f, day load;
— -gebühr f, day rate;
— -reichweite f, day range R.

tafeln, to rig.

Takt m, cadence T;
— geben, to cadence;
— -geber m, time tapper, cadence tapper;
— -gebung f, cadence;
— -zeichen n, cadence signal.

Tal n einer Welle, wave trough.

Talg m, tallow.

Talk m, talc.

Tandem-ämter pl, tandem offices;
— -betrieb m, tandem operation (von Verbindungsleitungen, of junctions).

Tangens m, Tangente f, tangent, ab: tan;
hyperbolischer —, hyperbolical tangent, ab: tanh.

Tangentenbussole f, tangent galvanometer.

tangential, tangential (to).
Tank *m*, tank;
  Kabel- —, cable tank.
  -**Anker** *m*, **Doppel**-, shuttle armature, H-armature.
Tanne *f*, deal.
Tannin *n*, tannic acid, tannin;
  mit — getränkt, tanned.
Tantal *n*, tantalum (Ta).
T-Antenne *f*, T-antenna;
  verlängerte — - —, extended T-antenna.
Tarif *m*, tariff;
  Einzelgebühren- —, measured rate tariff;
  Pausch- —, bulk tariff, flat rate tariff;
  Staffel- —, gratuated tariff;
  Zonen- —, zone tariff;
  — -**einheit** *f*, tariff unit;
  — -**system** *n*, tariff system.
Taschenlampe *f*, pocket lamp, flash lamp;
  — n-**batterie** *f*, flash lamp (dry cell) battery.
Tastatur *f*, keyboard;
  Universal- —, universal keyboard.
Taste *f*, key, throw key; **Druckknopf**: push key, press key; **Schreibmaschine** usw.: (character) key; **Morse**: manipulating or sending or operating key;
  eigene —, home key *T*;
  Anschlag *m* einer —, touch or depression of a key;
  Hinter-(Vorder)schiene der —, back (front) stop of key;
  Körper *m*, der —, centre of key;
  Abschalte- —, cut-out key;
  Abstands- —, blank key, space key;
  Amts- —, office key;
  Auslöse- —, release key;
  — - —, **Haupt**-, master release key;
  Blank- —, blank key;
  — - —, **Buchstaben**-, letter blank key;
  — - —, **Zahlen**-, figure blank key;
  Dienstleitungs- —, order wire key;
  Doppelstrom- —, double current key *T*;
  Haupt- —, master key;
  Irrungs- —, erase key *T*;
  Kabel- —, reversing key;
  Klopfer- —, sounder key;
  Kontroll- —, check key;
  Morse- —, Morse key;
  — - — mit selbsttätiger **Punktgebung**, vibroplex key;
  Rückruf- —, ring back key;
  Rückstell- —, resetting key, release key;
  Rückzug- —, back-spacing key *T*;
  Ruf- —, ringing key;
  Sende- —, sending key;
  Sprech- —, press key (am **Sprechhörer**, of the handset);
  Stromschluß- —, connection key;
  Überweisungs- —, assignment key;
  Umkehr- —, reversing key;
  Unterbrechungs- —, break key;
  Verbindungs- —, connection key;
  Zähl- —, meter key.
Tast-einrichtung *f*, keying device.
tasten, to key, to manipulate.
Tasten-feld *n*, keyboard;
  Universal- — - —, universal keyboard;
  — - — mit **Sperre**, locked keyboard;
  — -**geber** *m*, keyboard transmitter *T*;
  Fünf- — - —, five-key transmitter *T*;

Taften
- =**hebel** *m*, key lever;
- =**Impulsgeber** *m*, impulse sending key *A*;
- =**knopf** *m*, key button;
- =**locher** *m*, keyboard perforator;
- =**Nummerngeber** *m*, key (-set call) sender *A*;
- =**reihe** *f*, row of keys, bank of keys;
- =**sah** *m*, key set; Zehn= – = –, ten-button key set;
- =**sperre** *f*, keyboard lock.

Taft-leitung *f*, keying circuit;
- =**relais** *n*, key(ing) relay, contactor, relay key, magnetic key.

Taftung *f*, control, keying;
Erregerkreis= –, control of excitation;
Fern= –, remote control;
Gitter= –, grid control.

Tau *n*, rope, line;
Hanf= –, hemp rope;
Schlepp= –, tow.

tauchen, to plunge.

Tauch-kern *m*, plunger;
– = – =**relais** *n*, plunger relay;
– = – =**spule** *f*, sucking solenoid;
– =**spulen** *pl*, telescoping coils *pl*;
– =**transformator** *m*, telescoping coil transformer.

Taufender-syftem *n*, three-figure system, three digit system, *A*;
– =**wahlstufe** *f*, thousands digit *A*.

Taxquabrat *n*, telephone (trunk) zone.

Teakholz *n*, teak.

Technik *f*, art, engineering, technics *pl*, technique;
Fernmelde= –, communicating art;

Fernsprech= –, telephone engineering;
Funk= –, radio engineering;
Meß= –, measuring technique.

technisch, technical.

Teer *m*, tar;
Gas= –, gas tar;
Holz= –, wood tar;
Kohlen= –, coal tar;
– =**öl** *n*, coal tar oil.

teeren, to tar.

Teil *m*, part, portion, Abschnitt: division;
– =...., partial, component;
– =**amt** *n*, satellite exchange *A*;
– =**chen** *n*, particle.

teilen, to divide (durch by); sich –, to split.

Teiler *m*, divisor, divider.

Teilhaber *m*, shareholder.

Teilkreis *m*, divided circle, graduated circle, dial.

Teilnehmer *m*, subcriber, party; talker;
– **antwortet nicht**, there is no reply;
– **hängt an**, subscriber clears;
angerufener –, called party, required or wanted subscriber;
anrufender –, caller, calling party;
anrufenden –**s, Schleife** *f* **des**, calling loop;
verlangter –, called party, required or wanted party;
Gesprächsgebühren= –, measured rate subscriber;
Handamt= –, subcriber to a manual exchange;
Pauschgebühren= –, flat rate subscriber;
– =**anlage** *f*, substation plant;
– =**anrufzeichen** *n*, subscriber's line indicator;
– =**anschluß** *m*, subscriber's station; *am*: subset, substation;

Teilnehmer
— -apparat m, subscriber's set;
— -doppelleitung f, subscriber's loop;
— -gebühren pl, rates pl of subscription;
— -Hauptanschluß m, subscriber's main station;
— -klinke f, subscriber's jack;
— -leitung f, subscriber's line;
— -platz m, home position, answering position;
— -nebenanschluß m, — -nebenstelle f, subscriber's extension station;
— -schleife f, subscriber's loop;
— -sprechstelle f, subscriber's set, subscriber's station; am: subset, substation;
O. B. - — - —, subscriber's l. b. station;
Z. B. - — - —, subscriber's c. b. station;
— -verzeichnis n, telephone directory;
— -vielfach(feld) n, subscriber's multiple;
— -zentrale f, private branch exchange, ab: p. b. x.;
selbsttätige — - —, private automatic branch exchange, ab: p. a. b. x.

Teil-spannung f, component voltage;
— -strich m, (scale) division;
— -strom m, partial current, component current;
— - — -lauf m, circuit detail.

Teilung f, division; Skala: scale, dial; der Pole, Zähne: pitch;
mit einer — versehen, to scale, to divide;
Instrument mit vier —en, quadruple scale instrument;
gleichmäßige —, scale of equal divisions;

Kreis- —, divided circle, graduated circle, circular scale;
Pol- —, pole pitch.

Teil-welle f, partial wave;
— -zirkel m, dividers pl.

T-Eisen n, T-iron;
Doppel- — - —, I-iron, double T-iron.

Telautograph m, telautograph, telewriter.

Telautographie f, telautography.

telautographisch, telautographic(al).

Telegramm n, message, telegram;
— zu ermäßigten Gebühren, deferred (rate) telegram;
ein — abschreiben, to write up a message;
ein — annehmen, to accept a message;
ein — aufgeben, to hand in or to file a telegram;
ein — aufnehmen, to copy a message;
ein — prüfen, to check a message;
Amts- —, service message;
Auslands- —, foreign message;
Brief- —, night telegraph letter;
Chiffer- —, cipher(ed) message;
Dienst- —, service message;
Inlands- —, inland message;
Presse- —, press message, news message;
Privat- —, private message;
Sp- —, phonogram (message);
Zeitungs- —, press message, news message;
— -annahmestelle f, collecting office;
— -beförderung f, transmission of telegrams;
— -besteller m, messenger;

Telegramm
- -bestellung f, delivery of messages;
- -formular n, message blank;
- -kopf m, preamble, preface;
- -pult n, message desk;
- -reihe f, batch (of messages); in - = —n zu . . . ., in batches of . . . .;
- -schalter m, telegram counter;
- -verteilung f, distribution of messages;
- - —, mechanische, machine distribution of messages;
- -vordruck(blatt n) m, message form, message blank.

Telegraph m, telegraph;
- mit photographischem Zeichendruck, photo-printing telegraph;
- mit (un)gleich langen Zeichen, (un)equal letter telegraph;
optischer —, optical telegraph; semaphore;
Blattdruck= —, page printing telegraph;
Druck= —, printing telegraph;
Einfach= —, single-channel telegraph;
Einmann= —, single-operator telegraph;
Feuerwehr= —, fire-alarm telegraph;
Geh=Steh= —, start-stop telegraph;
Klippklapp= —, flip-flap telegraph, to-and-fro telegraph;
Kopier= —, copying telegraph;
Maschinen= —, automatic telegraph, machine telegraph;
Massen= —, high-capacity telegraph;
Mehrfach= —, multiple(x) telegraph, multi-channel telegraph, multiple-way telegraph;

- = — mit abgestimmten Wechselströmen, harmonic multiple telegraph;
- = — in Gabelschaltung, split multiplex telegraph;
- = — in Staffelschaltung echelon or series multiplex telegraph;
Nadel= —, needle telegraph;
- = —, Doppel=, double needle telegraph;
- = —, Ein=, single needle;
- = —, Mehr=, multiple needle telegraph;
Pendel= —, pendulum start-stop telegraph;
Photo= —, telephotograph;
Reihen= —, automatic telegraph, high-speed single channel telegraph;
Schnell= —, high-speed telegraph;
Schrittschalt= —, step-by-step telegraph;
Seitendruck= —, page-printing telegraph;
Streifendruck= —, tape-printing telegraph;
Typendruck= —, type-printing telegraph;
Vierfach= —, quadruple telegraph;
Zeiger= —, pointer telegraph;
Zweifach= —, double telegraph.

Telegraphen=alphabet n, telegraph code;
- = — mit (un)gleich langen Zeichen, (un)equal letter code;
- =amt n, telegraph office, telegraph station;
Haupt= = —, central telegraph office, ab: c. t. o.;
- =anlage f, telegraph plant;
Funk= = —, radio telegraph plant;
- =anstalt f, telegraph station;

Telegraphen
— -arbeiter m, line(s)man, wireman;
— -bote m, telegraph messenger;
— -gleichung f, telegraphic equation;
— -kabel n, telegraph cable;
— -kode m, telegraph code;
— -leitung f, telegraph line;
Simultan- - - —, telegraph superposed circuit;
— - — mit Sprechbetrieb, phonogram circuit;
— -schlüssel m, telegraph code;
— -truppe f, Signal Corps;
— -übertragung f, telegraph repeater (set);
— - — mit Berichtigung der Zeichenform, entzerrende — - —, regenerative repeater;
— - —s-amt n, repeater station, repeating telegraph station;
— -umgehungseinrichtung f (für Fernsprech-Zwischenverstärker), telegraph by-pass set (for telephone intermediate repeaters).

Telegraphie f, telegraphy;
Bild- —, image transmission, picture telegraphy;
Draht- —, wire telegraphy;
Funk- —, radio telegraphy, wireless telegraphy;
— - —, gerichtete, directional wireless telegraphy;
— - — mit gedämpften Wellen, spark telegraphy;
— - — mit ungedämpften Wellen, continuous wave telegraphy, c. w. telegraphy;
Gegensprech- —, duplex telegraphy;
— - —, einseitige, half duplex telegraphy;
Gleichstrom- —, d. c. telegraphy;
Mehrfach- —, multiple telegraphy;
— - — mit hochfrequenten Wechselströmen, h. f. multiple telegraphy.
Tonfrequenz- —, audio-frequency telegraphy, voice frequency telegraphy;
Wechselstrom- —, a. c. telegraphy.

telegraphieren, to telegraph, to wire.
Telegraphier-alphabet n, telegraph code;
— -fehler m, operator error;
— -frequenz f, signalling frequency, telegraphic frequency;
— -geräusch n, Morse thump, telegraph noise;
— -geschwindigkeit f, (line) speed;
— -grundfrequenz f, dot frequency;
— -leistung f, einzelne, telegraph transaction;
— -strom m, signal(ling) current;
— - — -schritt m, signal element, unit;
— - — -stärke f, signal strength;
— - -stoß m, kürzester, signal element, unit; [channel;
— -weg m, telegraph route;
— -zeichen n, telegraph signal.

Telegraphiesender m, (radio-) telegraphic transmitter.
telegraphisch, telegraphic(al).
Telegraphist m, telegraphist, telegraph operator.
Telegraphon n, telegraphone.
Telephon n, telephone;
Haus- —, house telephone, domestic telephone;
— - — -anlage f, house telephone plant;
Kondensator- —, condenser telephone;
Mono- —, monotelephone;

Telephon
Pan= —, pantelephone;
Thermo= —, thermo-telephone.

Telephonie *f*, telephony;
Draht= —, wire telephony;
Funk= —, radio telephony;
Mehrfach= —, multiple telephony;
— =sender *m*, (radio)telephonic transmitter.

telephonieren, to phone, to telephone.

telephonisch, telephonic(al).

Telephonograph *m*, telephonograph.

Teleskopmast *m*, telescopic mast.

Tellur *n*, tellurium (Te).

Temperatur *f*, temperature;
— =koeffizient *m*, temperature coefficient;

Temperguß *m*, malleable cast iron.

Tempo *n*, tempo, speed;
Hand= —, key speed TR.

Tenor *m*, tenor.

Terpentin *n*, (oil of) turpentine;
— =öl *n*, spirit of turpentine.

Tertiärkreis *m*, tertiary circuit.

T=förmig, tee-shaped;
— = —er Hebel *m*, tee lever.

Thallium *n*, thallium (Tl);
lichtempfindliche — =zelle *f*, thalofide cell.

Thermionen *pl*, thermions *pl*;
— =relais *n*, thermionic relay;
— =strom *m*, thermionic current.

thermionisch, thermionic.
thermisch, thermal.

Thermodetektor *m*, thermo-(electric) detector;
thermoelektrisch, thermo-electric;
— =er Strom *m*, thermo-current;
Thermo=elektrizität *f*, thermo-electricity;

— =element *n*, thermo-electric couple;
— =galvanometer *m*, thermo-galvanometer;
— =kreuz *n*, thermo-couple;
— =meter *n*, thermometer;
Fern= — = —, distance thermometer;
— =säule *f*, thermopile, thermoelectric pile;
— =strommesser *m*, thermo ammeter;
— =telephon *n*, thermo-telephone.

Thomson=kabel *n*, non-loaded (submarine) telegraph cable;
— =kurve *f*, Kelvin arrival curve T.

Thor *n*, **Thorium** *n*, thorium (Th).

thorhaltig, thoriated;
— er Wolframfaden *m*, thoriated tungsten filament.

thorieren, to thoriate.

Thoriumröhre *f*, thoriated filament valve.

ticken, to click.
Ticker *m*, ticker, tikker.
Tiefe *f*, depth.
Tiefsee *f*, deep sea;
— =kabel *n*, deep sea cable.

Tiegel *m*, pot;
Schmelz= —, melting tank, melting pot; Stahl: crucible;
— =stahl *m*, crucible steel.

tilgen, to amortize.
Tilgung *f*, amortization.
Tinte *f*, ink;
— n=schreiber *m*, inker *R*.

Tisch *m*, table;
Apparat= —, instrument table;
— =fernsprecher *m*, table telephone station;
Säulen= — = —, desk stand telephone, pedestal desk telephone;   [station.
— =gehäuse *n*, table telephone

Titan *n*, titanium (Ti).
Toleranz *f*, tolerance.
Ton *m*, tone, sound, note;
  hoher —, high-pitched note;
  musikalischer —, musical sound, musical tone;
  reiner —, pure note;
  tiefer —, low-pitched note;
  unreiner —, ragged note;
  Besetzt- —, busy (back) tone *FA*;
  Funken- —, spark note;
  Grund- —, fundamental note, fundamental tone;
  Kombinations- —, combination note;
  Kontroll- —, check tone;
  Schwebungs- —, beat note;
  Summer- —, buzzer tone, humming tone, buzzing sound.
Ton- ...., tonic, tone, note;
— -abstimmung *f*, tone tuning, note tuning.
tönen, to sound, to hum.
Tönen *n*, hum(ming), sounding;
  Selbst- —, squealing *V*;
— der Drähte, humming of wires.
tönend, sounding;
  voll- —, round;
  welch- —, sonorous;
— er Funken *m*, musical spark;
— er Funkensender *m*, musical spark transmitter.
Ton-erde *f*, argil, clay;
— -formstück *n*, tile, clay conduit;
  einzügiges — - —, single tile;
  mehrzügiges — - —, multiple tile;
— -frequenz *f*, audible frequency, audio-frequency, voice frequency;
— - —-telegraphie *f*, audio-frequency telegraphy, voice frequency telegraphy;
— - —-zerhacker *m*, audio-frequency chopper;

— -funkensender *m*, musical spark transmitter;
— -höhe *f*, pitch, note, pitch of tone;
  Schwebungs- — - —, pitch of the beat note;
— - — der Sprache, pitch of speech;
— -konstanz *f*, constancy of pitch;
— -minimum *n*, critical silence;
  auf — - — einstellen, to silence;
Tonne *f*, buoy.
Ton-prüfer *m*, tone tester;
— -quelle *f*, tone source, source of sound;
— -rad *n*, tone wheel *R*;
— -reinheit *f*, purity of tone;
— -senden *n*, modulated c. w. transmission;
— -spektrum *n*, tonic spectrum;
— -stärke *f*, intensity of tone.
tonüberlagert, modulated at audible frequencies.
Ton-überlagerung *f*, tonic sine modulation, modulation at audible frequency, *R*;
— -verstärker *m*, note amplifier, note magnifier;
— -verstärkung *f*, note magnification, note amplification.
Topf *m*, pot;
— -magnet *m*, pot-shaped magnet, iron-clad magnet;
  Klappe *f* mit — - —, tubular drop indicator.
Topographie *f*, topography.
topographisch, topographic.
Torfboden *m*, peat(y) soil.
Toroid *n*, toroid.
Torsion *f*, torsion.
Torsions- ...., torsion(al);
— -feder *f*, torsion spring;
— -festigkeit *f*, torsional strength;
— -kopf *m*, torsion head.
Torus *m*, torus.

tote Winbung *f*, idle turn, dead end.

Totpunkt *m*, dead point.

Touren-Konstanz *f*, constancy of speed;

— -zahl *f*, number of revolutions, speed;

— - -Belastungskennlinie *f*, speed-load characteristic.

— -zähler *m*, revolution counter.

Tragbalken *m*, girder.

tragbar, portable.

Tragbarkeit *f*, portability.

Tragdraht *m*, suspending wire, suspension wire.

träge, inert;

— antsprechend, sluggish in action.

Trage *f*, barrow;

Draht- —, drum barrow *B*.

tragen, to carry, to support, to bear.

Träger *m*, holder, support, bearer; Balken: girder; Strom: carrier;

Gitter- —, lattice girder;

Kabel- —, cable bearer;

— -frequenz *f*, carrier frequency;

örtlich überlagerte — - —, local carrier;

örtlich überlagerte asynchrone — - —, non-synchronous or asynchronous local carrier;

Empfang mit Überlagerung der — - —, zero-beat reception;

auf Überlagerung der — - —eingestellt, set for zero beat;

— - -überlagerer *m*, homodyne;

— -papier *n*, body paper;

— -strom *m*, carrier (current);

— - -telegraphie *f*, carrier current telegraphy;

— -welle *f*, carrier (wave);

unterdrückte — - —, suppressed or eliminated carrier;

Hochfrequenzbetrieb *m*, mit —r — - —, suppressed carrier operation;

Unterdrückung *f* der — - —, carrier suppression;

Wiedereinführung *f* der — - —, reintroduction of carrier;

— - -n-telegraphie *f*, carrier wave telegraphy;

— - -n-telephonie *f*, carrier telephony.

Trag-gerüst *n*, support, rack;

Kabel- — - —, cable rack;

— -gestell *n*, supporting structure.

Trägheit *f*, inertia, sluggishness;

—s-moment *n*, moment of inertia;

trägheitslos, inertialess.

Trag-kasten *m*, carrying case;

— -leiste *f*, mounting (strip);

— -platte *f*, mounting (plate);

— -ring *m* für Schaltdrähte, jumper ring;

— -schnur *f*, strain cord;

— -seil *n*, suspending wire, messenger wire, supporting strand, *B*;

— - — -klemme *f*, messenger wire clamp.

tränken, to impregnate, to inject, to soak;

mit Kupfervitriol —, to boucherize;

mit Teeröl —, to creosote;

mit Zinkchlorid —, to burnettize.

Tränk-gefäß *n*, impregnating tank;

— -kessel *m*, impregnating tank;

Tränken *n*, impregnating.

Tränkung *f*, impregnation, injection, soaking;

Kreosot- —, creosoting;

Kupfervitriol- —, boucherization;

Zinkchlorid- —, burnettization.

Transfiguration f, transfiguration.

Transformator m, transformer;
— mit geerdetem Kern, grounded core transformer, screened transformer;
— — geschlossenem Eisenkern, closed-core transformer;
— — offenem Eisenkern, open-core transformer;
mit —en gekoppelt, transformer-coupled V;
abgeschirmter —, screened transformer;
eisenloser —, air core transformer;
geschirmter —, shielded transformer;
luftgekühlter—, air-cooled transformer;
ruhender —, static transformer;
Abschluß- —, terminal transformer;
Abwärts- —, reducing transformer, step down transformer;
Anzapf- —, split transformer;
Aufwärts- —, step-up transformer;
Ausgangs- —, output transformer;
Ausgleichs- —, three-coil transformer, hybrid coil, V K;
Auto- —, auto-transformer;
Eingangs- —, input transformer;
Eisen- —, iron-core transformer;
Heiz- —, heating current transformer;
Hochfrequenz- —, h. f. transformer, oscillation transformer;
Igel- —, hedgehog transformer;
Kern- —, core transformer;
Klapp- —, hinged coil transformer, pair of rotatable coaxial coils;
Klingel- —, bell transformer;
Kopplungs- —, repeating transformer V; jigger R;
Luft- —, air-core transformer;
Mantel- —, ironclad transformer, shell transformer;
Öl- —, oil(-cooled) transformer;
Panzer- —, ironclad transformer;
Phasenschieber- —, phase-shifting transformer;
Reduktions- —, step-down transformer;
Regulier- —, regulating transformer; [former;
Resonanz- —, resonance transformer;
Rückkopplungs- —, reaction transformer;
Saug- —, booster transformer, suction transformer;
Schwingungs- —, oscillation transformer R;
Spar- —, auto-transformer;
Tauch- —, telescoping coil transformer;
Tesla- —, Tesla h. f. transformer, oscillation transformer;
Verstärker- —, amplifier transformer;
Wechselstrom- —, alternating current transformer;
Zwischen- —, intermediate transformer;
Zwischen(rohr)- —, intervalve transformer;
— **-gefäß** n, transformer tank.
transformieren, to transform;
abwärts —, to step down;
aufwärts —, to step up.
Transformierung f, transformation.

Transit *m*, transit.
transkontinental, transcontinental.
transozeanisch, trans-oceanic.
Transparenz *f*, transparency.
transponieren, to transpose.
Transponierungsempfang *m*, transposition reception.
Transport *m*, transport, transportation;
  Energie- —, transport of energy;
  — **-daumen** *m*, spacing cam *T*.
Transporteur *m*, protractor.
Transversalschwingungen *pl*, transverse vibrations *pl*.
transzendente Funktion *f*, transcendental function.
Trapez *n*, trapezoid.
trapezförmig, trapezoidal.
Traverse *f*, transom *B*.
treiben, to drive, to move, to impel; Sand, Schnee: to drift.
treibend, motive.
Treibriemen *m*, driving belt.
Trennbatterie *f*, spacing battery *T*.
trennen, to separate, to disconnect, to sever, to cut;
  ein Ortsgespräch für ein Ferngespräch —, to break a local call for a toll call;
  einen Teilnehmer —, to cut off a subscriber;
  eine Verbindung —, to disconnect, to take down a connection.
Trennen *n*, disconnecting, disconnection.
Trenn-klinke *f*, break jack, interrupt jack;
  fünfteilige — — —, five-point break jack;
  Doppel- — — —, double break jack;

— **-kontakt** *m*, break contact; spacing contact *T*;
— **-relais** *n*, cut-off relay;
  Doppel- — — —, double break relay;
— **-schalter** *m*, disconnecting switch, circuit breaker, disconnector;
  Öl- — — —, oil break switch;
— **-seite** *f*, spacing position *T*;
  Relais liegt auf — — —, the relay spaces *T*;
  Überwiegen *n* der — — —, spacing bias *T*;
— **-stöpsel** *m*, infinity plug;
— **-strom** *n*, spacing current *T*;
— **-senden**, to space;
— **-stück** *n*, separator.
Trennung *f*, disconnection, separation, severing, breaking;
  Fernamts- —, breaking of local calls for toll calls.
Trense *f*, worming, worm, filler;
  Garn- —, yarn worming.
trensen, to worm.
Trensenaderpaar *n*, worming pair.
Tresse *f*, lace.
Tret-dynamo *f*, pedal dynamo;
— **-schalter** *m*, foot switch.
treue Wiedergabe *f*, faithful reproduction.
Treue *f* **der Wiedergabe**, faithfulness of reproduction.
Trichter *m*, tun dish, funnel;
  Schalltrichter *m*, mouthpiece; Lautsprecher: horn, trumpet, großer projector;
  Sprech- —, mouthpiece;
— **-antenne** *f*, funnel-shaped aerial;
— **-klang** *m*, characteristic horn sound (der Lautsprecher, of loudspeakers);
— **-lautsprecher** *m*, horn type loudspeaker.

Trieb *n*, pinion;
Hohl= —, lantern pinion;
— =feder *f*, driving spring, main spring;
— =kraft *f*, motive power;
— =sand *m*, drifting sand;
— = — =grund *m*, shifting ground;
— =welle *f*, driving shaft;
— =werk *n*, gearing.
Trigonometrie *f*, trigonometry.
trigonometrische Funktion *f*, trigonometric function.
Tritt *m*, step;
in — kommen, to come in step;
— =brett *n*, treadle.
trocken, dry;
— e Räume *pl*, dry rooms *pl*.
Trocken=batterie *f*, dry cell battery;
— =element *n*, dry cell;
— =haltung *f*, keeping dry;
— =ofen *m*, drying stove.
trocknen, to dry, to desiccate;
Holz durch Lagern: to season.
Trockner *m*, desiccator.
Trocknung *f*, desiccation, drying-up;
— im Vakuum, desiccation under vacuum.
Trog *m*, trough(ing), tray;
— =element *n*, tray cell.
Trommel *f*, drum, barrel, reel;
Feder= —, spring drum, spring barrel;
Kabel= —, cable reel, cable drum;
— =anker *m*, drum armature;
— =länge *f*, drum length (des Kabels, of cable).
Trompete *f*, trumpet.
Tropen *pl*, tropics *pl*.
— =ausführung *f*, tropical finish.
tropisch, tropic(al).

Tropf=glas *n*, pipette;
— =ring *m*, drip ring.
trübe, dull, dim.
Trupp *m*, gang;
Bau= —, construction gang;
Störungs= —, repair gang.
Tuba *f*, tuba.
Tülle *f*, nozzle.
Tumblerschalter *m*, tumbler switch.
Tunnel *m*, tunnel.
Turbinen=gebläse *n*, turbine blower;
— =unterbrecher *m*, turbine break, turbine interrupter.
Tür=kontakt *m*, door push;
— =schalter *m*, door switch.
Turm *m*, tower, pylon;
freitragender —, self-supporting tower;
Funk= —, radio tower, radio mast.
Tusche *f*, ink;
mit — ausziehen, to ink.
Typ *m*, type.
Type *f*, type, letter;
erhabene —, raised type;
Typendruck *m*, type printing;
— =drucker *m*, type printer;
— =drucktelegraph *m*, type printing telegraph;
— =fläche *f*, type surface;
— =hebel *m*, type bar;
— = — =übersetzer *m*, type bar translator or printer;
— =korb *m*, type basket (der Schreibmaschine, of the typewriter);
— =rad *n*, type wheel;
— = — =achse *f*, type wheel shaft;
— = — =übersetzer *m*, type wheel translator.

U.

üben, (sich), to practice.
überbrücken, to bridge (across); to span.
Überbrückung f, bridging;
— =draht m, jumper;
— =klemme f, bridge connector;
— =kondensator m, by-pass condenser, bridging condenser.
überdrehen, das Gewinde, to strip the thread.
übereinandergreifen, to overlap.
Übereinandergreifen n, overlap(ping) (der Gruppen of groups A).
überentladen, to run down (Sammler, storage cells).
übererregen, to overexcite.
Übererregung f, overexcitation.
Überführung f, transition;
— =isolator m, terminal insulator;
— = — mit Vergußkammer, pothead insulator;
— =säule f, —=stange f, terminal pole; für Teilnehmerkabel: distributing pole F.
überfüllt, congested.
Überfüllung f, congestion.
Übergang m, passage, transition;
Übergangs=stelle f, junction;
— =verlust m, zwischen zwei Stromkreisen, loss at a junction, transition loss; zwischen Kontakten: contact loss;
— =vorgang m, transient;
— =widerstand m, contact resistance;
— =zeit f, transition period;
— =zustand m, transient state.
überhängend, overhung.
Überheizung f, overheating.
überhitzen, to overheat.
Überhitzung f, overheating.

überholen, to overhaul.
überhörfrequent, ultra-audible.
Überhörfrequenz f, ultra-audible frequency, ultra-audio frequency.
überkompensieren, to overcompensate.
Überkompensierung f, overcompensation.
überkompoundieren, to overcompound.
Überkreuzung f, crossing-over.
überladen, to overcharge.
Überladung f, overcharge, overcharging.
Überlagerer m, heterodyne, (heterodyne) local oscillator;
Fremd= —, separate heterodyne local oscillator;
Selbst= —, self-heterodyne or autodyne local oscillator;
Trägerfrequenz=—, homodyne;
— =frequenz f, local oscillation frequency.
überlagern, to superpose, to superimpose; eine abweichende Schwingung: to heterodyne R.
Überlagerung f, superposition, super(im)posing, superimposition;
mit — empfangen, to heterodyne;
Ton= —, (zerhackte), (interrupted) sine modulation;
Überlagerungs=.... heterodyne;
— =empfang m, heterodyne reception;
— = — mit Selbsterregung, self-heterodyne reception, autodyne reception;
— = — — Fremderregung, separate heterodyne reception;

Überlagerungs-
— = **empfang mit Überhörfrequenz**, supertonic heterodyne reception;
— =**empfänger** *m*, heterodyne receiver;
— = — **mit Selbsterregung**, autodyne receiver, self-heterodyne receiver;
— = — — **Fremderregung**, separate heterodyne receiver;
— =**schaltung** *f*, superimposed connection *F T*;
Arbeiten *n* **in** — = —, superimposed working *F T*;
— =**schwingungen** *pl*, heterodyne oscillations, local oscillations, *pl R*.
Überland-Fernkabel *n*, overland l. d. cable;
— =**zentrale** *f*, h. t. power plant.
überlappen, to lap (over), to overlap.
Überlappen *n*, lapping(-over), overlap(ping);
— **der Wellenzüge**, overlap of wave trains.
Überlast *f*, overload.
überlasten, to overload.
übermitteln, to transmit.
Übermittlung *f*, transmission.
übersättigen, to supersaturate.
Übersättigung *f*, supersaturation.
überschieben, to slip over.
Überschlag *m*, estimate, computation; **Funken**: flashover; **Funken-** —, spark-over, flashover, breakdown.
überschlagen, to compute, to estimate; **Funken**: to flash over, to spark over.
Überschlagspannung *f*, breakdown voltage, spark-over voltage.
Überschneidung *f*, (point of) intersection.
überseeisch, transoceanic.

Überseeverbindung *f*, transoceanic communication.
übersetzen, to translate.
Übersetzer *m*, translator, translating device, *T*;
— =**rechen** *m*, combiner comb *T*;
— =**scheibe** *f*, combiner disc *T*.
Übersetzung *f*, translation;
— =**verhältnis** *n*, **Transformator**:. transformation ratio, **Räder**: gear(ing) ratio;
— =**— 1:1**, unity transformation ratio.
Überspannung *f*, overtension, excess(ive) voltage, overvoltage;
— =**schutz** *m*, protection from overtension; surge arrester;
— =**sicherung** *f*, excess voltage cut-out.
Übersprechdämpfung *f*, crosstalk transmission equivalent, *F K*.
Übersprechen *n*, crosstalk *F K*.
Übersprech-kopplung *f*, crosstalk path, side-to-side unbalance, *F K*;
— =**weg** *m*, crosstalk path.
Überspringen *n*, flashing-over; side-flashing (**des Blitzes**, of the lightning).
übersteuern, to overmodulate *R*.
Übersteuerung *f*, overmodulation *R*.
überstreifen, to slip over.
Überstrom *m*, excess current;
— =**ausschalter** *m*, overload circuit breaker;
— =**relais** *n*, overload relay.
übertragen, to transfer; **Strom usw**: to convey, to transmit, to transport; **mit Relais** to relay.
Übertrager *m*, repeater *T F*; **Transformator**: transformer repeating coil;·

Übertrager
- mit hohem Umsetzungsverhältnis, high-ratio transformer;
- — zwei Wicklungen, two-winding transformer;
geschirmter —, shielded transformer;
Abschluß= —, terminal transformer;
Ausgangs= —, output transformer, outlet transformer;
Ausgleichs= —, hybrid coil, three-coil transformer, balanced differential transformer $F\ V$;
Differential= —, differential repeating coil, differential transformer;
Eingangs= —, input transformer;
Fernsprech= —, telephone transformer;
Gabel= —, forked repeater T;
Impuls= —, impulse repeater A;
Mithör= —, monitoring coil;
Nach= —, output transformer, outlet transformer, V;
Ring= —, toroidal repeating coil, ring transformer, F;
— = —, Doppelsprech=, phantom repeating coil;
Stromstoß= —, impulse repeater A;
Telegraphen= —, telegraph repeater (set);
Vor= —, input transformer V;
— =amt n, repeater or repeating station T;
— =gestell n, repeating coil rack F;
— =spule f, repeating coil.

Übertragung f, transmission; zwischen zwei Stromkreisen: transfer, transference; Fortleitung: conduction, conveyance, transport(ation); mit Relais, Verstärkern: repetition (into); Zession: assignment; Telegraphen= —: (through) repeater, für Doppelleitung: metallic repeater;
entzerrende —, regenerative or rectifying repeater T;
umlaufende —, rotary repeater T;
— — mit Berichtigung der Zeichenform, regenerative rotary repeater;
Energie= —, über Leitungen: transport of energy; durch Kopplungen: transfer of energy; [peater T;
Gegensprech= —, duplex re-
Kegelrad= —, bevel gear;
— = — 1:1, equal ratio bevel gear;
Kraft= —, power transmission;
Nachrichten= —, transmission of intelligence;
Räder= —, gear(ing);
Rück= —, retransfer;
Telegraphen= —, telegraph repeater (set);
— = — mit Anrufer, alarm repeater;

Übertragungs=amt n, repeating telegraph station;
— =beamter m, relay clerk;
— =bereich m, transmission range;
— =filter n, transmission filter circuit;
— =geschwindigkeit f, speed of transmission;
— =kenngröße f, transmission characteristics $pl\ L$;
— =klopfer m, relaying sounder, uprighting sounder;
— =leitung f, transmission line;
— =maß n, transmission unit, transmission measure, transmission equivalent;

Übertragungs=
— = maß in Meilen Standard=kabel, transmission equivalent in m. s. c. (= miles of standard cable);
— = —, Gesamt=, total transmission equivalent, transmission efficiency FL;
— = —, zulässiges Gesamt=, total permissible transmission equivalent;
— =messung f, transmission measurement F;
— =mittel n, transmitting medium;
— =niveau n, transmission level FL;
— =normal n, transmission standard;
— =relais n, translating relay, repeating relay;
— =vergleichssystem n, transmission reference system;
— =verlust m, transmission loss;
— =wirkungsgrad m, transmission efficiency;
— =zeit f, duration of transmission, transit time.

überwiegend, einseitig, bias.

überwachen, to watch, to observe, to supervise.

Überwachung f, supervision, observation;
laufende —, routining (der Leitungen usw, of lines);
Netz= —, transmission maintenance work;

Überwachungs= —, supervisory;
— =kreis m, monitoring circuit R;
— =lampe f, supervisory lamp F; pilot signal;
— = — des rufenden (verlangten) Teilnehmers, answering (calling) supervisory lamp;
— =messung f, maintenance test;
— = — regelmäßige — = —, routine maintenance test;
— =platz m, (Dienst)=, observation desk F;
— =relais n, pilot relay; supervisory relay F;
— =zeichen n, supervisory (signal), pilot signal.

Überweisungs=taste f, assignment key;
— =wähler m, allotting switch A.

überwiegen, to preponderate.

Überwiegen n, preponderance; bias;
— nach der Zeichen=(Trenn)seite, marking (spacing) bias T.

überziehen, to coat.

Überzug m, coat, coating, serving.

Übung f, practice; Fertigkeit: skill.

U=Eisen n, U-iron, channel iron.

Ufer n, shore, beach.

Uhr f, clock;
Haupt= —, master clock;
Neben= —, auxiliary clock;
Sand= —, sand glass;
Stopp= —, stop watch;
— en=zeichen n, time (signal);
— =werk n, clockwork;
— = — =antrieb m, clockwork train;
— =zeigersinn, im, clockwise;
— = —, entgegen dem, counterclockwise, anti-clockwise.

ultra=rot, ultra-red;
— =violett, ultra-violet.

umändern, to alter, to change, to vary.

Umänderung f, alteration, change, variation.

Umdrehung f, revolution, turn;
— en in der Minute, revolutions per minute, ab: r. p. m.
— s=anzeiger m, speed indicator;
— s=zahl f, number of revolutions, speed.

Umfang f, volume; Kreis: periphery, circumference;
Kreis= —, circumference, periphery;
Verkehrs= —, traffic volume;
Umfangs= — ..., peripheral, circumferential;
— =geschwindigkeit f, peripheral speed, circumferential speed.
umfassen, to embrace.
umflechten, to braid.
Umflechtmaschine f, braider.
umformen, to transform, to convert.
Umformer, converter;
Einanker= —, rotary converter;
Frequenz= —, frequency transformer;
— = —, ruhender, static frequency transformer;
Kaskaden= —, cascade converter;
Pendel= —, vibrating rectifier.
Umformung f, transformation;
Frequenz= —, frequency transformation;
— =verhältnis n, transformation ratio;
— =wirkungsgrad m, efficiency of transformation.
umgeben, to surround, to encircle.
umgebrochen, broken down, laid flat, B.
umgehen, to by-pass.
Umgehungsschaltung f, by-pass connection.
umgekehrt, inverse;
— es Verhältnis n, inverse ratio.
Umgitterung f, barrier, enclosure.
Umgrenzung f, definition; einer Kurve: envelope;
— =linie f, envelope.
umhüllen, to wrap (up), to lap, to serve.

Umhüllung f, serving, lapping, wrapping;
Papier= —, wrapping of paper.
Umkehr f, reversal;
Phasen= —, phase reversal;
Polaritäts= —, reversal of polarity.
umkehrbar, reversible;
nicht —, irreversible, non-return;
— e Permeabilität f, reversible permeability.
Umkehrbarkeit f, reversibility.
umkehren, to reverse, to revert; Formeln: to invert.
Umkehr-schalter m, reversing switch, poling switch;
— =taste f, reversing key.
Umkehrung f, reversal, inversion.
umklöppeln, to braid.
Umklöppelung f, braiding.
äußere (innere), outer (inner) braiding;
Glanzgarn= —, glazed cotton braiding.
Umkreis m, circumference.
umlappen, to lap (round), to wrap; mit Band: to tape.
Umlappung f, lapping, wrapping;
mit Band: taping;
Messingband= —, brass taping.
Umlauf m, Fluß: circulation; Umdrehung: rotation, revolution;
umlaufen, to rotate, to revolve; to circulate.
umlaufend, rotary.
Umlauf-geschwindigkeit f, rotational speed;
— =regler m, speed governor, speed controlling device;
— =zahl f, number of revolutions, speed;
— =zähler m, revolution counter, cyclometer.

umlegen, to reverse, Kippen: to tilt; Schalter: to throw (a key).

Umlegen *n,* reversal, tilting; Schalter: throwing.

umleiten, to divert, to deviate.

Umleiter *m,* director, translator, controller, *A*;

— =system *n,* director system *A*.

Umleitung *f,* diversion, deviation;

— s=wähler *m,* director selector *A*.

Umlötung *f,* wiring change.

ummagnetisieren, to reverse the magnetism.

Ummagnetisierung *f,* magnetic reversal.

Umpol(arisier)ung *f,* reversal of polarity.

Umrahmung *f,* framing.

umrechnen, to translate *A*.

Umrechner *m,* translating device *A*;

— =feld *n,* translation field.

Umrechnung *f,* translation *A*;

— s=faktor *m,* reduction factor, conversion factor;

— s=tafel *f,* reduction table.

Umriß *m,* contour, outline.

umschalten, to switch, to change over.

Umschaltegestell *n,* distributing frame.

Umschalter *m,* (change-over) switch, commutator, double throw switch;

— mit zwei Stellungen, throw-over switch, two-position switch;

mehrwegiger —, multiple-way switch;

Doppelhebel= —, double lever switch;

Gabel= —, cradle switch *F*;

Haken= —, hookswitch, switch hook, *F*;

Hebel= —, lever switch;

Klinken= —, (line) switchboard, jack switchboard;

Kurbel= —, lever switch;

Linien= —, line switchboard;

Luftdraht= —, aerial (change-over) switch;

Nebenstellen= —, substation switchboard *F*;

Platz= —, position switching key, coupling key, *F*;

Sitz= —, socket contact *F*;

Stöpsel= —, plug commutator, plug switch;

Verstimmungs= —, wavelength changing switch, change-tune switch;

Voltmeter= —, voltmeter switch;

Wellen= —, wavelength changing switch, change tune switch;

Zentral= —, intercommunication switch *T*;

Zwischenstellen= —, inter-through switch *F*.

Umschalterelais *n,* auto(matic) switch.

Umschaltung *f,* switching, commutation, changing-over, cross, wiring change; Typenrad: shift, inversion;

einpolige —, single commutation;

doppelpolige —, double commutation;

Buchstaben= —, letter shift *T*;

Leitungs= —, line change;

Zahlen= —, figure shift *T*.

Umschlag *m,* transit, travel, (des Relaisankers); Hülle: envelope;

Anker= —, armature travel;

Fenster= —, window envelope.

— s=zeit *f,* transit time.

umschließen, to encircle, to enclose.

umschnüren, to lace, to serve with thread.

Umschnürung f, serving of thread.

umsetzen, to translate, to convert, to transpose;
elektrischen Strom in Sprach=energie —, to convert electric current to speech.

Umsetzer m, translator, transposer, translating device.

Umsetzung f, translation, transposition, conversion;
— s=verhältnis n, conversion ratio.

Umspanner m, transformer.

umspannen, Strom: to transform; to span.

umspinnen, to braid, to cover, to whip.

Umspinnung f, braiding, covering, whipping;
(Eisen= —, iron whipping, wrapping of iron;
Krarup= —, iron whipping, Krarup winding.

umsponnen, covered;
einfach (doppelt, dreifach) —er Draht, single (double, triple) covered wire.

umsteuern, to reverse.

Umsteuerung f, reversal. [mit.

umtelegraphieren, to retrans-

Umtelegraphierung f, retransmission, (additional) transit.

umwandeln, to transform, to transpose, to convert (eine gewählte Nummer in eine solche von anderer Stellen=zahl, a number dialled in into a number of either more or less digits A).

Umwandlung f, conversion, transformation;
— s=faktor m, conversion factor;
— s=temperatur f, magnetische, magnetic transition temperature;

— s=verhältnis n, transformation ratio.

umwickeln, to lap, to whip (round) mit Band: to tape;
eine Spule neu wickeln: to rewind;
mit Isolierband —, to serve with insulating tape.

Umwicklung f, lapping, serving, whipping, taping.

Umzäunung f, enclosure.

unabgeschirmt, unscreened.

Unabhängigkeit f, independence;
Frequenz= —, independence of frequency.

unabhörbar, untapped.

unangreifbar, non-corrosive.

unausführbare Verbindung f, unobtainable number F;
Summerzeichen n zur Kenn=zeichnung —r —en, number unobtainable tone, ab: n. u. tone A.

unausgeglichen, unbalanced;
— er Zustand m, out-of-balance condition.

unbeantworteter Anruf m, no-reply call.

unbearbeitet, raw B.

unbedientes Amt n, unattended exchange A.

unbeeinflußt, unaffected (by).

unbegrenzt, unlimited.

unbekannt, unknown;
— e Größe f, unknown quantity.

unbelastet, unloaded, nonloaded, K.

unbesetzt, clear, free, disengaged, idle, F A.

unbeständig, instable, unsteady, inconstant.

Unbeständigkeit f, instability, inconstancy.

unbestimmt, indeterminate.

unbiegsam, rigid.

18*

undeutlich, indistinct; inarticulate F; Zeichen: illegible, indefinite T;
— e **Aussprache** f, inarticulateness.

Undeutlichkeit f, indistinctness; inarticulateness; illegibility, [indefinition.

undicht, leaky;
— **sein**, to leak.

Undichtheit f, leak(iness).

undulieren, to undulate;
— **d,** undulating, undulatory.

unelastisch, unelastic(al).

unelektrisch, unelectric.

unempfindlich, insensitive (gegen, to); gegen Beschädigung: robust;
gegen Ströme unter 10 mA — **gemacht**, biased against currents of 10 ma.

Unempfindlichkeit f, insensitiveness; robustness.

unendlich, infinite;
— **werden**, to approach infinity;
— **viele**, an infinity of;
— **klein**, infinitesimal, indefinitely minute;
— e **Reihe** f, infinite series.

unentflammbar, flame-proof, non-inflammable;
— **gemacht**, flame-proofed.

unerregt, unexcited.

unerreichbar, unobtainable.

Unfall m, accident;
— -**bericht** m, accident report;
— -**verhütung** f, accident prevention;
— - —**s-vorrichtung** f, safety device;
— - —**s** -**vorschriften** pl, safety rules pl;
— -**versicherung** f, accident insurance.

ungedämpft, undamped;
— e **Schwingungen** pl, undamped or sustained or persistent oscillations pl;

— e **Wellen** pl, continuous waves, ab: c. w., undamped waves, type A waves, pl;
— e —, **zerhackte** oder **unterbrochene**, interrupted continuous waves, ab: i. c. w.;
— e **Zeichen** pl, undamped wave signals pl.

ungeerdet, ungrounded.

ungekreuzte Doppelleitung f, non-transposed metallic circuit.

ungeladen, Sammler: discharged uncharged; Leitung: non-loaded.

ungerade, Zahl: odd; uneven;
— **s Vielfaches** n, odd multiple.

ungeradzahlig, odd-numbered.

ungerichtet, non-directional, non-directive, Luftleiter auch: equi-radial.

ungestöpselt, unplugged.

ungiltig, non-valid, void.

unglasiert, unglazed.

ungleich, unequal, unbalanced;
— -**förmig**, discontinuous, non-uniform, irregular, asymmetrical.

Ungleichförmigkeit f, discontinuity, irregularity, asymmetry;
— **des Wellenwiderstandes**, impedance irregularity.

Ungleichheit f, inequality, unbalance, diversity.

ungleichmäßig, discontinuous, uneven, non-uniform.

Ungleichmäßigkeit f, discontinuity, non-uniformity.

ungleichnamig, unlike, opposite.

Ungleichung f, inequality M.

Unglück n, emergency, accident.

unhörbar, inaudible.

Unhörbarkeit f, inaudibility.

unhygroskopisch, non-hygroscopical.

unifilar, unifilar.

unionisiert, un-ionized.

unipolar, unipolar, single-polar;
— e **Leitung** *f*, uni-directional conductance.

Unipolardynamo *f*, homopolar dynamo, unipolar dynamo.

unisoliert, uninsulated.

Unkosten *pl*, cost, expense; **laufende** —, running cost.

unlegiert, unalloyed.

unleserlich, unreadable.

unmagnetisch, unmagnetized, non-magnetic.

unmoduliert, unmodulated.

Unordnung, in, out of order, (*ab*: o. o. o.), tied up, out of gear.

unpassend, misfitting;
— **sein**, to misfit;
— **er Streifen** *m*, misfitting slip *T*.

unpolarisiert, non-poled, non-polarized.

unregelmäßig, irregular.

Unregelmäßigkeit *f*, irregularity.

unrein, impure.

Unreinheit *f*, impureness, impurity.

Unruhe *f*, balance wheel (des Uhrwerks, of the clockwork).

unsauber, smudgy;
— **er Abdruck** *m*, smudgy impression *T*.

unscharf, flat;
— e **Abstimmung** *f*, flat tuning.

unschmelzbar, infusible.

unsichtbar, invisible;
— e **Irrung** *f*, invisible correction *T*.

unstabil, unstable, instable, unsteady.

Unstabilität *f*, instability.

unstetig, uneven, non-uniform, discontinous.

Unstetigkeit *f*, discontinuity, non-uniformity.

Unsymmetrie *f*, asymmetry, dissymmetry, unbalance;

Wirkung *f* der —, asymmetrical effect;
— **einer Leitung gegen Erde**, wire-to-earth unbalance;
— **zwischen den Drähten einer Doppelleitung**, wire-to-wire unbalance.

unsymmetrisch, asymmetrical, dissymmetrical, (zu, with regard to), unbalanced.

Unteramt *n*, sub-office, branch exchange, sub-exchange.

Unterbau *m*, substructure.

unterbrechen, to interrupt, to break, to disconnect, to open, to open-circuit; to intermit, to stop.

Unterbrecher *m*, break, cut-out; interrupter, buzzer, chopper;
elektrolytischer —, electrolytic interrupter;
schwingender —, buzzer, vibrating contact;
umlaufender —, rotary interrupter;
Hammer= —, vibrating break, hammer break;
Hochfrequenz= —, high-frequency interrupter;
Kommutator= —, commutator break, commutator interrupter;
Schnell= —, ticker *R*;
Selbst= —, self-interrupter;
Stimmgabel= —, tuning fork interrupter, vibrator;
Summer= —, buzzer (interrupter);
Turbinen= —, turbine break, turbine interrupter;
Wehnelt= —, electrolytic interrupter;
Zahnrad= —, crown wheel commutator;
Zungen= —, vibrating reed interrupter; [*A*.
— =**maschine** *f*, impulse machine

Unterbrechung *f*, disconnection, interruption, break, open(ing), stoppage;
funkenfreie —, clean break;
selbsttätige —, automatic break;
zeitweilige —, intermittent disconnection, intermittency;
Unterbrechungs-funke *m*, spark at break;
— -klinke *f*, break jack;
fünfteilige — - —, five-point break jack;
Doppel- — - —, double break jack;
Vielfach- — - —, series multiple jack;
— -kontakt *m*, break contact;
— -relais *n*, break relay;
Doppel- — - —, double break relay;
— -taste *f*, break key.
unterbrochen sein oder werden, to discontinue.
unterdrücken, to suppress; to damp out, to choke out.
Unterdrückung *f*, suppression; choking-out.
untere Reihe *f*, bottom row;
— -s Ende *n*, bottom.
unterfahren, to underrun *B*.
Untergrund *m*, subsoil.
Untergruppe *f*, subgroup *A*.
Unterhalt *m*, maintenance, upkeep.
unterhalten, to maintain.
Unterhaltung *f*, upkeep, maintenance;
regelmäßige —, routine maintenance;
— -s-kosten *pl*, cost of upkeep, maintenance cost.
Unter-Hörfrequenz *f*, sub-audio frequency.
unterirdisch, underground.

Unterlage *f*, base, underlayer, support;
erschütterungsfreie —, resilient support;
Filz- —, felt underlayer.
Unterlagscheibe *f*, washer.
Untersalpetersäure *f*, nitrogen dioxide (NO_2).
Untersatz *m*, stand.
unterschnitten, undercut.
Unterseite *f*, underside, bottom side.
Unterstation *f*, substation.
unterstopfen, to tamp, to pack up.
untersuchen, to test (auf, for), to examine, to research (into), to investigate;
auf Berührung (Kurzschluß, Erdschluß) —, to test for contact (short-circuit, earth);
nochmals —, to re-test.
Untersuchung *f*, test(ing), examination; research, investigation; study; chem., analysis;
bei nochmaliger —, on re-test.
Untersuchungs-abschnitt *m*, testing section;
— -amt *n*, testing office;
— -kasten *m*, test box, test case;
Stangen- — - —, pole test box;
— -platz *m*, test position;
— -säule *f*, pillar test box;
— -stange *f*, pole test box;
— -stelle *f*, testing point;
Kasten- — - —, test case;
— -tisch *m*, test desk.
Unterteil *n*, foot, bottom; base.
unterteilen, to subdivide, to section(alize).
unterteilt, subdivided;
fein —, finely subdivided.
Unterteilung *f*, subdivision.
Unter-Vermittlungsstelle *f*, sub-exchange;

Unter
— -Verteilerstelle *f.* subsidiary distributing point.
Unterwasser-antenne *f*, underwater antenna;
— -schallempfänger *m*, subaqueous sound receiver.
unterweisen, to inform, to instruct.
Unterweisung *f*, information, instruction.
unterzeichnen, to sign.
unüberwachtes Amt *n*, unattended office *A*.
ununterbrochen, continuous;
— belastet, continuously loaded.
unveränderlich, invariable.
unverbrennbar, incombustible.
Unverbrennbarkeit *f*, incombustibility.
unverstärkt, unamplified.
unvertauschbar, non-interchangeable.
Unvertauschbarkeit *f*, non-interchangeability.
unverwechselbar, non-interchangeable;
— er Stecker *m*, non-interchangeable plug.
unverzerrt, undistorted.

unwandelbar, invariable; fest: rigid.
Unwetter-bericht *m*, weather report;
— -warnung *f*, weather warnings *pl.*
unwirksam, ineffective, inefficient, inoperative.
Unwirksamkeit *f*, ineffectiveness, inefficiency.
Unze *f*, ounce, *ab*: oz., (= 28,3495 g).
unzersetzbar, indecomposable.
Unzersetzbarkeit *f*, indecomposability.
unzerstörbar, non-corrodible, non-corrosive.
unzubereitete Stange *f*, untreated or plain pole *B*.
unzugeteilte Nummer *f*, unallotted number *F*.
unzulässig, undue;
— e Beanspruchung *f*, undue strain.
Uran *n*, uranium (U).
Urkunde *f*, letter.
Ursprung *m*, origin;
— s-land *n*, country of origin;
— s-verkehr *m*, originating traffic.

V.

Vagabondierende Ströme *pl*, leakage or stray currents, vagabond currents, *pl*.
Vakuum *n*, vacuum;
Güte *f* des —s, degree of vacuum;
— -reiniger *m*, vacuum cleaner;
— -röhre *f*, vacuum tube;
Hoch- - —, high vacuum tube.
Vanadium *n*, vanadium (V).
variabel, variable.

Variable *f*, variable.
Variokoppler *m*, variocoupler.
Variometer *n*, variometer, syntonizing inductance;
Klapp- —, hinged coil variometer, pair of rotable coaxial coils;
Kugel- —, ball variometer;
— -spule *f*, variometer coil;
drehbare — - —, variometer rotor;
feste — - —, variometer stator.

Vaselin n, Vaseline f, vaseline, petroleum jelly;
— -öl n, vaseline oil.

Vektor m, vector;
— -analysis f, vector analysis;
— -diagramm n, vector diagram;
— -größe f, vector quantity;
— -summe f, vector sum.

vektoriell, vectorial;
— addieren, to add vectorially;
— darstellen, to represent vectorially;
— e Darstellung f, vector representation.

Ventil n, valve;
Auslaß- —, escape valve;
Kolben- —, piston valve;
Schieber- —, slide valve;
Sicherheits- —, safety valve.

Ventilation f, ventilation.

Ventilator m, fan, motor fan, groß: blower, ventilator.

ventilieren, to ventilate.

Ventil-röhre f, valve, mit zwei Elektroden auch: diode, diode valve;
— -wirkung f, valve action;
— -zelle f, valve, cell;
elektrolytische — - —, electrolytic valve.

verallgemeinern, to generalize.

Verallgemeinerung f, generalization.

Veralten n, obsolescence.

veraltet, obsolescent, obsolete.

veränderlich, variable, varying; differential.

Veränderliche f, variable;
(un)abhängige —, (in)dependent variable.

verändern (sich), to vary, to change, to alter.

Veränderung f, variation, change, alteration.

verankern m, to anchor, Stange: to stay, to guy.

verankerte Stange f, pole and stay, stayed pole.

Verankerung f, anchoring, stay(ing).

veranschlagen, to rate, to estimate.

Veranschlagung f, estimation.

verarmen, to impoverish.

Verarmung f, impoverishment (an Säure of acid). [box;

Verband-kasten m, ambulance
— -zeug n, first aid outfit.

verbessern, to improve.

Verbesserung f, improvement.

verbinden, to join, to connect;
zwei Orte: to interconnect, to connect up, to join up; zwei Klemmen usw., to common together; durch Schaltdrähte: to jumper (mit, to); durch Laschen: to bond, to strap (together), to tie; zwei Leitungen: to extend a line to another line; durch Löten, Schweißen usw: to join, to joint; verspleißen: to joint (mit, to); chemisch: sich —, to combine;
durch- — to connect through, to cut through, to put through;

Verbinder m, binder B;
Hülsen- —, jointing sleeve B.

Verbindung f, connection; zweier Orte: communication, interconnection; durch Schaltdraht: jumpering; Verspleißung, Verlötung usw: joint; Zusammenstellung: combination;
— mit angeschärften Enden, scarfed joint B;
— en pl zwischen den Ämtern (inter-office) trunks A, junctions F;
eine — herstellen, to set up a connection, to complete a call F;

Verbindung
in — stehen, to communicate, to be connected up;
Herstellung f einer —, completion of a call F;
abgehende —, outlet, outgoing trunk, A; outgoing junction F;
biegsame —, flexible coupling;
drahtlose —, radio communication;
freie —, idle trunk A;
gemeinsame —, common connection;
stumpfe —, butt joint B;
unausführbare —, unobtainable number F A;
Doppel= —, double connection F;
Falsch= —, wrong connection F;
Fern= —, l. d. conversation F;
Funk= —, radio communication;
Hülsen= —, sleeve joint B;
Kreuz= —, cross-connection;
Muffen= —, sleeve joint, spigot (and socket) joint (von Röhren, of tubes);
Niet= —, rivet joint;
Orts= —, local call;
Quer= —, cross(-connection); tie line F B;
Schraub= —, bolted connection;
Übersee= —, transoceanic communication;
Vororts= —, suburban connection;
Würge= —, twist(ed) joint B;
Verbindungs=aufbau m, trunking scheme A;
— =bolzen m, tie bolt;
— =draht m, connection wire;
— =gestell n, connecting rack;
— =hülse f, (Kupfer=), (copper) jointing sleeve;
— =kabel n, junction cable F, trunk (line) cable A;
Quer= — = —, tie cable;
— =leitung f, junction (line), trunk (am.); trunk, trunking circuit, A; zwischen Wählern eines Amts: link;
abgehende — = —, out(going) junction;
ankommende — = —, in(coming) junction;
— — = —, Wähler m für, injunction selector;
besetzte — = —, busy junction F, busy trunk A;
doppeltgerichtete — = —, (für Wechselverkehr) both-way junction, two-way trunk circuit;
freie — = —, idle junction F, idle trunk A;
Orts= — = —, junction, trunk;
Vororts= — = —, suburban junction;
— = — für Dienstleitungsbetrieb, order wire junction;
— = — — Tandembetrieb, tandem junction;
— = — — Wählerauslösung, selector release trunk;
— =leitungs=betrieb m, trunking;
— = — =bündel n, trunk group, trunk line bundle;
— = — =kabel n, junction cable; trunk (line) cable A;
— = — =klappe f, junction indicator;
— = — =klinke f, junction jack;
abgehende — = — = —, outjunction jack;
— = — =netz n, junction network;
— = — =schrank m, junction board;
— = — =stöpsel m, junction plug;
— = — =sucher m, junction finder;
— = — =verkehr m, trunking;

Verbindungs
— -leitungsvielfachfeld n, junction multiple;
— -mittel n, communication;
— -schnur f, (connecting) cord;
— -stelle f, zweier Stromkreise: junction; zweier Körper: joint;
— -stöpsel m, calling plug, ringing plug;
— -streifen m, connection strip;
— -stück n, connector; strap, bond, tie, link.
verbleien, to lead, to plumb.
verbogen, buckled.
verbolzen, to bolt.
Verbrauch m, consumption;
Strom- —, current consumption.
verbrauchen, to consume; aufzehren: to dissipate.
Verbraucher m, (Apparat usw.) receiver, load;
— -kreis m, receiver circuit;
— -seite f, receiver end.
verbrennen, to burn.
Verbrennung f, burning, combustion;
— -maschine f, —-motor m, (internal) combustion engine.
Verbund-dynamo f, compound (-wound) dynamo;
Gegen- — - —, differential compound-wound dynamo, differentially wound dynamo;
— -erregung f, compound excitation;
— -motor m, compound-wound motor;
Gegen- — - —, differentially wound motor.
verdampfen, to vaporize, to volatilize, to evaporate.
Verdampfung f, vaporization, volatilization evaporation;

— -wärme f, heat of vaporization, (latent) heat of evaporation;
verdecken, to cover; Laute: to cloud, to mask.
verdeckte Leitungsführung f, concealed wiring.
Verdeckung f, covering; Laute: clouding, masking.
verdichten, to condense.
Verdichter m, condenser.
Verdichtung f, condensation.
verdoppeln, to double, to duplicate.
Verdopplung f, doubling;
Frequenz- —, doubling of frequency.
verdrahten, to wire.
Verdrahtung f, wiring;
feste —, permanent wiring;
Amts- —, internal wiring;
Kontaktsatz- —, bank wiring A;
Vielfach- —, multiple wiring.
verdrallen, to twist; verlitzen: to strand.
verdrallt, twisted; stranded;
— e und abgeschirmte Zuführungen pl, twisted and screened leads.
— e Doppelader f, twisted pair.
Verdrallung f, twisting;
— -schema n, (symmetrisches), (symmetrical) twist system B;
verdrängen, to displace; durch Neues ersetzen: to supersede.
Verdrängung f, displacement; supersession;
Strom- —, skin effect.
Verdrehung f, contortion.
verdreifachen, to treble.
verdrillen, to twist; verlitzen: to strand.
verdunkeln, to darken, to obscure.
verdünnen, to dilute, to thin; Gase: to rarefy.

Verdünnung *f*, dilution; Gase: rarefaction.

verdunsten, to evaporate, to vaporize.

Verdunstung *f*, evaporation, vaporization.

vereinfachen, to simplify.

Vereinfachung *f*, simplification.

Verengung *f*, stricture, constriction.

Verfahren *n*, method, practice; process, operation.

verfallen, I to decay; II void;
— es Patent *n*, void patent.

verfaulen, to rot.

verflechten, to plait.

verfügbar, available.

verflüchtigen, sich, to volatilize.

vergänglich, transient.

Vergänglichkeit *f*, transientness.

vergeuden, to waste, to dissipate.

Vergeudung *f*, waste, wastage, dissipation.

vergipsen, to plaster.

verglast, vitrified.

Vergleich *m*, comparison.

vergleichbar, comparative.

vergleichen, to compare;
— b, comparative.

Vergleichs-apparat *m*, reference instrument;
— -grundlage *f*, basis of comparison;
— -leitung *f*, reference circuit;
— -methode *f*, comparison method;
— -stromkreis *m*, Fernsprech-, standard reference telephone circuit;
— verfahren *n*, comparison method;
— -widerstand *m*, standard resistance, reference resistance;
— -zahlen *pl*, comparative figures *pl*.

vergoldet, gilt.

vergrößern, to increase, to enlarge, to magnify;

vergrößerte Ansicht *f*, enlarged view.

Vergrößerung *f*, increase, magnification.

Verguß-kammer *f*, sealing chamber;
— -masse *f*, sealing compound.

vergüten, to remunerate.

Vergütung *f*, remuneration.

verhalten, sich, to behave.

Verhalten *n*, behaviour.

Verhältnis *n*, proportion, ratio, Satz: rate;
— 1:1, unity ratio;
gerades —, direct ratio;
umgekehrtes —, inverse ratio;
Umformungs- —, transformation ratio;
— -arme *pl*, ratio arms *pl* (einer Wheatstoneschen Brücke, of a Wheatstone bridge).

Verhinderungsschaltung *f*, (für Privatnebenstellen) exchange prohibitory circuit, prohibiting equipment.

verhüllen, to cloud, to mask.

Verhüllen *n*, clouding, masking, (der Zeichen of signals).

verjüngt, thinned, taper.

Verjüngung *f*, tapering.

verkabeln, to cable.

Verkabelung *f*, cabling, wiring;
Amts- —, office wiring;
Kontaktsatz- —, bank-to-bank cabling;
— -s-system *n*, cabling system;
offenes — —, (Zweigsystem) tapering cabling system $F B$.

Verkehr *m*, traffic; service;
in — stehen, to communicate, to correspond;
abgehender —, outgoing traffic; [traffic;
ankommender —, incoming

Verkehr
doppelseitiger —, doppeltgerichteter —, both-way traffic, two-way or duplex traffic;
einseitiger —, one-way traffic;
gemischter —, (Amts- und Privatnebenstellen) mixed service;
schwacher —, slack traffic;
starker —, heavy traffic;
A-B- —, junction service, trunk service;
Abgangs- —, outgoing traffic;
Ankunfts- —, incoming traffic;
Durchgangs- —, transit traffic, through traffic;
Fern- —, long-distance traffic;
— - —, Nah-, short haul toll traffic (am);
Fernsprech- —, telephone traffic; [(am);
Nah- —, short haul traffic
Orts- —, local traffic;
Ortsfernsprech- —, local telephone traffic;
Schnell- —, no-delay traffic, no-delay service;
Spitzen- —, peak traffic;
Ursprungs- —, originating traffic;
Vororts- —, suburban traffic;
Wechsel- —, intercommunication;
Weit- —, toll traffic, long-distance traffic;
Verkehrs-andrang, rush (of traffic);
— -anhäufung f, accumulation of traffic;
— -bedürfnis n, traffic requirements pl;
— -belastung f, traffic load;
— -beobachtung f, traffic observation;
Maschine f zur — -, traffic recording machine F;
— -einheit f, traffic unit F;
— -leistung f, traffic load;
mögliche — - —, traffic-carrying capacity;
— -schreiber m, telephone traffic recorder.
verkehrsschwache Zeit f, slack period;
Verkehrsspitze f, traffic peak;
verkehrsstarke Zeit f, busy period; busy hours, rush hours, pl;
Verkehrs-statistik f, traffic statistics pl;
— -umfang m, volume of traffic;
— -weg m, channel;
— -wert m, telephone traffic unit;
— -zahlen pl, traffic figures pl.
verkehren, to communicate, to intercommunicate.
verketten, to link, to interlink.
verkettet sein, to link, to intermesh, (mit, with).
Verkettung f, interlinking, (inter-) linkage.
— -s-spannung f, interlinked voltage.
verkitten, to putty.
verkleiden, to cover, to sheath.
Verkleidung f, covering, sheathing.
verkrusten, to incrust.
Verkrustung f, incrustation.
verkupfern, to copper, to copperplate.
verkupfert, coppered, copperplate.
verkürzen, to shorten, to clip.
Verkürzung f, shortening;
— -s -kondensator m, aerial series condenser, shortwave or shortening condenser, R.
verlagern, to shift, to unbalance, to displace.
Verlagerung f, displacement.
verlängern, to lengthen, to prolong, to extend, to elongate;

verlängern
  ein Gespräch über drei Minuten —, to extend a call beyond three minutes;
  eine Leitung —, to extend a circuit. (nach, to).
Verlängerung f, prolongation, extension, elongation;
  Leitungs- —, artificial extension circuit, excess network, K; extension of a line.
— s-gebühr f, renewal fee;
— s-leitung f, extension circuit, excess network, K;
— s-spule f, Luftdraht-, aerial loading inductance, antenna load coil.
verlangsamen, to slow down.
verlangter Teilnehmer m, wanted subscriber, called subscriber.
verlaschen, to lash, to bond, (mit, to).
Verlauf, course; Fortschreiten: progress;
  Resonanz- —, resonance curve.
verlegen, to lay; Drähte: to string (wires);
  in die Erde —, to bury;
  oberirdisch —, to run (a wire) overhead.
Verlegung f, Kabel: laying.
verletzen, to infringe (Patentrechte, upon patents).
Verletzung f, infringement.
verlitzen, to strand.
Verlitzmaschine f, stranding machine.
verlöten, to solder up, mit Blei: to plumb.
Verlust m, loss;
— e pl bei Sprechfrequenz, telephonic frequency losses pl;
  innere —e, internal losses;
  Ableitungs- —, leakage loss, leakance loss;
  Eigen- —e, internal losses;
  Eisen- —e, iron losses; core losses;
  End- —, terminal loss L;
  Erwärmungs- —, Joulean loss;
  Gesamt- —, total loss;
  Glimm- —e, corona(l) losses;
  Hysteresis- —, hysteretic loss;
  Kern- —e, core losses;
  Korona- —e, corona(l) losses;
  Kupfer- —, copper loss;
  Leitungs- —, line loss;
  Luftreibungs- —, windage loss;
  Nachwirkungs- —, hysteretic loss;
  Reibungs- —, friction load, friction loss;
  Spiegelungs- —, loss at a junction, reflection loss;
  Strahlungs- —, corona(l) loss;
  Strom- —, current loss;
  Übergangs- —, contact loss, transition loss; an Stoßstellen: loss at a junction;
  Übertragungs- —, transmission loss;
  Wechselstrom- —e, alternating current losses, a. c. losses;
— -dämpfung f, loss damping;
verlustfrei, loss free, free of losses;
verlustlos, free of losses, non-dissipative;
— e Leitung f, line of no loss L;
Verlust-widerstand m, loss resistance
  Reihen- — eines Kondensators, equivalent series resistance;
— -winkel m, dielektrischer, phase (angle) difference of a condense.
vermerken, to record.
vermieten, to lease.
Vermittlungs-amt n, exchange, central office (am.);
  Hand- — —, manual exchange;

Vermittlungs=amt
O. B.= = =, local battery exchange, l. b. exchange, magneto exchange;
S. A.= — = —, automatic exchange;
Z. B.= — = —, c. b. exchange, common battery exchange;
— =leitung f, Fern=, toll switching trunk;
— =schrank m, switchboard;
Glühlampen= — = —, lamp switchboard;
Wand= — = —, wall pattern switchboard;
— =stelle f, exchange, central office; [ge;
Haupt= — = —, main exchan-
Neben= — = —, Unter= — = —, sub-exchange;
— = — n, Fernsprechanlage f, mit mehreren, multi-office exchange.

vernachlässigen, to neglect, to disregard.

Vernachlässigung f, neglect(ion).

vernichten, to nullify.

vernickeln, to nickel (-plate).

vernickelt, nickel-plate(d).

vernieten, to rivet.

verpachten, to lease.

Verputz m, plaster.

verputzen, to plaster.

verquicken, to amalgamate.

Verquickung f, amalgamation.

verriegeln, to bolt, to latch, to block;
wieder —, to re-latch.

verringern, to diminish, to decrease.

Verringerung f, diminution, decrease.

verrosten, to rust.

Verrosten n, rusting.

versagen, to fail; ablehnen: to refuse.

Versagen n, failure; refusal.

verschalten, to cross.

Verschaltung f, wrong connection.

verschiebbar, sliding.

verschieben, to shift, to displace;
in der Phase —, to dephase.

Verschiebung f, shift, shifting, displacement;
— um 180^0, opposition;
dielektrisch —, electric displacement;
Phasen= —, phase displacement;
— s=strom m, displacement current.

verschiedenartig, heterogeneous.

Verschiedenartigkeit f, heterogeneity.

Verschlag m, partition.

verschlechtern, to deteriorate, to impair.

Verschlechterung f, deterioration, impairment.

Verschleiß m, wear.

verschleißen, to wear (out).

verschließen, to seal; to lock (up).

Verschluß m, seal, Schloß, Sperre: lock;
luftdichter —, hermetic seal.

verschmelzen mit, to burn to (Sammlerplatten, storage cell plates).

verschoben, displaced (um, by);
—e Phase f, displaced phase.

verschränken, to interconnect A; to traverse.

Verschränkung f, interconnecting A.

verschrauben, to bolt.

Verschraubung f, bolting.

verschütten, to spill (Säure, acid).

verschwinden, to fade, to vanish, to disappear.

Verschwinden, n, fading, vanishing; zum —bringen, to damp out.

verſeilen, to twist, to strand; zu zweien —, to twin.

Verſeilmaſchine *f*, stranding machine.

Verſeilung *f*, twisting; Dieſelhorſt=Martin=—, D. M.=—, multiple twin formation; Stern= —, spiral quad formation; Vierer= —, quad formation.

verſenken, to sink; eintauchen: to submerge; Schrauben: to counter-sink.

Verſenken *n*, sinking; immersion.

Verſenker *m*, rose bit, countersink.

verſenkte Linie *f*, underground line, covered line; —r Schraubenkopf *m*, countersunk screw head.

verſetzen, to shift, to displace; ſtaffeln: to stagger.

Verſetzung *f*, shift, displacement; staggering.

verſichern, to insure (gegen, against).

verſilbern, to silver.

verſorgen, to supply; to serve.

Verſorgung *f*, supply; Strom= —, current supply.

erſpannen, to guy, to span, to stay.

Verſpannung *f*, guying, staying; Anker: guy, stay.

verſpleißen, to splice, t joint.

Verſpleißung *f*, splicing, jointing.

verſtändlich, intelligible.

Verſtändlichkeit *f*, intelligibility; gute —, articulation.

verſtärken, to intensify; to amplify, to magnify, *V*; mechaniſch: to strengthen, to reinforce; eine Batterie: to boost; ein Geſtänge: to truss.

Verſtärker *m*, amplifier, magnifier; repeater *F*;

— mit magnetiſcher (Widerſtands=, induktiver) Kopplung, transformer- (resistance-, inductance-) repeating amplifier;

— — zwei gegeneinander geſchalteten Röhren, push-pull amplifier;

mit —n verſehen, relayed, repeatered, *F*;

mehrſtufiger —, multi-stage amplifier;

zweiſtufiger —, two-stage amplifier;

Audion= —, amplifying detector;

Doppelrohr= —, two-valve repeater *F*;

Druck=Zug= —, push-pull amplifier;

Einrohr= —, single-valve repeater;

Einwege= —, one-way repeater, simplex repeater, *F*;

End= —, terminal repeater *F*;

Fernſprech= —, telephone repeater, telephonic repeater

Hochfrequenz= —, h. f. amplifier, radio frequency amplifier;

Hörfrequenz= —, note magnifier, audio-frequency amplifier;

Kaskaden= —, cascade amplifier;

Laut= —, sound magnifier;

Mehrfach= —, multiple amplifier;

Mehrröhren= —, multi-valve amplifier;

Mikrophon= —, receiver-transmitter amplifier;

Niederfrequenz= —, low frequency amplifier, note magnifier;

Reflex= —, dual amplifier, reflex amplifier *R*;

Röhren= —, valve amplifier;

Verstärker
 Schnur= —, cord circuit repeater F;
 Schwebungs= —, heterodyne amplifier;
 Sende= —, sending amplifier;
 Sprach= —, speech amplifier;
 Ton= —, note amplifier;
 Vor= —, input amplifier;
 Zweifach= —, two-stage amplifier;
 Zweiröhren= —, two-valve repeater, two-valve amplifier;
 — = — mit aperiodischer Rückkopplung, kallirotron;
 Zweiwege= —, two-way repeater, duplex repeater, F;
 Zwischen= —, intermediate repeater F;
 — = —, Fernsprech=, telephone intermediate repeater;
 — = —, Zweidraht=Doppelrohr=, two-valve two-wire intermediate repeater F;
 — =abschnitt m, repeater section F;
 — =abstand m, repeater spacing F;
 — =amt n, repeater station, amplifying relay station, F;
 — =betrieb m, Fernsprech=, telephonic repeater operation;
 — =bucht f, repeater bay F;
 — =einheit f, repeater unit;
 — =feld n, repeater section F;
 — =gestell n, repeater rack;
 — =lampe f, valve, (amplifier) lamp, amplifier valve, runde: amplifier bulb;
 — =röhre f, amplifier valve, amplifying tube, amplifier triode, seltener: strengthening tube;
 Fernsprech= — = —, telephone repeater tube or valve;
 — =satz m, (basic) repeater unit;
 — =schaltung f, repeater circuit;
 — =transformator m, amplifier transformer.

Verstärkung f, amplification, magnification, V; intensification, mechanisch reinforcement, strengthening; Gegensatz von Dämpfung: gain, repeater gain, F;
 40fache —, 40fold amplification;
 mehrstufige —, multi-stage amplification;
 zweistufige —, two-stage amplification;
 End= —, terminal amplification;
 Hochfrequenz= —, h(igh) f(requency) amplification;
 Hörfrequenz= —, audio-frequency amplification;
 Kaskaden= —, cascade amplification;
 Leistungs= —, power amplification, power magnification;
 Mehrfach= —, multiple amplification;
 Niederfrequenz= —, l(ow) f(requency) amplification;
 Reflex= —, dual amplification, reflex amplification, R;
 Rückkopplungs= —, regenerative amplification, regeneration;
 Spannungs= —, voltage amplification.

Verstärkungs=faktor m (reziproker Durchgriff) amplification factor, magnification factor;
 — =grad m, power amplification ratio;
 — =konstante f, amplification constant;
 — =kurve f, amplification curve;
 — =maß n, repeater gain (equivalent) F;
 — =messer m, repeater gain measuring set;

Verstärkungs
- **-regelung** f, gain control F;
- **-regler** m, gain controller F;
- **-stufe** f, amplification stage;
- **-überschuß** m, (repeater) gain F;
- **-verhältnis** n, (Leistungs-), power amplification ratio;
- **-zahl** f, **- -ziffer** f, amplification coefficient, amplification factor.

versteifen, to stiffen; **verstreben**: to strut, to prop.

Versteifung f, stiffening; strutting, propping; stiffening piece.

verstellen, to shift; **die Bürsten —**, to shift the brushes.

Verstellen n, shifting.
Verstellung f, shift.
Verstellvorrichtung f, shifter.
verstiften, to pin (auf, to).
verstimmen, to detune.
Verstimmen n, detuning.
Verstimmung f, detuning;
ungedämpftes Senden n **mit —**, compensated c. w. (= continuous wave) transmission;
- **-s-schalter** m, wave(length) changing switch;
- **-s-welle** f, compensation wave, spacing wave.

verstopft, congested.
Verstopfung f, congestion.
verstreben, to prop, to strut.
Verstreben n, propping, strutting.
Verstrebung f, prop, strut.
verstümmeln, to mutilate, to alter, T.
Verstümmelung f, mutilation T.
Versuch m, test, trial, experiment, experimental test;
- **e anstellen**, to experiment;
vergleichender —, comparative test;

Betriebs- —, field trial;
Laboratoriums- —, laboratory test;
Strecken- —, field trial;
Versuchs- ...., experimental;
- **-erlaubnis** f, experimenter's licence;
- **-gestell** n, test rack, test stand;
- **-lizenz** f, experimenter's license;
- **-raum** m, laboratory;
- **-stadium** n, experimental stage.

versuchen, to try; **proben**: to test, to try.

vertauschen, to interchange, to permute; **zwei Leitungen**: to cross, to transpose.

Vertauschung f, permutation, interchange; crossing, transposition.

Vertakelung f, rigging.
verteilen, to distribute; **zuteilen**: to allot; **zerstreuen**: to disperse; **sortieren**: to sort out; **von neuem —**, to redistribute.

Verteiler m, distributor TA;
Telegramme, Zettel: check table; **Gestell**: distributing frame F T;
korrigierender —, correcting or controlling distributor;
korrigierter —, corrected or controlled distributor;
mehrwegiger —, multi-channel distributor;
umlaufender —, rotary distributor;
Empfangs- —, receiving distributor; [butor T;
Geh-Steh- —, start-stop distributor;
Haupt- —, main distribution frame, ab: MDF;
Mehrfach- —, multiplex distributor, multichannel distributor;

Verteiler
  Sende- —, sending or transmitting distributor;
  Zwischen- —, intermediate distribution frame, *ab*: IDF;
  — -bürsten *pl*, distributor brushes *pl*;
  Korrektion *f* der — - —, durch Rückwärtsdrehen, (Baudot) shift-the-hands correction;
  — -einrichtung *f*, selbsttätige, automatic distributing facilities *pl*;
  — -gestell *n*, distributing frame *FT*;
  — -relais *n*, switching relay *T*;
  — -ring *m*, distributor ring, crown, *T*;
  — -scheibe *f*, distributor plate or face or head or plateau, *T*;
    bewegliche — - —, movable distributor plate;
    feste — - —, fixed distributor plate;
    hintere — - —, rear distributor plate;
    vordere — - —, front distributor plate;
  — -segmente *pl*, distributor segments *pl*;
  — -seite *f*, jumper side;
  — -stelle *f*, distributing point *TF*
verteilt, distributed;
  gleichmäßig —, evenly or continuously distributed;
  punktförmig —, distributed in lumps, lumped;
  — —e Ladung *f*, Kapazität: lumped capacity, concentrated capacity; Induktivität: lump-loading, lumped loading *L*;
  stetig —, evenly or uniformly distributed;
  — e Induktivität *f*, distributed
  — inductance.

Verteilung *f*, distribution; Zuteilung: allotment; Zerstreuung: dispersion;
  Wieder- —, redistribution;
  — der Anrufe, call distribution *F*;
  Einrichtung *f* zur selbsttätigen — — —, automatic distributing device *F*;
  Feld- —, field distribution;
  Spannungs- —, distribution of voltage;
  Strom- —, current distribution;
Verteilungs-kasten *m*, distribution box or case *B*;
  — -mast *m*, distributing pole *B*;
  — -punkt *m*, distributing point;
  — -schrank *m*, distributing cabinet;
  — -tafel *f*, distributing board.
Vertiefung *f*, pit, cavity.
vertikal, vertical.
Vertikalantenne *f*, vertical wire aerial.
Vertikale *f*, vertical.
Vertikal-führung *f*, vertical guide;
  — -intensität *f*, vertical force, earth's vertical field, vertical component of earth's magnetic field. [ment.
Vertrag *m*, contract, agreeeinen — schließen, to contract, to agree.
Verunreinigung *f*, impurity.
vervielfältigen, to multiply, to manifold.
verwerfen, to reject;
Verwerfung *f*, rejection.
verwickelt, complicated, complex.
verwischte Sprache *f*, blurred voice.
verwürfeln, to jumble.
Verwürfelungsschlüssel *m*, jumble code *T*.

verwürgen, to twist (together) B.

Verzahnung f, teeth pl; **Innen-** —, internal teeth pl.

verzehren, to dissipate; **— b**, dissipative.

Verzehrung f, dissipation; **Energie-** —, energy dissipation.

Verzeichnis n, index, schedule.

verzerren, to distort, to contort, to deform. **— b**, distortional.

verzerrt, distorted.

Verzerrung f, distortion, contortion, deformation;
— I Art, amplitude distortion L;
— II Art, frequency distortion L;
— durch Ein- und Ausschwingen, transient distortion L;
stationäre —, stationary distortion;
Amplituden- —, amplitude distortion;
Feld- —, field deformation, field distortion;
Frequenz- —, frequency distortion;
Gesamt- —, $\left(\text{Längenmaß}, \dfrac{R}{2Z} - \dfrac{GZ}{2}\right)$ total distortion of a line, T;
Impuls- —, impulse distortion A;

Verzerrungsfaktor m, distortion factor $\left(\dfrac{Z_1 + Z_2}{Z_1 - Z_2}\right)$.

verzerrungsfrei, distortionless, non-distorting;
— e Leitung f, distortionless circuit L.

Verzerrungs-freiheit f, absence of distortion; [stant;
— -konstante f, distortion con-
— -messer m, distortion measuring system.

verziehen, sich, to buckle, Holz: to warp.

Verziehen n, buckling, warping.

verzinken, to galvanize.

Verzinkung f, Feuer- —, galvanizing, galvanization.

verzinnen, to tin.

verzinnt, tinned; feuer- —, fire-tinned.

verzogen, warped.

Verzögerer m, slug; retarder; — -kette f, delay network K.

verzögern, to retard, to delay T.

Verzögerung f, retardation, delay T;
Leitungs- —, line lag;
Phasen- —, lagging of phase;

Verzögerungs-relais n, slow-acting relay; copper collar relay, coppered relay, F;
— -widerstand m, timing resistance, retardation coil T;
— -winkel m, retardation angle L.

verzweigen, to branch.

Verzweiger m, distribution box B;
End- —, terminal block;
Kabel- —, cable connection box, cable distribution case.

verzweigte Ströme pl, branched currents pl.

Verzweigung f, branching (-off);
Verzweigungs-muffe f, cable distribution plug;
— -s-punkt m, split (point), branching point.

Vibration f, vibration;
Vibrations-galvanometer n, vibration galvanometer;
— -relais n, vibrating relay.

Vibrator m, vibrator.

vibrieren, to vibrate; — b, vibrating, vibratory.

Vibrieren n, vibration.

vielabrig, multicore;
— es Kabel n, multicore cable.
Vieleck n, polygon.
vieleckig, polygonal.
vielfach, multiple;
— schalten, to multiple (to).
Vielfach n (m), multiple F A.
Deladen= —, Höhenschritt= —,
level multiple A.
Vielfaches n, multiple;
ganzes —, integral multiple;
gerades —, even multiple;
ungerades —, odd multiple.
Vielfach=Abstimmvorrichtung f,
multiple tuner;
— =antenne f, multiple antenna;
— =anschluß m, private branch
exchange (or p. b. x.) junction,
F; multiple connection;
— = — =Leitungswähler m, private branch exchange final
selector, p. b. x. final selector, A;
— =feld n, multiple, multiple
field F; bank multiple A;
Klinken= —, jack multiple;
Leitungswähler= — =—, final
selector (bank) multiple A;
Teilnehmer = —= —, subscribers
multiple;
Verbindungsleitungs= — = —,
junction multiple;
— = — = drähte pl, bank wires
pl A;
— = — =kabel n, multiple cable;
— =funkenstrecke f, multiple spark
discharger;
— = — für die Erzeugung ungedämpfter Schwingungen, timed
spark discharger;
vielfachgeschaltet, multiple-connected, connected in multiple, multipled (to);
Vielfachkabel n, multiple cable;
bank cable A;
— =klinke f, multiple jack;

— = —n=feld n, (jacks) multiple;
— =kontakt m, multiple contact;
— = — =bank f, contact bank A;
an die — = — =bänke der Leitungswähler geführt, multipled
to the connector banks A;
— =kontaktfeld n, multiple (bank
contact).
— =Parallelklinke f, parallel multiple jack;
— =schaltung f, multiple connection;
— =schrank m, multiple switchboard;
— =Unterbrechungsklinke f, series
multiple jack;
— =verdrahtung f, multiple wires
pl, bank wiring A;
— =verkabelung f, multiple cabling;
— =Zwillingsverseilung f, multiple twin formation;
— =Zwillingskabel n, multiple
twin cable F.
vielgliedriger Kettenleiter m,
multi-mesh network.
vielpaarig, multi-pair.
vielpolig, multipolar.
vielstufiger Schalter m, multipoint switch.
vierabrig, four-wire.
Vierdraht=betrieb m, four-wire
operation;
— =leitung f, four-wire circuit;
— =schaltung f, four-wire connection.
Vierelektrodenröhre f, four-electrode valve, tetrode.
Vierer m, four-wire unit, Vierer=
bündel: quad, K;
zum — schalten, to phantom;
zum — verseilen, to quad;
Unsymmetrie f zwischen Stamm
und —, phantom-to-side unbalance K;
Dieselhorst=Martin= —, D. M.=
—, two pair core;

Vierer
  Kern= —, central quad;
  Spiral= —, spiral quad;
  Stern= —, spiral(led) four, spiral quad;
  — =**Abzweigübertrager** m, combining transformer;
  — =**belastung** f, phantom loading;
  — =**betrieb** m, phantom circuit operation, duplex operation;
  — =**bündel** n, four-wire core, quad;
  — =**kapazität** f, pair-to-pair capacity, side-to-side capacity;
  — =**kreis** m, phantom or duplex circuit, superposed or combined circuit, plus circuit;
  — =**leitung** f, phantom circuit, phantom pair;
  — =**pupinisierung** f, phantom loading, composite loading, superposed loading;
viererpupinisiert, composite loaded, phantom loaded;
Vierer=schaltung f, phantom connection;
  — =**Schleifenkapazität** f, phantom capacity, side-to-side capacity, pair-to-pair capacity;
  — =**spule** f, phantom coil, superimposed circuit coil;
  — = —**satz** m, phantom coil set;
  — =**verbindung** f, phantom circuit, plus circuit;
viererverseilt, quadded;
  — es **Kabel** n, duplex cable, phantom cable;
Viererverseilung f, quad formation;
  Dieselhorst=Martin= —, D.M.= —, multiple twin formation;
  Stern= —, spiral(led) four formation, spiral quad formation.
vierfach, quadruple.
Vierfach=telegraph m, quadruple telegraph (system);

  — =**Zwilling** m, quadruple twin K;
  — = —**S=kabel** n, quadruple pair cable.
Vierkant n, square;
  — =**kopf** m, square head.
vierphasig, quarter-phase.
Vierpol m, ideal artificial line (with two input and two output terminals), transducer (am).
vierpolig, four-pole, four-polar.
Vierteldrehung f, quarter turn.
Vierung f, quadrature.
Viola f, alto;
Violine f, violin; [viol;
Violincello (violon)cello, bass
virtuell, virtual.
visieren, to sight.
viskos, viscous.
Vokal m, vowel (sound).
Vollamt n, main office.
Vollast f, full load.
vollautomatisch, **vollselbsttätig** full automatic(al).
Volt n, (ab: V), volt;
  Kilo= —, (ab: kV), kilovolt;
  Mikro= —, (ab: μV), microvolt;
  Milli= — (ab: mV), millivolt;
  — =**ampere** n (ab: VA), volt-ampere;
  Kilo= — = —, (ab: kVA), kilovolt-ampere;
  — =**meter** n, voltmeter;
  Phasen= — = —, phase voltmeter;
  Röhren= — = —, amplifying voltmeter;
  — = —**umschalter** m,. voltmeter switch;
  — = — = **vorschaltwiderstand** m, multiplier.
Volumen n, volume.
Volumprozent n, percent by volume.
Voranschlag m, estimate.

Vorarbeiter *m*, foreman.
vorausberechnen, to predetermine.
Vorausberechnung *f*, predetermination.
voraussetzen, to (pre)suppose.
Voraussetzung *f*, (pre)supposition, Annahme: assumption.
vorausbezahlen, to prepay.
Vorausbezahlung *f*, prepayment.
Vorbenutzung *f*, prior use.
vorbereiten, to prepare.
Vorbereitung *f*, preparation.
Vorder=ansicht *f*, front view, front elevation.
— =seite *f*, front; face;
— =teil *n*, head.
Vordruck *m*, — =blatt *n*, form, paper blank.
voreilen, to run fast, to advance, Phase: to lead (um ....⁰, by ....⁰).
Voreilen *n*, **Voreilung** *f*, leading, lead, advance;
Phasen= —, leading of phase;
Winkel= —, advance angle;
— =winkel *m*, angle of lead, advance angle.
Vorgang *m*, process, act, operation;
flüchtiger —, transient;
periodischer —, cyclic operation;
Druck= —, act of printing *T*.
vorgehen, zu schnell sein: to run fast; Phase: to lead.
vorgespannt, bias(s)ed;
elektrisch —, electrically biased.
vorherbestimmen, to predetermine.
Vorhersage *f*, forecast;
Wetter= —, weather forecast.
Vorherrschen *n*, predomination, prevailing, preponderance.
vorherrschende Winde *pl*, prevailing winds *pl*.

vorkommen, to occur.
vorläufig, provisional.
vormagnetisieren, to polarize, to bias.
Vormagnetisierung *f*, magnetic bias, superposed magnetization, polarization;
— =strom *m*, biasing current;
vorn, in front.
Vorort *m*, suburb;
Vororts=gebiet *n*, suburban area;
— =gespräch *n*, suburban call;
— =leitung *f*, suburban junction;
— =platz *m*, suburban position;
— =schrank *m*, suburban switchboard;
— =verbindung *f*, suburban connection;
— =verkehr *m*, suburban service.
Vorrang *m*, priority.
Vorrat *m*, stock, store;
auf —, in stock;
— =kondensator *m*, tank condenser, reservoir condenser.
vorrätig, in stock; available.
Vorrichtung *f*, device; für Zusammenbau: assembling jig, Lehre, Futter: fixture.
vorrücken, to feed (den Streifen, the slip); fortschreiten: to progress;
zu weit —, to overfeed.
Vorschalteleitung *f*, trunk junction circuit.
Vorschaltwiderstand *m*, series resistance, für Spanungsmesser: multiplier.
Vorschlag *m*, proposition.
vorschlagen, to propose.
Vorschrift *f*, instruction, rule, specification;
Bau= —, specification;
Betriebs= —, rules *pl* of operation;
Dienst= —, service rules *pl*.
— **en=bereich** *m*, range;

Vorschriftenbereich (außer)europäischer Dienst- —, (extra)-European range.

Vorschub *m*, feed;
Blatt- —, page feed, **Einzelblätter:** cut page feed, *T*;
Buchstaben- —, letter feed;
Differential- —, differential feed (der **Wheatstone-Tastenlocher,** of Wheatstone keyboard perforators);
Papier- —, paper feed(ing);
Seiten- —, page feed;
Streifen- —, paper feed, tape feed;
— **-daumen** *m*, spacing cam *T*;
— **-einrichtung** *f*, feed, feeding device;
— **-klinke** *f*, feed pawl;
— **-magnet** *m*, feeding magnet, spacing magnet *T*;
— **-rad** *n*, feed wheel.

Vorspannung *f*, priming or initial or biasing potential, bias, **Feder:** tension;
eine — erteilen, to bias;
einseitige —, bias;
magnetische —, magnetic polarization, magnetic bias;
negative —, negative bias;
Gitter- —, grid bias, biasing or priming or initial grid volt.

vorspringen, to project; [age.
— **d,** salient, projecting.

Vorsprung *m*, projection, (projecting) lug, shoulder.

Vorstadt, suburb;
Vorstadt-...., **vorstädtisch,** suburban.

vorstechen, to pierce.

Vorstecher *m*, bradawl.

Vortexring *m*, vortex, *pl* vortices.

vorübergehend, transient;
— **er Vorgang** *m*, transient *L*;
— **er Zustand** *m*, transient state *L*.

Vorverstärker *m*, input amplifier.

Vorübertrager *m*, input transformer, input repeating coil.

Vorwahl *f*, preselection, finding action, *A*;
doppelte —, tandem preselection.

Vorwähler *m*, preselector, lineswitch, *A*;
erster —, first preselector *(engl.)*, primary lineswitch *(am.)*;
zweiter —, second preselector, secondary lineswitch, trunk hunting switch;
zehnteiliger —, ten-point preselector;
Dreh- —, rotary lineswitch;
Relais- —, relay preselector;
— **-antrieb** *m*, master switch;
— **-gestell** *n*, lineswitchboard.

Vorwärts-auslösung *f*, calling party release *A*;
— **-und Rückwärtsauslösung** *f*, first party release *A*.

vorwärtsschalten, to step on, to step up (to).

Vorzeichen *n*, sense, sign, *M*;
von gleichem —, of the same sign;
von verschiedenem —, of different sign;
Wechsel *m* **des —s,** change of sign.

vorzeitig, premature.

Vulkanfiber *f*, vulcanized fibre.

vulkanisieren, to vulcanize.

Vulkanisierkessel *m*, vulcanizing pan.

Vulkanisierung *f*, vulcanization.

W.

Waben-spule f, honeycomb coil, duo-lateral coil;
— **-wicklung** f, honeycomb winding, duo-lateral winding.

Wachs n, wax;
mit — überziehen, to wax;
Bienen- —, beeswax;
Erd- —, ozokerite;

wachsartig, waxy;

Wachs-draht m, waxed wire;
— **-papier** n, waxed paper.

wachsen, to grow, to swell;
Sammlerplatten: to fan out, to tree; mit Wachs behandeln: to wax.

Wachsen n, growth, swelling; fanning-out, treeing; waxing.

Wackelkontakt m, defective contact, loose or variable connection.

Wage f, balance;
Coulombsche —, Coulomb's balance;
magnetische —, magnetic balance.

Wagen m, car, carriage;
Anhänge- —, truck;
Greifer- —, pick-up carrier;
Hänge- —, cableway carriage (für Luftkabel, for aereal cables) B;
Kraft- —, **Motor-** —, motor car, automobile;
Last- —, truck;
Lastkraft- —, motor truck;
Roll- —, truck, lorry;
— **-park** m, fleet.

wagerecht, horizontal, level.

Wahl f, selection;
freie —, **Frei-** —, hunting, hunting operation;
Fern- —, toll switching, l. d. selection;
Gruppen- —, group selection;

Nummern- —, impulse-stepping, impulse action;
Vor- —, preselection, finding action;
— **-anruf** m, selective ringing or signalling; selector calling T;
abgestimmter — **-** —, harmonic selective ringing.

wählen, to select, to dial A;
Einer (Zehner) —, to dial units (tens) digit A;
durch- —, to dial through;
frei —, to hunt (for).

Wählen n, selection, dialling A;
Durch- —, through-dialling;
Frei- —, hunting action, finding action.

Wähler m, selector, (selective or automatic) switch, auto-switch;
zehnteiliger —, — mit 10 Ausgängen oder Richtungen, ten point selector;
großer —, major switch;
kleiner —, minor switch (**Vorwähler, Steuerschalter,** pre-selectors, master switches);
— **mit freier Wahl**, selector, hunting switch;
— **für ankommende Verbindungsleitungen**, injunction switch;
Amtsnamen- —, (1st, 2nd) code switch;
Dienst- —, service connector;
Dreh- —, spindle switch, rotary selector;
Fernleitungs- —, F L- —, l. d. connector;
Gruppen- —, group selector, trunking switch, intermediate selector;
— **-** — **in Millionennetzen, erster**, (mit Leitbuchstaben) code selec-tor; [selector;
— **-** —, **II., III,** usw., tandem

Wähler

Gruppen- —mit Stromstoßübertrager, selector repeater;
Heb-Dreh- —, vertical and rotary selector, Strowger switch;
Leitungs- —, connector, final selector, final switch;
— - — mit Frequenzwahl für Gesellschaftsleitungen, frequency selecting connector;
— - —, Mehrfach-, private branch exchange final selector, p. b. x. final selector;
Linien- —, final selector, connector, A; commutator F;
— - —, Stöpsel-, plug switch;
Misch- —, load distributing switch;
Mitlauf- —, companion work selector, simultaneous movement selector;
Nummern- —, numerical switch (Gegensatz: GW für Leitbuchstaben);
Relais- —, relay selector;
Rückruf- —, reverting call switch;
Schrittschalt- —, step-by-step selector;
Strowger- —, Strowger switch;
Überweisungs- —, allotting switch; [lector;
Umleitungs- —, director se-
Vielfachanschluß- —, private branch exchange final selector, p. b. x. final selector;
Zuteilungs- —, allotting switch;
— -auslösung f, selector release;
Verbindungsleitung f für —-—, selector release trunk;
— -arm m, wiper A;
— -bank f, selector bank;
— -gestell n, selector rack, switch frame, autoswitch rack; [T;
— -kamm m permutation plate

— -raum m, auto-room A;
— -scheibe ~ f, Nummernscheibe: dial switch A; permutation disc, selector plate, T;
— -schiene f, permutation bar, combination bar, code bar, selector bar, T;
— -stufe f, rank of switches;
— -system n, selector system; automatic telephone system;
— -vielfach n (m), bank multiple;
— - — -kabel n, bank cable.

Wähl-impuls m, dialling impulse;
— -magnet m, selecting magnet T;
— -mechanismus m, selecting mechanism;
— -stromstoß m, dialling impulse;
— -vorgang m, selective process;
— -werk n, selective or selecting mechanism.

Wahl-schalter m, selector switch;
— -stufe f, digit A;
— -vorgang m, selection, selecting operation;
freier — - —, hunting operation.

wahlweise, selective;
— rufen, to call selectively.

wahrnehmbar, perceptible, perceivable.

wahrnehmen, to perceive, to observe.

Wahrnehmung f, perception, observation.

Wahrscheinlichkeitsrechnung, f, probability theory.

Waldhorn n, French horn.

Walzblei n, rolled lead;
— -draht m, rolled wire.

Walze f, roller, wheel; Schreibmaschine: platen.

Walzeisen n, rolled iron.

walzen, wälzen, to roll.

Walzen-lager n, roller bearing;

Walzen
— -mikrophon *n*, pencil transmitter;
— -schalter *m*, barrel switch.
Walz-profil *n*, rolled section;
— -werk *n*, rolling mill, calender.
Wand *f*, wall;
Schirm- —, screened wall.
Wanderfeld *n*, travelling field, moving field.
wandern, Feld, Welle: to travel, to move; Jonen: to migrate, Nullpunkt: to shift, to wander.
Wandern *n*, travelling; migration; shifting.
Wanderwelle *f*, surge, travelling wave transient wave;
Spannungs- —, voltage surge.
Wandfernsprecher *m*, wall telephone station.
Wandler *m* transformer;
Frequenz- —, ruhender oder statischer, static frequency transformer.
Wandlung *f*, transformation;
Frequenz- —, frequency transformation;
— -verhältnis *n*, transformation ratio.
Wand-stärke *f*, thickness;
— -Vermittlungsschrank *m*, wall pattern switchboard.
Wandung *f*, wall(s *pl*).
Wange *f*, side wall, cheek, side plate.
warm, warm, hot;
rot- —, red hot;
weiß- —, white hot;
— werden, to heat up;
— -brüchig, hot-short brittle;
— -laufen, sich, to run hot.
Wärme *f*, heat; temperature;
Verdampfungs- —, heat of vaporization, (latent) heat of evaporation;
— -(aus)strahlung *f*, heat radiation;

— -leiter *m*, heat conductor;
— -leitfähigkeit *f*, heat conductivity, thermal conductivity;
— -wirkung *f*, thermal effect.
Warm-laufen *n*, heating-up;
— -werden *n*, heating (-up).
Warnungs-tafel *n*, danger or caution or warning board;
— -zeichen *n*, warning sign.
warten, to wait; beaufsichtigen: to attend (to).
Wärter *m*, attendant (to),
Batterie- —, battery attendant.
Wartezeit *f*, delay, wait, waiting time, *F*;
— für ein Gespräch, delay on a call *F*;
— wahrscheinlich .... Minuten, delay likely to be ....minutes *F*.
Wartung *f*, attendance (einer Anlage, to a plant); Pflege: maintenance;
— -s-kosten *pl*, cost of attendance.
Warze *f*, pimple.
Wasser *n*, water.
— zersetzen, to split up water;
angesäuertes —, acidulated water;
destilliertes —, distilled water;
Leitungs- —, tap water.
wasserdicht, waterproof, watertight.
Wasserdruck *m*, hydraulic pressure;
— -presse *f*, hydraulic press.
wasser-frei, anhydrous;
— -gekühlt, water-cooled.
Wasserglas *n*, water-glass (K_4SiO_4, Na_4SiO_4).
wasserhaltig, aqueous.
Wasser-kühlung *f*, water-cooling;
— -leitung *f*, water pipes *pl*, water conduit;

Waffer
— =leitungš=ḩaḩn *m*, water tap, water spout.

wafferlöšlich, water soluble.

Waffer=ftoff *m*, hydrogen (H);
— =ftraḩl *m*, water jet;
— =temperatur *f*, water temperature.
— =wage (bubble) level.

Watt *n* (*ab*: W), watt;
Silo= —, (*ab*: kW), kilowatt, kw;
Mikro= —, (*ab*: μW), microwatt;
Milli= — (*ab*: mW), milliwatt, mw;
— =komponente *f*, watt component.

wattloš, wattless;
— e Komponente *f*, wattless component;

Watt=meter *n*, wattmeter;
— =ftunde *f* (*ab*: Wh), watt hour;
— =jaḩl *f*, wattage.

Wechfel *m*, change; inversion, shift *T*; Strom: alternation;
— *pl*, reversals *T*;
Buchstaben= —, letter shift, unshift;
Figuren= —, inversion, shift (signal), *T*;
Strom= — *pl*, reversals *pl T*;
Zaḩlen= —, shift, figure shift, *T*;
— =feld *m*, alternating field;
magnetisches — = —, alternating magnetic field;
schnelles — = —, oscillatory field;
— =fluß *m*, alternating flux;
— =geschwindigkeit *f*, angular velocity, frequency in radians;
— =kontakt *m*, make-and-break contact.

wechseln, to alternate, to change; to shift *T*.

wechselpolig, heteropolar.

Wechselspannung *f*, alternating potential, a. c. potential;
— š=komponente *f*, alternating component of voltage.

Wechselstrom *m*, alternating current, a. c.;
schneller —, oscillatory current, undulating current;
— =erzeuger *m*, alternator, alternating current generator;
500 periodiger — = —, 500 cycle alternator;
— =generator *m*, alternator;
— = — mit umlaufender Funkenstrecke, alternator disc set *R*;
— =komponente *f*, a. c. component, alternating component of current;
— =lichtbogen *m*, alternating arc;
— =Mehrfachtelegraphie *f*, carrier current multiple telegraphy, alternating current multiple telegraphy;
— =meßbrücke *f*, a. c. bridge;
— =messer *m*, a. c. ammeter;
— =messung *f*, a. c. measurement;
— =quelle *f*, a. c. source;
— =relais *n*, a. c. relay;
— =transformator *m*, alternating current transformer;
— =verluste *pl*, a. c. losses *pl*;
— =wecker *m*, magneto bell, a. c. bell;
— = — mit Ankerumlegefeder, biased magneto bell;
— =widerstand *m*, alternating current resistance.

Wechsel=taste *f*, shift key *T*;
— =verkehr *m*, intercommunication;
— =winkel *m*, alternate angle;
— =wirkung *f*, interaction;
in — = — stehen, to reciprocate;
— =jaḩl *f*, frequency;

Wechsel
— -zeichen n, shift signal T.
wecken, to ring.
Wecken n, ringing.
Wecker m (call) bell, (electro-magnetic) alarm, ringer;
einschaliger —, single dome bell;
zweischaliger —, double dome bell;
zweiter —, extension bell F;
Anruf- —, call bell;
Außen- —, extension bell;
Dosen- —, circular bell;
Einbruch- —, burglar alarm;
Einschlag- —, single stroke bell;
Fallscheiben- —, indicator bell;
Fortschell- —, continuous(ly) ringing bell;
Gleichstrom- —, trembler bell, trembling bell;
Gleichstrom-Dosen- —, circular trembler;
Gruben- —, mining bell;
Haus- —, domestic electric bell;
Nacht- —, night bell, night alarm;
Schalmei- —, gong bell;
Schnarr- —, buzzer;
Starkstrom- —, power bell;
Wechselstrom- —, magneto bell, (a. c.) ringer;
— — mit Ankerumlegefeder, biased magneto bell;
— -stromkreis m, bell circuit.
Weckstrom m, ringing current;
— -kreis m, ringing circuit, ringing loop.
Weg m, path; Straße: road;
gemeinsamer —, common path;
geschlossener —, closed path;
öffentlicher —, public road;
Absatz- —, channel T F;
Anker- —, armature travel;

Leit- —, (telegraph) route;
Telegraphier- —, telegraph route;
Verkehrs- —, channel;
Wegerecht n, wayleave, rights-of-way pl.
wegheben, sich, to cancel out M.
Wehnelt-Kathode f, Wehnelt cathode;
— -röhre f, Wehnelt valve;
— -unterbrecher m, electrolytic interrupter.
weich, soft; Holz: sappy;
— e Röhre f, soft valve V.
Weich-blei n, soft lead;
— -eisen-anker m, soft iron armature;
— - —blech n, lackiertes, ferrotype;
— - —instrument n, moving iron instrument;
— - —kern m, soft iron core;
— - —membran f, lackierte, ferrotype diaphragm;
— - —strommesser m, soft-iron vane ammeter, moving iron ammeter;
— -gummi n (m), soft rubber;
— -lot n, tin solder, soft solder;
— -porzellan n, soft porcelain.
Weiche f, elektrische, separating filter.
weiß, white, blank;
— -glühend, — -warm, white hot, incandescent; bright V.
Weiß n, blank, spacing signal, T;
Buchstaben- —, letter blank;
Zahlen- —, figure blank;
— -blech n, tinned sheet iron, tin plate, tin;
— -bleierz n, cerusite ($PbCO_3$);
— -buche f, white beech;
— -glut f, white heat, incandescence;
— -strick m, pipe yarn B.

weit, wide; far.
Weit= ...., long-range ....R, long-distance .... F.
Weite f, width, opening.
weiterbefördern, to retransmit.
Weiterbeförderung f, retransmission T;
weiter=drehen, die Schaltarme, to step round the wipers (auf, to) A;
— =bewegen, to feed;
— =führen, eine Leitung, to extend a line (nach, to);
Weitergabe f, retransmission T;
selbsttätige —, automatic retransmission;
— mit der Hand, manual retransmission;
weitergeben, to retransmit, mit Relais: to repeat (on, into), T;
Weitergeber m, retransmitter T;
weiter=schalten, to step up, to step on (auf, to);
— =senden, to retransmit;
Weitersendung f, retransmission.
weitmaschig, wide-meshed;
— es Gitter n, open grid.
Weitverkehr m, long-distance traffic;
Fernsprech= —, long-distance telephony.
Wellblech n, corrugated sheet iron.
Welle f, wave; Achse: shaft(ing), axle, arbor;
auf — 600 arbeiten, to operate on a wavelength of 600 m;
kleine —n pl ripple(s pl);
— —n bilden, to ripple;
aperiodische —, aperiodic wave;
ausbreitende —, sich, travelling wave, proceeding wave;
Raum: propagating wave;
ausgesandte —, transmitted wave;
einfallende —, oncoming wave;

elektromagnetische —, electromagnetic wave;
empfangene — received wave;
fortschreitende —, advancing wave;
gedämpfte —n pl, damped waves, type B waves, discontinuous waves, pl;
gleichstromüberlagerte —, pulsating wave, wave superposed on direct current;
hinlaufende —, main wave L;
hin= und hergehende —, rock shaft;
modulierte —, modulated wave;
pulsierende —, pulsating wave;
reflektierte —, reflected wave L;
rücklaufende —, reflected wave L;
stehende —n, stationary waves, standing waves;
ungedämpfte —n, continuous waves, c. w., type A waves, undamped waves;
— —n, getastete, key controlled continuous waves, ab: C. W., type A 1 waves;
— —n, getastete tonüberlagerte, key controlled continuous waves modulated at audiofrequency, type A 2 waves, ab: I. C. W.;
— n — n, Senden n mit, cut-in c. w. transmission;
Ankernuten= —n, slot ripple;
Antriebs= —, driving shaft, driver shaft;
Äther= —, ether wave;
Eigen= —, natural wavelength;
Elementar= —, elementary wavelet R;
Empfangs= —, received wave;
Geh=Steh= —, start-stop spindle;
Grund= —, fundamental wave;

Welle
Halb= —, half-wave;
Hohl= —, tubular shaft;
Kommutierungs= —n, commutator ripple;
Kopplungs= —n, coupling waves, partial waves;
Kugel= —, spherical wave;
Modulations= —, wave of modulation;
Nocken= —, cam spindle;
Ober= —, overtone, harmonic (wave);
Raum= —, spherical wave;
Schall= —, sound wave;
— = —, sinusförmige, sine wave of sound;
Sende= —, transmitted wave;
Sinus= —, sine wave, harmonic wave;
— = —, reine, pure sine wave;
— = —, zusammengesetzte, complex sine wave;
Spannungs= —, voltage wave;
Strom= —, current wave;
— = —n des Gleichstroms, current ripple(s);
Teil= —, partial wave;
Träger= —, carrier (wave);
Trieb= —, driving shaft;
Verstimmungs= —, compensation wave, spacing wave;
Wander= —, surge, transient wave, travelling wave;
— = —, Spannungs= , voltage surge;
Zeichen= —, marking wave, signal wave;
Zwischenzeichen= —, spacing wave;

Wellen=anzeiger m, oscillation detector, wave detector, cymoscope;
— =ausbreitung f, wave propagation;
— =band n, wave band;
— =bauch m, bulge, loop, antinode;
— =bereich m, wave range, wave band;
— =berg m, wave crest;
— =bewegung f, wave motion; undulation,
— =empfänger m, wave receiver;
— =erzeuger m, wave generator;
— =filter n, wave filter;
— =form f, wave shape, wave form;
— =fortpflanzung f, wave propagation;
— =front f, wave front T;
geneigte — = —, tilted wave front T;
steile — = —, steep wave front T;
— =generator m, wave generator;
— =geschwindigkeit f, wave velocity;
— =gestalt f, wave shape;
— =konstante f, line angle, (γl) L;
— =kontakt m, shaft contact(s pl) A;
— =länge f, wavelength;
Betriebs= — = —, operating wavelength;
Eigen= — = —, natural wavelength; unloaded wavelength (des Luftdrahts of an aerial);
— = —n =konstante f, wave length constant, phase shift constant, L;
— = —n-konstanz f, steadiness of the wave, R;
— =linie f, wave line;
— = —n-schreiber m, wave line recorder, ondograph;
— =messer m, wavemeter, cymometer;
Normal= — = —, standard wavemeter;
— = — mit Summererregung, buzzer-driven wavemeter;

Wellen
— **-schlucker** m, wave trap, frequency sifter, frequency trap, rejective circuit;
— **-schreiber** m, ondograph; undulator T;
— **-schwanz** m, wave tail T;
— **-stirn** f, wave front T;
geneigte — - —, tilted wave front T;
steile — - —, steep wave front T;
— **-tal** n, wave trough, trough, hollow;
— **-telegraphie** f längs Leitungen, wired wireless telegraphy;
— **-umschalter** m, change-tune switch, wavelength changing switch;
— **-widerstand** m, characteristic impedance, surge impedance;
Ungleichförmigkeit f oder Schwankungen pl des — - —es, impedance irregularities pl;
Verlauf m des — - —es in Abhängigkeit von der Frequenz, impedance-frequency characteristic;
— - — einer mit halber Spule beginnenden Leitung, mid-load characteristic impedance K;
— **-zug** m, wave train, train of waves, ungedämpft auch: beat;
— - **-frequenz** f, group frequency, wave train frequency.

wellig, wavy;
kurz- —, short-wave . . . ;
lang- —, long-wave . . . ;
— e **Gleichspannung** f, ripple voltage;
— er **Strom** m, ripple current.

Welligkeit f, ripple;
— von n % ripple of n percent;
Ankernuten- —, slot ripple;

Kommutierungs- —, commutation ripple;
— **-s-frequenz** f, ripple frequency.

Wellplattenkondensator m, corrugated plate condenser.

Wendepol- m, reversing pole, inter-pole;
— **-punkt** m einer Kurve, cusp;
— **-schalter** m reversing key.

werfen, Lichtstrahl: to project (auf, on to);
sich —, to warp (Holz, wood).

Werg n, tow.

Werk n, Arbeit: labour, work;
Fabrik: works pl, factory;
Mechanismus: mechanism, gear;
Regler- —, governing mechanism;
— **-bank** f (work) bench;
— **-statt** f, workshop, shop, große: factory;
— **-stoff** m, material;
— **-zeug** n, tool(s pl);
— - — **kasten** m, tool box;
— - — **-maschine** f, machine tool.

Wert m, value;
— eins, unity;
kritischer —, critical value;
Augenblicks- —, instantaneous value;
Dauer- —, steady state value;
Durchschnitts- —, average value;
Effektiv- —, virtual value, r. m. s. (= root mean squares) value, effective value;
Erfahrungs- —, empirical value;
Garantie- —, guaranteed value;
Grenz- —, limiting value;
Höchst- —, **Maximal-** —, maximum value;
Meß- —, measured value;

Wert
 Mittel= —, mean value, average value;
 — = —, quadratischer, r. m. s. value, virtual value;
 Momentan= —, instantaneous value;
 Nenn= —, nominal value;
 Pflicht= —, specification value, contract value;
 Prüf= —, test value;
 Regel= —, average value;
 Sättigungs= —, saturation value;
 Scheitel= —, amplitude, crest;
 Spitzen= —, peak value;
 Verkehrs= —, telephone traffic unit F;
 Zeit= —, standing value.
Western=(3W=)Schaltung *f*, repeating coil c. b. system.
Wetter=bericht *m*, weather report;
 — =berührung *f*, weather contact;
 — =dienst *m*, weather signals *pl*;
 — =nebenschluß *m*, weather leakage;
 — =vorhersage *f*, weather forecast.
Wickel *m*, reel; **Wicklung:** binder, binding;
 Kondensator= —, condenser reel;
 — =draht *m*, taping wire B;
 — =kondensator *m*, roll type condenser;
 — =lötstelle *f*, Britannia joint B;
 — =maschine *f*, winding machine.
wickeln, to wind;
 be= —, to wrap, to tape, B;
 neu —, to rewind.
Wicklung *f*, winding;
 hochohmige —, high-resistance winding;
 niederohmige —, low-resistance winding; [ter-winding;
 zwischen den —en wirkend, in-

Ausgleichs= —, compensation winding (des Differentialrelais, of the differential relay) T;
Beschleunigungs= —, acceleration winding (des Gulstabrelais, of the Gulstad relay);
Differential= —, differential winding;
Erreger= —, exciting winding;
Erst= —, primary winding;
Feld= —, field coil, field winding;
Halte= —, holding coil;
Hilfs= —, auxiliary winding;
Jute= —, wrapping of jute;
Leitungs= —, line winding (des Differentialrelais, of the differential relay) T;
Nuten= —, slot winding;
— = —, mit, slot wound;
Primär= —, primary (winding);
Sekundär= —, secondary (winding);
Stufen= —, bank(ed) winding R;
Zweit= —, secondary (winding);
Wicklungs=querschnitt *m*, cross-sectional area of winding;
 — =raum *m*, winding space volume of winding;
 — =verhältnis *n*, turns ratio.
Widerlager *n*, abutment.
Widerstand *m*, resistance; rheostat, resistor;
 — ausschalten, to cut out resistance;
 — einschalten, to insert or switch in resistance; im Stöpselrheostat: to unplug resistance;
 — entgegensetzen, to impede, to offer a resistance;
 mit — behaftet, resistive;
 äquivalenter —, equivalent resistance; [ance;
 äußerer —, external resist-

Widerstand
effektiver —, effective resistance; [sistance;
gemeinsamer —, common re-
gerichteter —, Kristall: asymmetrical resistance; **Reaktanz:** reactive resistance;
induktionsfreier —, plain or non-inductive resistance;
induktiver —, inductive resistance;
innerer —, internal resistance; output resistance V;
kombinierter —, joint or combined resistance;
kritischer —, critical resistance;
magnetischer —, magnetic resistance, reluctance;
negativer —, negative resistance, third-class resistance;
Ohmscher —, ohmic resistance, steady current resistance;
resultierender —, resultant resistance;
scheinbarer —, apparent resistance;
spezifischer —, specific resistance, resistivity;
— — in Mikrohm/cm³, volume resistivity;
— — — Ohm/m,g, mass resistivity;
wahrer —, true resistance;
Ableitungs —, der Leitung: leakage resistance; **der Röhre:** resistance leak;
Abschluß —, terminal resistance;
Abzweig —, leak coil, leak resistance (**der Telegraphenübertragung,** of telegraph repeaters);
Anodenkreis —, output resistance V; [sistance;
Ausbreitungs —, diffusion re-

Ausgleichs —, balancing resistence, compensating resistance;
Ballast —, loading resistance, ballast resistance;
Batterie —, battery resistance; earthing resistance T;
Begrenzungs —, limiting resistance;
Belastungs —, loading resistance;
Blind —, reactance;
mit — - behaftet, reactive;
Dekaden —, decade resistance box, decimal resistance;
Doppelleitungs —, loop resistance;
Eisen —, iron resistance, iron filament ballast lamp;
End —, terminal resistance;
Erdungs —, ground resistance; earthing resistance;
Faden —, filament resistance;
Fehler —, fault resistance;
Feld —, field rheostat, field resistance;
Flüssigkeits —, water resistance;
Gesamt —, total resistance;
Gitter —, grid leak, grid resistance;
Gitterkreis —, (internal) input resistance;
Gleichstrom —, steady (current) resistance, d. c. (= direct current) or c. c. (= continuous current) resistance;
Grenz —, critical resistance;
Heiz —, filament rheostat, heating resistance;
Innen —, internal resistance; outpout resistance V;
Isolations —, insulation (resistance), dielectric resistance;

Widerstand
Kapazitäts- —, capacitance;
Kopplungs- —, repeating resistance;
Lampen- —, lamp resistance;
Leit- —, conduction resistance;
Leiter- —, conductor resistance;
Luft- —, air resistance;
Luftdraht- —, aerial resistance;
Nebenschluß- —, shunt resistance, leak resistance; resistance leak V;
Normal- —, standard resistance;
Nutz- —, useful resistance;
Oberflächen- —, surface resistance;
Parallel- —, parallel resistance;
Potentiometer- —, potentiometer resistance;
Quer- —, stem, shunt element (der Kettenleiter, of networks);
Regler- —, rheostat;
Reihen- —, series resistance;
Röhren- —, tube resistance;
Schein- —, impedance;
— - —$, **Blindkomponente** f **des**, reactive (component of) impedance;
— - —$, **Wirkkomponente** f **des**, dissipative (component of) impedance;
Schieber- —, slide rheostat;
Schleifen- —, (conductor) loop resistance;
Schutz- —, protective resistance;
Schwächungs- —, gain controller, potentiometer, gain regulator, $K V$;
Strahlungs- —, radiation resistance; characteristic impedance L;

Übergangs- —, contact resistance;
Vergleichs —, reference resistance, standard resistance;
Verlust —, loss resistance;
— - —, **Reihen-**, equivalent series resistance (eines Kondensators, of a condenser);
Verzögerungs —, retardation or timing resistance;
Vorschalt- —, series resistance, reductor; für Voltmeter: multiplier;
Wasser- —, water resistance;
Wechselstrom- —, alternating current or a. c. resistance;
Wellen- —, characteristic impedance, surge impedance;
— - —es, reziproker Wert m des, characteristic admittance;
— - —es, Verlauf des, in Abhängigkeit von der Frequenz, frequency-impedance characteristic curve;
— - —es, Unregelmäßigkeiten pl im Verlauf des, impedance irregularities pl K;
— - —, Abschließung f einer Leitung durch ihren, termination of a line in its own impedance;
— - — einer mit halber Spule beginnenden Pupinleitung, mid-load characteristic impedance;
— - — — — einem halben Spulenfeld beginnenden Pupinleitung, mid-series characteristic impedance;
Wirk- —, non-reactive resistance, dissipative resistance;
Widerstands-änderung f, resistance variation;
— -äquivalent n, equivalent resistance; [lance;
— -ausgleich m, resistance ba-

Widerstands
— **-dämpfung** f, resistance loss;
— **-draht** m, resistance wire;
— **-erhöhung** f, (scheinbare), (apparent) increase of resistance;
— **-fähigkeit** f, strength;
dielektrische — - —, dielectric strength, selten: elastance;
widerstandsgekoppelt, resistance-coupled;
Widerstands-kasten m, resistance box; [coefficient;
— **-koeffizient** m, resistance
— **-komponente** f, resistance component;
— **-kopplung** f, resistance or resistive coupling;
Verstärker m **mit** — - —, resistance-repeating amplifier;
— **-lampe** f, resistance lamp;
widerstandslos, resistanceless;
Widerstands-messer m, ohmmeter;
— **-messung** f, resistance test;
— **-normal** n, resistance standard;
— **-schwankung** f, resistance variation;
— **-spule** f, resistance coil;
Einer- (**Zehner-**, **Hunderter-**), units (tens, hundreds) resistance coil;
— **-stufe** f, resistance step;
— **-symmetrie** f, resistance balance (beider Spulenhälften, of the two halves of the coil K;)
— **-verlust** m, resistance loss.
wieder aufladen, to recharge.
Wiederaufladung f, recharge.
wiederausstrahlen, to re-radiate.
Wiederausstrahlung f, re-radiation.
wiedereinführen, to reintroduce.
Wiedergabe f, reproduction;
genaue —, faithful reproduction;

Treue f **der** —, faithfulness of reproduction;
Zeichen- —, signal reproduction.
wieder-geben, to reproduce.
— **-herstellen**, to restore, to rebuild, to re-establish.
Wiederherstellung f, restoration, re-establishment.
Wiederholungs-klinke f, ancillary jack F;
— **-lampe** f, ancillary lamp F.
wiederinstandsetzen, to repair, to reinstate.
Wiederinstandsetzung f, reinstatement.
Wiederkehr f, **regelmäßige**, periodicity.
wiederkehrend, regelmäßig, periodic, (periodic) recurrent.
wiederverriegeln, to relatch.
Wiederverriegelung f, relatching.
wiederverstärken, to reamplify.
wiederverteilen, to redistribute.
Wiederverteilung f, redistribution.
wiederzünden, to reignite.
Wiederzündung f, re-ignition.
Wiege f, cradle; **Wippe**: rocker.
wiegen, to weigh.
Wind m, wind;
— **-belastung** f, wind load;
— **-druck** m, wind pressure;
— **-eisen** n, twisting pliers pl;
— **-fang** m, fan;
— - — **-regler** m, fan governor;
— **-öffnung** f, air hole;
— **-rose** f, card.

Winde f, winch, (lifting-) jack, hoist;
Hand- —, hand winch;
Kabel- —, cable winch;
Motor- —, motor winch, power-driven winch.
winden, to wind, hoch: to hoist.
windschief, warped.

Windung f, winding, einzelne: turn, convolution;
tote —, idle turn;
Steigung f der —, pitch of winding;
Anker= —, armature coil;
Gegen= —, opposing winding;
Windungs-ebene f, winding plane;
— =fläche f, turn area;
— =kapazität f, internal capacity (of a coil);
— =verhältnis n, turns ratio;
— =zahl f, number of turns.
Winkel m, angle, corner;
rechter —, right angle;
im rechten —, at right angles;
spitzer —, acute angle;
stumpfer —, obtuse angle;
Gegen= —, opposite angle;
Komplement= —, complementary angle;
Neben= —, adjacent or adjoining angle;
Neigungs= —, angle of slope;
Phasen= —, phase angle;
— = —, negativer (positiver), negative (positive) impedance angle;
Steigungs= —, angle of slope;
Wechsel= —, alternate angle;
Winkel- ...., angular;
— =ablenkung f, angular deflection;
— =bewegung f, angular motion;
— =eisen n, angle iron;
— =geschwindigkeit f, angular velocity, frequency in radians;
Einheit f der — = —, radian (= $360° : 2\pi$);
— =hebel m, bell crank (lever), angle lever, crank; am Baubotübersetzer: pointsman;
winkelhebelartig, bell crank ....
Winkel-komplement n, complement of angle;

— =maß n, impedance angle L;
— = — je Längeneinheit, wavelength constant L;
— =messer m, goniometer;
— =punkt m, Kurve, Leitung: inflection point;
— =stange f, angle pole B;
— =stellung f, angular position;
— =verzahnung f, double-helical teeth pl;
Getriebe n mit — = —, double-helical gearing.
Wippe f, rocker, rocking beam.
Wirbel m, eddy, whirl; Vortexring: vortex pl vortices.
wirbeln, to whirl, to eddy.
Wirbelströme pl, eddy currents, Foucault currents pl.
wirken, to function, to operate, to act (auf, on).
Wirkdämpfung f, transmission efficiency.
Wirken n, functioning, operation, performance, action.
Wirk-komponente f, energy component, active component, watt component;
— = — des Scheinwiderstandes, dissipative impedance;
— =leistung f, real power;
— =leitwert m, conductance.
wirksam, active, effective, efficient;
— e Oberfläche f, active surface.
Wirksamkeit f, activity, effectiveness, efficiency.
Wirkstrom m, energy current, active current;
— =komponente f, energy component of current.
Wirkung f, action, effect;
schlechte —, inefficiency;
Außen= —, external effect;
Nutz= —, useful effect.
Wirkungsgrad m, efficiency;
maximaler —, maximum efficiency;

Wirkungsgrad
Gesamt- —, total efficiency, overall efficiency;
Gesamt- — einer Anlage, commercial efficiency;
Kupfer- —, copper efficiency;
Strahlungs- —, efficiency of radiation.

wirtschaftlich, economical; commercial.

Wirtschaftlichkeit f, economy;
— s-frage f, question of economics.

Wirtschaftszentrum n, commercial centre.

wischen, to wipe.

Wismuth n, bismuth (Bi);
— -spirale f, bismuth coil.

Wohngegend f, residential district.

Wölbung f, curvature.

Wolfram n, tungsten (W);
— -faden m, tungsten filament; thorhaltiger oder thorierter — - —, thoriated tungsten filament;
— -lampe f, tungsten lamp.

Wollastondraht m, Wollaston wire.

Wolle f, wool.

Woodsches Metall n, Wood's alloy (25 Pb, 12,5 Sn, 50 Bi, 12,5 Cd, 73°).

Wörter pl **in der Minute**, words pl per minute, w. p. m.

Wulst m, torus.

Würgeverbindung f, twist(ed) joint B;
Kupferröhren- —, twisted sleeve joint, copper sleeve joint.

Wurzel f, root;
die — ziehen, to extract the root (aus, of);
zweite —, Quadrat- —, square root;
dritte —, Kubik- —, third power root, cube root;
— -zeichen n, root sign, radical.

3.

zähflüssig, viscous.
Zahl f, number; Ziffer, figure;
— **en und Zeichen** pl, figures, lower-case characters, pl T;
ganze —, integer, integral number;
gerade —, even number;
ungerade —, odd number;
Kubik- —, cube;
Quadrat- —, square number.

Zähl-ader f, pilot wire, marked wire;
— - —n-paar n, key pair, pilot pair, marked pair;
— -einrichtung f, counting device, counter; [ing device T;
Buchstaben- — - —, letter-count-

— -relais n, meter(ing) relay F;
— -taste f, meter key F;
— -vorrichtung f, counting mechanism.

zahlen, to pay.
Zahlen- ...., numeric(al);
— -blank n, figure space, figure blank, T;
— -folge f, **(in der)**, (in the) numerical order;

zahlenmäßig, numerical;
Zahlen-umschaltung f, figure shift T;
— -wechsel m, figure shift, shift, T;
— -weiß n, figure blank, figure space, T.

zählen, to meter to record (on the market), *F*; to count.
Zähler *m*, numerator *M*; meter *F*; counter;
Gesprächs= —, (service) meter;
Platz= —, position meter *F*;
Touren= —, Umlauf= —, revolution counter, cyclometer;
— =ablesung *f*, meter reading;
— =batterie *f*, meter battery;
— =gestell *n*, (service) meter rack;
— =kontrollampe *f*, meter lamp;
— =kontrollzeichen *n*, meter indicator.
Zählung *f*, counting; metering *F*;
Gesprächs= —, metering;
Zonen= —, zone metering.
Zahn *m*, tooth (*pl* teeth);
eingesetzter —, cog;
mit Zähnen versehen, toothed; studded;
Magnet=zahn, field projection;
Pol= —, pole tooth, spoke;
Sperr= —, ratchet tooth, ratchet step; pawl, detent;
— = — kranz *m*, ratchet drum;
— =breite *f*, tooth pitch;
eine halbe — = — auseinander, half a tooth pitch apart;
— =induktion *f*, tooth induction;
— =länge *f*, tooth pitch;
— =lücke *f*, tooth gap;
— =rad *n*, tooth(ed) wheel;
— = — antrieb *m*, gear drive;
— =rädergetriebe *n*, toothed wheel gearing, gear;
— = — mit gekreuzten Wellen, skew gearing;
— =radunterbrecher *m*, toothed wheel circuit breaker, crown wheel commutator;
— =scheibe *f*, studded disc *R*;
— = —n=funkenstrecke *f*, studded disc discharger;
— =segment *n*, segmental rack;
— =stange *f*, (toothed) rack, mit Sperrzähnen: ratch;
— =teilung *f*, tooth pitch.
Zange *f*, tongs, pliers, *pl*.
Beiß= —, cutting pliers, nippers, *pl*;
Biege= —, bending pliers, bending tongs *pl*; [tongs;
— = —, Rohr=, pipe bending
Flach= —, flat nose(d) pliers;
Rohr= —, pipe wrench;
Rund= —, round nose(d) pliers.
Zapfen *m*, faucet, trunnion;
Angel: pivot; der Achse: journal; des Fasses: spigot;
in — lagern, to pivot (on);
Dreh= —, pivot;
— =lager *n*, journal bearing.
Zapfstelle *f*, tap, tapping (point), tap connection.
Zaun *m*, fence.
Z. B., common battery, c. b., central energy (*am.*);
— =Fernhörer *m*, c. b. receiver;
— =Fernsprecher *m*, c. b. telephone station;
— =System *n*, common battery system, c. b. system;
Ericsson= — = —, bridged impedance c. b. system;
Western= — = —, repeating coil c. b. system.
Zeder *f*, cedar.
rote virginische —, red cedar.
zedieren, to assign (to).
Zehnerstufe *f*, tens digit *A*.
Zehntastensatz *m*, ten button key set *A*.
Zeichen *n*, signal, mark;
die — brechen, the marks split *T*;
die — laufen zusammen, the signals run together *T*;
durch Störer verdeckte —, swamped or clouded signals;

Zeichen
  gebämpfte —, spark signals R;
  richtige —, straight signals (Gegensatz: umgekehrte —, reversed signals) T;
  schwache —, weak signals;
  starke —, strong signals;
  ungebämpfte —, continuous wave signals;
  Besetzt- —, busy tone, busy back tone FA;
  Glocken- —, bell signal;
  Licht- —, luminous signal;
  Melde- —, signal;
  — - —, Gruppen-, pilot signal;
  Not- —, distress signal R;
  Schau- —, visual signal;
  Summer- —, humming sound, buzzer tone;
  Telegraphier- —, telegraph signal;
  Überwachungs- —, supervisory (signal), pilot signal;
  Warnungs- —, warning (signal);
  Zeit- —, time signal;
  — -abstand m, figure space T;
  — -batterie f, marking battery T;
  — -frequenz f, signal frequency R;
  — -front f, signal front, signal head, T;
  — -gabe f, — -gebung f, signalling, transmission of signals;
  — -intensität f, signal intensity;
  — -kontakt m, marking contact, marking stop, T;
  — -kopf m, signal head, signal front L;
  — -loch n, signal hole (des Sendelochstreifens, of the perforated tape) T;
  — -material n, drawing materials pl;
  — -papier n, drawing paper;
  — -saal m, drafting room;
  — -seite f, marking side;
  Relais n liegt auf der — - —, the relay marks T;
  Überhang m nach der — - —, marking bias T;
  — -stärke f, signal strength, signal intensity;
  — -stirn f, signal head TL;
  — -strom m, marking current T;
  — - — geben, to mark T;
  Zwischen- — - —, spacing current T;
  — -ton m, signal note R;
  — - — -höhe f, pitch of the signal note;
  — -welle f, signal wave R;
  Zwischen- — - —, spacing wave R;
  — -wiedergabe f, signal reproduction.
zeichnen, to draw; unterzeichnen: to sign;
  eine Kurve —, to plot a curve.
Zeichner m, draughtsman.
Zeichnung f, drawing, sketch, picture;
  schematische —, skeleton sketch,
  Schnitt- —, sectional drawing.
Zeiger m, index, hand, pointer;
  Licht- —, spot of light;
  Merk- —, adjustable index, indicator needle;
  Minuten- —, minute hand;
  — -galvanometer n, pointer galvanometer;
  — -telegraph m, pointer telegraph.
Zeile f, line;
  Druck- —, line of print;
  -n-magnet m, line-feed magnet;
  -n-vorschub m, line feed.
Z-Eisen n, Z-iron.
Zeit f, time; Zeitdauer: duration;
  zur — Null, at zero time;

Zeit
nach der — abmessen, einteilen, to time;
Abfertigungs- —, Abwicklungs- handling time;
Anmelde- —, booking time F;
Anspruch- —, operating time;
Aufgabe- —, code time, time of acceptance, T;
Beförderungs- —, time of transmission;
Lauf- —, time of transit;
Leitungs- —, line time, circuit time, FT;
Meß- —, testing time;
Übertragungs- —, duration of transmission;
Warte- —, wait(ing) time, delay, F;
— -achse f, time axis;
— -ball m, time ball;
— -dauer f, duration;
— -einheit f, unit (of) time;
in der — — —, per unit time;
— -gebühr f, measured rate;
— -konstante f, time constant;
— -maß n, tempo, rhythm;
— — —stab m, scale of time;
— -messung f, timing;
— -relais n, time-delay relay;
— — — mit Bremszylinder, dashpot relay;
— -schalter m, time switch;
— -schreiber m, calculagraph F;
— -signal n, time signal;
— -stempel m, time stamp;
— -teilchen n, small interval of time;
— -unterschied m, difference of time, Nacheilung: time lag, Voreilung: time lead;
— -vergeudung f, waste of time;
— -verlust m, lost time;
— -wert m, standing value (einer Anlage, of a plant);
— -zeichen n, time signal;

— - — -geber m, time signal transmitter; chronopher.
zeiten, to time.
Zeitungs-dienst m, news work;
— -leitung f, news circuit;
— - — mit mehreren Empfangsstellen, Y Q-circuit (engl.), way circuit (am.);
— -telegramm n, news message, press message.
zeitweilig, temporary.
Zelle f, cell; Schrank: cabinet;
gegengeschaltete —, counter-cell;
lichtelektrische —, photo-electric cell;
lichtempfindliche —, light-reactive cell;
schalldichte —, silence cabinet F;
Fernsprech- —, telephone cabin, silence cabinet;
Polarisations- —, polarization cell;
Sammler- —, storage cell;
Ventil- —, valve;
— - —, elektrolytische, electrolytic valve;
Zersetzungs- —, decomposition cell.
zellenartig, cellular.
Zellenschalter m, cell switch, battery (cell) switch;
Doppel- —, double cell switch.
Zellhorn n, Zelluloid n, celluloid.
Zelt n, tent;
Löter- —, wireman's tent.
Zement m, cement;
in — einschwemmen, to float in cement;
Portland- —, portland cement;
— -formstück n, concrete block;
einzügiges — - —, single-duct concrete block;
mehrzügiges — - —, multiple-duct concrete block;

Zement
— = formstückkanal m, concrete block conduit;
— =fußboden m, concrete floor;
— =mörtel m, cement (mortar);
— =rohr n, concrete pipe.
zementieren, to cement.
Zentesimal= ...., centesimal.
Zentimeter n, (ab: cm), centimetre (= 0,3937 inch);
Quadrat= —, (ab: cm², qcm) square centimetre (= 0.15501 squ. in.);
Kubik= —, (ab: cm³, ccm), cubic centimetre (= 0.061026 cub. in.);
— =würfel n, centimetre cube.
Zentner m, hundredweight, ab: cwt. (1 cwt. = 112 lbs. = 50,80 kg).
Zentralanrufschrank m, concentration switchboard, concentrator.
Zentrale f, central office, exchange;
Nebenstellen= —, private branch exchange, ab: p. b. x.;
— = zu 6 Amtsleitungen und 50 Nebenstellen, 50 line 6 trunk private branch exchange;
— = —, Selbstanschluß=, private automatic branch exchange, ab: p. a. b. x.;
Privat= —, private exchange, ab: p. x.;
— = —, Selbstanschluß=, private automatic exchange, ab: p.a.x.;   [zentrale;
Teilnehmer= — = Nebenstellen=
Zentral=schrank m, concentrator, concentrating switchboard;
in einem = — vereinigte Leitungen pl, concentrated trunks pl F T;
— =umschalter m, concentrator (für 20 Leitungen: for 20 lines) F T;

Nacht= —=—, night concentrator;
— =— für Telegraphenleitungen, intercommunication switch.
zentralisieren, to centralize.
Zentralisierung f, centralization.
zentrieren, to centre.
Zentrieren n, centering.
Zentrierfeder f, centering spring.
Zentrifugal=kraft f, centrifugal force;
— =regler m, centrifugal governor.
zentripetal, centripetal.
zentrisch, concentric.
Zentrumbohrer m, centre bit.
zerbrechen, to break, to fracture, to rupture.
Zerbrechen n, breaking, fracture.
zerbrechlich, fragile.
zerdrücken, to crush.
Zerfall m, decay, disintegration.
zerfallen, to decay, to disintegrate.
zerhacken, to chop (Strom, current).
Zerhacker m, chopper;
Tonfrequenz= —, audio frequency chopper.
zerlegen, to split up, to analyze.
zerreißen, to break, to disrupt.
Zerreißen n, rupture, disruption, breaking.
Zerreißfestigkeit f, tearing strength.
zersetzbar, decomposable.
zersetzen, (sich), to decompose, to disintegrate.
Zersetzung f, decomposition, disintegration;
— =zelle f, decomposition cell.
zerspringen, to burst.
zerstören, to destroy; to corrode.
zerstört werden, to decay.
Zerstörung f, corrosion, decay; destruction.

zerstreuen, to disperse; Energie: to dissipate.
Zerstreuung f, dispersion; dissipation.
Zession, assignment (to).
Zessionar m, assignee.
Zettel m, ticket;
— =rohrpost f, pneumatic ticket carrier;
— =verteiler m, ticket distribution position;
Rohrpost= — = —, pneumatic ticket distribution desk.
Zickzacklinie f, zig-zag line.
Ziegel m, brick;
— =mauerwerk n, brick work;
— =stein m, brick;
— = — =schicht f, course of bricks.
Ziehband n, clamp, strap.
Ziehbank f, drawing bench.
ziehen, to pull, to draw, to haul; Drähte herstellen: to draw; Drähte verlegen: to string (wires); Röhrensender: to draw out.
Ziehen n, pull, drawing; stringing; instability, drawing-out, R.
Zieh-feder f, drawing pen;
— =strumpf m, cable grip, wire grip;
— =vorgänge pl, drawing-out, instability, R. [B.
Ziersockel m, ornamental sleeve
Ziffer f, figure; Faktor: coefficient, figure;
— =blatt n, dial;
— n=rolle f, counter;
— n=scheibe f, figure dial.
Zimmer=antenne f, indoor aerial;
— =leitung f, office wiring, office cable; internal wiring;
— =temperatur f, (normal) room temperature.
Zink n, zinc (Zn);
schwefelsaures —, sulphate of zinc ($ZnSO_4$);

— =amalgam n, zinc amalgam;
— =becher m, — =behälter m, zinc container, zinc containing vessel;
— =blech n, sheet zinc;
— =chlorid n, chloride of zinc ($ZnCl_2$);
mit — = — tränken, to burnettize B;
Tränkung f mit — = —, burnettization B;
— =platte f, zinc plate;
— =pol m, zinc pole, zinc terminal, ab: Z;
— =sulphat n, sulphate of zinc, white vitriol, ($ZnSO_4$);
— =weiß n, zinc white (ZnO);
— =vitriol n, white vitriol, sulphate of zinc ($ZnSO_4$).
Zinke f, prong, tine;
Stimmgabel= —, tuning fork
Zinn n, tin (Sn); [tine.
— =folie f, tin foil;
— =oxyd n, — =säure f, — =stein m, cassiterite, tin dioxide, tinstone, (SnO_2).
Zinsen pl, interest.
Zinseszins m, compound interest
Zins=fuß m, — =satz m, (rate of) interest.
Zirkulation f, circulation.
zirkulieren, to circulate.
zischen, to hiss, to sizzle.
Zischen n, hisses pl.
Zischlaut m, sibilant sound, hissing sound.
Zone f, zone; region, district;
neutrale —, neutral zone;
Fern= —, telephone trunk zone;
Fernsprech= —, telephone zone;
Indifferenz= —, neutral zone;
Nachbar= —, adjacent zone;
Zonen=gebühr f, zone rate;
— =hauptort m, — =mittelpunkt m, (telephone) zone centre;
zweiter — —, sub-zone centre;

Zonen
— -system n, repeating centre system, zone system, T.
— -tarif m, zone tariff.
Zopfende n, top end B;
Zubehör n, accessories pl.
zubereitet, prepared, treated;
— e Stange f, treated pole B.
zuführen, to supply (to), to convey.
Zuführung f, Strom: supply, conveyance; Draht: lead;
lose —, wandering lead;
verdrallte und abgeschirmte — en pl, twisted and screened leads pl;
Batterie= —, battery lead;
Luftdraht= —, downleads pl of an aerial.
Zug m, (tractional) pull (auf, on); Beanspruchung: stress, strain; Spannung: tension; Eisenbahn, Wellen: train; Ziehen: traction;
seitlicher —, lateral pull, transverse stress;
Draht= —, pull of wire;
Seiten= —, lateral stress;
Wellen= —, wave train;
— -abfertigungsdienst m, train dispatch service;
— -balken m, balk;
— -beanspruchung f, tensile stress;
— -deckung f, train blocking;
— - -s-system n, train blocking system;
— -dienstleiter m, train dispatcher;
— -festigkeit f, tensile strength;
— -kraft f, pull; Magnet: lifting power;
— -schalter m, pull switch;
— -seilchen n, draw wire B;
— -spannung f, tensile stress;
— -stange f, pull rod;
— -vorrichtung f, train.
Zugang m, entrance; access, A.

zugänglich, accessible.
Zugänglichkeit f, accessibility.
zugeführte Leistung f, (power) input.
Zuhaltung f, tumbler.
Zuhörerschaft f, audience.
zulässig, permissible;
— er Heizstrom m, safe filament current;
— e Stromstärke f, rated current.
Zuleitung f, lead; [pl;
Apparat= —, instrument leads
Luftdraht= —, aerial feeder;
— -s-draht m, lead-in wire;
— -s-platz m, B-position F.
Zunahme f, increase, increment;
prozentuale —, percentage increase.
zünden, to ignite (Lichtbogen: arc);
wieder —, to re-ignite.
Zunder m, scale.
Zünd-kerze f, spark(ing) plug;
— -spannung f, Lichtbogen: ignition voltage; Funkenstrecke: breakdown voltage;
— -störungen pl, ignition interference (durch Explosionsmotoren, from internal combustion engines).
Zündung f, ignition; breakdown;
Lichtbogen= —, arc ignition;
Neu= —, Wieder= —, reignition.
zunehmen, to grow, to increase.
Zunge f, tongue; reed;
schwingende —, vibrating reed;
— n-pfeife f, reed pipe;
— n-summer m, reed hummer;
— n-unterbrecher m, vibrating reed break.
zuordnen, to assign (to).
zurichten, to trim.
zurück-behalten, to retain;
— -bewegen, to move back; in die frühere Lage: to unshift;

zurück
— **-gehen**, to return;
— **-halten**, to retain, hemmen: to retard;
— **-kehren**, to return; in die Ruhelage — - —, to return to normal;
— **-leiten**, to lead back, to return; rückkoppeln: to feed back;
— **-rufen**, to recall, to ring back, F;
— **-schnellen**, to jump back;
— **-stellen**, to reset, to release;
— **-werfen**, to throw back; Schall: to reverberate;
— **-ziehen**, to retract, to pull back, to withdraw.

Zurückziehung f, retraction, withdrawal.

zusammen-arbeiten, to interwork (with);
— **-backen**, to agglomerate.

Zusammenbacken n, agglomeration, packing.

zusammenballen, to bunch (Mikrophonkohlen, transmitter carbons).

Zusammenbau m, assembly;
— **-lehre** f, assembling jig, fixture.

zusammen-bauen, to assemble.
— **-brechen**, to break down; to collapse.

Zusammenbruch m, breakdown, collapse.

zusammen-drehen, to twist together.
— **-drücken**, to compress, zerdrücken: to crush.
— **-fallen**, magn. Feld usw: to collapse; übereinstimmen: to coincide.

Zusammenfallen n, collapse; coincidence.

zusammen-fallend, coincident (mit, with).

Zusammenfassung f, centralization.

zusammen-gesetzt, gemischt: compound; composite, composed (aus, of); Welle: complex; resultierend: resultant;
— **-klappbar**, collapsable, collapsible (am.);
— **-laufen**, to run together (Morsezeichen, Morse signals);
— **-legbar**, collapsable, collapsible (am.);
— **-legen**, to centralize, to concentrate.

Zusammenlegung f, centralization, concentration.

zusammen-löten, to solder together;
— **-pressen**, to compress;
— **-rechnen**, to add up; to compute;
— **-schalten**, to join up; bündeln: to bunch; Schleife: to loop;
— **-schrauben**, to bolt together;
— **-setzen**, to compose, to assemble; (aus, of), mischen: to compound.

Zusammen-setzung f, composition;
— **-stellung** f, combination, assemblage (b. Zusammengestellte: assembly); Liste: list;
— **-stoß** m, collision;

zusammen-stoßen, to collide;
— **-wirken**, to interact.

Zusammenwirken n, interaction.
zusammenziehen, to contract.
Zusammenziehung f, contraction.
Zusatz m, addition;
— **-batterie** f, booster battery;
— **-dynamo** f, booster (dynamo);
— **-gestell** n, additional rack;
— **-patent** n, addition (to), additional patent;
— **-spannung** f, additional voltage, boosting voltage.

zusätzlich, additional, incremental;

zufäglich
— c **Permeabilität** f, incremental permeability.
Zufchlag m, addition;
— für **Abtrieb**, slack B.
zufchmelzen, to seal (off) V.
zufpitzen, to tip, to point.
zufprechen, to telephone (ein **Telegramm**, a message).
Zuftand m, state, order, condition;
Anfangs- —, initial state;
Ausgangs- —, initial conditions pl;
Betriebs- —, working or operating order, service order;
Dauer- —, steady state;
End- —, final state;
Erhaltungs- —, maintenance standard;
— - —, guter, high maintenance standard.
zuftöpfeln, to plug up.
zuteilen, to allot, to preassign, to appropriate, to allocate.
Zuteilung f, allotment, appropriation, allocation, assignment;
— **s-wähler** m, allotting switch.
zuweifen, to assign, to appropriate.
Zuweifung f, assignment, appropriation, assignation.
zuverläffig, reliable.
Zuverläffigkeit f, reliability;
— **s-probe** f, — **s-prüfung** f, reliability test.
zweiadrig, twin ...., twin leader ...., bifilar, double conductor ....;
— es **Kabel** n, bifilar cable, twin core cable.
zweiarmig, two-armed;
— er **Hebel** m, two-armed lever.
Zweidraht-betrieb m, two-wire opération F K;

— **-Doppelrohr-Zwischenverstärker** m, two-wire two-valve intermediate repeater F K;
— **-leitung** f, two-wire circuit F K.
zweidrähtiger Luftleiter m, two-wire aerial R.
Zweielektrodenröhre f, two-electrode valve, diode.
Zweifach- ...., double;
— **-telegraph** m, double telegraph set;
— **-verftärker** m, two-stage amplifier.
zweifädig, bifilar.
Zweig m, branch; leg;
a- — geerdet, a-leg earthed;
— **-amt** n, branch exchange;
— **-kabel** n, branch cable;
— **-linie** f, branch line, kurze: spur;
— **-fchaltung** f, parallel connection;
— **-ftrom** m, branch current;
— - — **-kreis** m, branch circuit;
— **-fyftem**, n, tapering cabling system B.
Zweigitterröhre f, double grid valve.
zweigleifige Bahn f, double track railway.
Zweikreis-empfang m, secondary reception R;
— **-empfänger** m, double circuit receiver R.
Zweileiterkabel n, twin core cable.
zweipaariges Kabel n, two pair core cable.
zweiphafig, two-phase, biphase, diphase.
zweipolig, bipolar, two-polar.
zweireihig, in two rows.
Zweirohrverftärker m, two-valve repeater, double relay repeater, F K.
zweifchenklig, two-legged.

Zweischnur=Klappenschrank *m*, double-cord switchboard;
— =**system** *n*, double cord system.
zweispitzige Kurve *f*, double-peaked curve.
zweispulig, double spool....
zweistufig, two-stage.
Zweit=kreis *m*, secondary circuit;
— =**wicklung** *f*, secondary (winding).
Zweiwegeverstärker *m*, two-way repeater, duplex repeater.
zweiwertige Zeichen *pl*, two-power signals *pl T*.
zweizinkig, double-pronged.
Zwilling *m*, twin;
Zwillings=antenne *f*, pair of aerials, twin aerial;
— =**kabel** *n*, twin cable;
  Mehrfach= — = —, **Vielfach=** — = —, multiple twin cable, *ab*: m. t. cable (Dieselhorst= Martin=Verseilung);
  Vierfach= — = —, quadruple pair cable (Achterverseilung);
— =**klinke** *f*, pair of jacks;
— =**stecker** *m*, biplug, pair of plugs.
Zwinge *f*, ferrule.
Zwirn *m*, twine, yarn.
Zwischen=amt *n*, intermediate station, waystation, intermediate office;
— =**boden** *m*, — =**decke** *f*, false floor *B*;
— =**frequenzempfänger** *m*, transposition receiver, *R*;
— =**glied** *n*, link;
— =**kabel** *n*, intermediate cable;
  Fernleitungs= — = —, toll intermediate cable;
— =**kreis** *m*, intermediate circuit, link circuit;
  abgestimmter — = —, tuned intermediate circuit;
  aperiodischer — = —, intermediate aperiodic circuit;

— =**lage** *f*, intermediate layer, separator; [parator.
  Isolier= — = —, insulating se=
zwischenliegend, intermediate;
  zwischen.... **und**.... **liegend**, intermediate of.... and....
Zwischen=pol *m*, inter-pole;
— =**raum** *m*, interstice; gap, space; bes. zeitlich: interval; Abstand: clearance, distance; freier —, clearance;
— =**röhrtransformator** *m*, inter-valve transformer;
— =**satzstück** *n*, adapter.
zwischenschalten, to interpose, to interpolate (in).
Zwischen=schaltung *f*, interposition, interpolation;
— =**sender** *m*, retransmitter *T*; repeater or repeating station *R*;
  Rundfunk= — = —, broadcast repeating station, remotely controlled broadcast transmitter.
zwischensetzen, to interpose.
Zwischen=stecker *m*, adaptor, adapter;
— = —**für Röhren**, valve adapter, socket adapter;
— =**stelle** *f*, intermediate telephone set;
— =**stellenumschalter** *m*, inter-through switch *F*;
— =**transformator** *m*, intermediate transformer;
— =**verstärker** *m*, **Fernsprech=**, telephone intermediate repeater, through line repeater;
  Zweidraht=Zweirohr= — = —, two-valve two-wire intermediate repeater;
— =**verteiler** *m*, intermediate distributing frame, *ab*: IDF, cross-connection field, cross-connecting board;

Zwischen
— =wand *n*, partition;
— =zeichen=strom *m*, spacing current *T*;
— = — =welle *f*, spacing wave, compensation wave, *R*.

Zwitschern *n*, birdies *pl R*.

zyklisch, cyclic(al).

Zyklus *m*, cycle;
 magnetischer —, magnetic cycle.

Zylinder *m*, cylinder;
 Halb= —, semi-cylinder;
— =lager *n*, journal bearing.

zylindrisch, cylindrical.

Verlag von Julius Springer in Berlin W 9

Englisch-Deutsches und Deutsch-Englisches Wörterbuch der Elektrischen Nachrichtentechnik
von
O. Sattelberg
im Telegraphentechnischen Reichsamt Berlin

Erster Teil

Englisch-Deutsch

(292 S.) 1925. Gebunden RM. 9.—

Es ist zu begrüßen, daß sich ein Fachmann gefunden hat, der das technische Englisch in gute deutsche Wortverbindungen gefaßt hat. — Das vorliegende Werkchen ist mit viel Übersicht und Sorgfalt zusammengestellt, so daß es ohne weiteres in der Lage ist, bei der Verdeutschung englischer und amerikanischer Fachliteratur eine zuverlässige Hilfe zu sein. Druck, Format, Anordnung und notwendige Wortanalysen sind ebenfalls dem praktischen Gebrauch entsprechend getroffen. (Radio-Umschau.)

Der Fernsprechverkehr als Massenerscheinung mit starken Schwankungen. Von Dr. G. Rückle und Dr.-Ing. F. Lubberger. Mit 19 Abb. und auf 1 Tafel. (155 S.) 1924. RM. 11.—; gebunden RM. 12.—

Der Poulsen-Lichtbogen-Generator. Von C. F. Elwell. Deutsch von Dr. A. Semm, Postrat im Telegraphen-Techn. Reichsamt, und Dr. F. Gerth. Mit 149 Textabbild. (189 S.) 1926. RM. 12.—; gebunden RM. 13.50

Die Grundlagen der Hochfrequenztechnik. Eine Einführung in die Theorie von Dr.-Ing. Franz Ollendorf, Charlottenburg. Mit 379 Abbildungen im Text und 3 Tafeln. (656 S.) 1926. Gebunden RM. 36.—

Hochfrequenzmeßtechnik. Ihre wissenschaftlichen und praktischen Grundlagen. Von Dr.-Ing. August Hund, Beratender Ingenieur. Mit 150 Textabbildungen. (340 S.) 1922. Gebunden RM. 11.—

Der Radio-Amateur (Radio-Telephonie). Ein Lehr- und Hilfsbuch für die Radio-Amateure aller Länder. Von Dr. Eugen Nesper. Sechste, bedeutend vermehrte und verbesserte Auflage. Mit 955 Textabbildungen. (886 S.) 1925. Gebunden RM. 27.—

Verlag von Julius Springer in Berlin W 9

Bibliothek des Radio=Amateurs. Herausgegeben von Dr. **Eugen Nesper.**

Fertig liegen vor:

1. Band: **Meßtechnik für Radio=Amateure.** Von Dr. **Eugen Nesper.** Dritte Auflage. Mit 48 Textabbildungen. (56 S.) 1925. RM. 0.90
2. Band: **Die physikalischen Grundlagen der Radiotechnik.** Von Dr. **Wilhelm Spreen.** Dritte, verbesserte und vermehrte Auflage. Mit 127 Textabbildungen. (162 S.) 1925. RM. 2.70
3. Band: **Schaltungsbuch für Radio=Amateure.** Von **Karl Treyse.** Neudruck der zweiten, vervollständigten Auflage. (19.—23. Tausend.) Mit 141 Textabbildungen. (60 S.) 1925. RM. 1.20
4. Band: **Die Röhre und ihre Anwendung.** Von **Hellmuth C. Riepka.** Dritte, verbesserte und vermehrte Auflage. Erscheint im Februar 1926.
5. Band: **Praktischer Rahmen=Empfang.** Von Ing. **Max Baumgart.** Zweite, vermehrte und verbesserte Auflage. Mit 51 Textabbildungen. (82 S.) 1925. RM. 1.80
6. Band: **Stromquellen für den Röhrenempfang** (Batterien und Akkumulatoren). Von Dr. **Wilhelm Spreen.** Mit 61 Textabbildungen. (76 S.) 1924. RM. 1.50
7. Band: **Wie baue ich einen einfachen Detektor=Empfänger?** Von Dr. **Eugen Nesper.** Zweite, vermehrte Auflage. Mit 31 Abbildungen im Text und auf einer Tafel. (60 S.) 1925. RM. 1.35
8. Band: **Nomographische Tafeln für den Gebrauch in der Radiotechnik.** Von Dr. **Ludwig Bergmann.** Mit 53 Textabbildungen und zwei Tafeln. Zweite, vermehrte Auflage. (94 S.) 1926. RM. 2.70
9. Band: **Der Neutrodyne=Empfänger.** Von O. **Schöpflin** und **Carl Eichelberger.** (Zweite Auflage des Buches „Der Neutrodyne=Empfänger" von Dr. Rosa Horsky.) In Vorbereitung.
10. Band: **Wie lernt man morsen?** Von Studienrat **Julius Albrecht.** Mit 7 Textabbildungen. Zweite Auflage. (44 S.) 1925. RM. 1.35
11. Band: **Der Niederfrequenz=Verstärker.** Von Ing. **O. Kappelmayer.** Zweite, verbesserte Auflage. Mit 57 Textabbildungen. (112 S.) 1925. RM. 1.80
12. Band: **Formeln und Tabellen** aus dem Gebiete der Funktechnik. Von Dr. **Wilhelm Spreen.** Mit 34 Textabbildungen. (80 S.) 1925. RM. 1.65

Fortsetzung siehe nächste Seite.

Verlag von Julius Springer in Berlin W 9

Bibliothek des Radio-Amateurs. Herausgegeben von Dr. **Eugen Nesper.**
13. Band: **Wie baue ich einen einfachen Röhrenempfänger?** Von Karl Treyse. Mit 28 Textabbildungen. (55 S.) 1925. RM. 1.35
15. Band: **Innen-Antenne und Rahmen-Antenne.** Von Dipl.-Ing. **Friedrich Dietsche.** Mit 25 Textabbildungen. (67 S.) 1925. RM. 1.35
16. Band: **Baumaterialien für Radio-Amateure.** Von **Felix Cremers.** Mit 10 Textabbildungen. (101 S.) 1925. RM. 1.80
17. Band: **Reflex-Empfänger.** Von Radio-Ingenieur **Paul Adorján.** Mit 60 Textabbildungen. (61 S.) 1925. RM. 2.10
18. Band: **Das Fehlerbuch des Radio-Amateurs.** Von Ing. **Siegmund Strauß.** Mit 75 Textabbildungen. (86 S.) 1925. RM. 2.10
19. Band: **Rufzeichen-Liste für Radio-Amateure.** Von **Erwin Meißner.** (140 S.) 1925. RM. 3.—
20. Band: **Lautsprecher.** Von Dr. **Eugen Nesper.** Mit 159 Textabbildungen. (145 S.) 1925. RM. 3.80; geb. RM. 4.20
21. Band: **Funktechnische Aufgaben und Zahlenbeispiele.** Von Dr.-Ing. **Karl Mühlbrett.** Mit 46 Textabbildungen. (97 S.) 1925. RM. 2.10
22. Band: **Ladevorrichtungen und Regenerier-Einrichtungen** der Betriebsbatterien für den Röhren-Empfang. Von Dipl.-Ing. **Friedrich Dietsche.** Mit 56 Textabbildungen. (62 S.) 1925. RM. 2.10
23. Band: **Kettenleiter und Sperrkreise** in Theorie u. Praxis. Von Elektro-Ingenieur **C. Eichelberger.** Mit 120 Textabbildungen und einer Rechentafel. (99 S.) 1925. RM. 3.—
24. Band: **Hochfrequenz-Verstärker.** Von Dipl.-Ing. Dr. **Arth. Hamm.** Mit 106 Textabbildungen. (133 S.) 1926. RM. 3.90
27. Band: **Superheterodyne-Empfänger.** Von Ing. **E. F. Medinger.** Mit 49 Textabbildungen. (74 S.) 1926. RM. 2.70
28. Band: **Die Methode der graphischen Darstellung und ihre Anwendung in Theorie und Praxis der Radiotechnik.** Von Dipl.-Ing. **O. Herold.** Mit 74 Textabbildungen. (87 S.) 1925. RM. 2.70

Die Vakuumröhren und ihre Schaltungen für den Radio-Amateur. Von J. Scott Taggart. Deutsche Bearbeitung von Dr. **Siegm. Loewe** und Dr. **Eugen Nesper.** Mit 136 Textabbildungen. (188 S.) 1925. RM. 13.50

Verlag von Julius Springer in Berlin W 9

Drahtlose Telegraphie und Telephonie. Ein Leitfaden für Ingenieure und Studierende von **L. B. Turner**. Ins Deutsche übersetzt von Dipl.-Ing. **W. Glitsch**, Darmstadt. Mit 143 Textabb. (229 S.) 1925. Geb. RM. 10.50

Radio-Technik für Amateure. Anleitungen und Anregungen für die Selbstherstellung von Radio-Apparaturen, ihren Einzelteilen und ihren Nebenapparaten. Von Dr. **Ernst Kadisch**. Mit 216 Textabbildungen. (216 S.) 1925.
Gebunden RM. 5.10

Lehrbuch für Radio-Amateure. Leichtverständliche Darstellung der drahtlosen Telegraphie und Telephonie unter besonderer Berücksichtigung der Röhren-Empfänger. Von **H. E. Riepka**, Mitglied des Hauptprüfungsausschusses des Deutschen Radio-Clubs e. V., Berlin. Mit 151 Textabbildungen. (159 S.) 1925. Gebunden RM. 4.50

Grundversuche mit Detektor und Röhre. Von Dr. **Adolf Semiller**, Studienrat am Askanischen Gymnasium und Realgymnasium zu Berlin. Mit 28 Textabbildungen. (48 S.) 1925. RM. 2.10

Kalender der Deutschen Funkfreunde 1926. Herausgegeben im Auftrage des Deutschen Funktechnischen Verbandes e. V., Berlin von Dr.-Ing. **Karl Mühlbrett**, Techn. Staatslehranstalten Hamburg und Ziviling. **Friedr. Schmidt**, Generalsekretär Hamburg. Mit einem Geleitwort von Prof. Dr. **A. Esau**, Physikalisches Institut Jena, Präsident des Deutschen Funktechnischen Verbandes e. V. Zweiter Jahrgang. (216 S.) Gebunden RM. 3.60
Vorzugspreis für Mitglieder des Deutschen Funktechnischen Verbandes und der ihm angeschl. Vereine. RM. 2.70

Verlag von Julius Springer und M. Krayn in Berlin W 9

Der Radio-Amateur. Zeitschrift für Freunde der drahtlosen Telephonie und Telegraphie. Organ des Deutschen Radio-Clubs. Unter ständiger Mitarbeit von Dr. **Walther Burstyn**-Berlin, Dr. **Peter Lertes**-Frankfurt a. M., Dr. **Siegmund Loewe**-Berlin und Dr. **Georg Seibt**-Berlin u. a. m. Herausgegeben von Dr. **Eugen Nesper**-Berlin und Dr. **Paul Gehne**-Berlin. Erscheint wöchentlich im Umfange von je 20—24 Seiten.
Monatlich RM. 2.40 / Einzelheft RM. 0.60
(Die Auslieferung erfolgt vom Verlag Julius Springer, Berlin W 9)

MIX
Papier aus verantwortungsvollen Quellen
Paper from responsible sources
FSC® C105338

If you have any concerns about our products,
you can contact us on
ProductSafety@springernature.com

In case Publisher is established outside the EU,
the EU authorized representative is:
**Springer Nature Customer Service Center GmbH
Europaplatz 3, 69115 Heidelberg, Germany**

Printed by Libri Plureos GmbH
in Hamburg, Germany